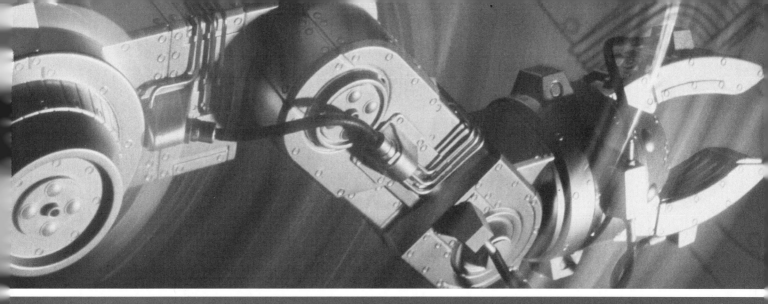

MACHINE DESIGN

Timothy H. Wentzell, P.E.

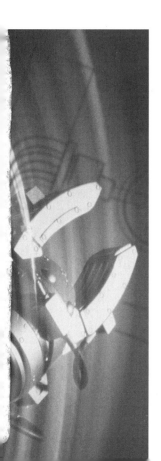

THOMSON
DELMAR LEARNING

Australia Canada Mexico Singapore Spain United Kingdon United States

MACHINE DESIGN
Timothy H. Wentzell, P.E.

VP, Tech/Trades SBU:
Alar Elken

Marketing Director:
Cyndi Eichelman

Production Manager:
Andrew Crouth

Editorial Director:
Sandy Clark

Channel Manager:
Fair Huntoon

Production Editor:
Stacy Masucci

Acquisitions Editor:
Alison Weintraub

Production Director:
Mary Ellen Black

Editorial Assistant:
Jennifer Luck

Development:
Dawn Daugherty

COPYRIGHT 2004 by Delmar Learning, a division of Thomson Learning, Inc. Thomson Learning' is a trademark used herein under license.

Printed in the United States of America
2 3 4 5 XX 07 06 05 04 03

For more information contact
Delmar Learning
Executive Woods
5 Maxwell Drive, PO Box 8007,
Clifton Park, NY 12065-8007
Or find us on the World Wide Web at
http://www.delmar.com

ALL RIGHTS RESERVED. No part of this work covered by the copyright hereon may be reproduced in any form or by any means—graphic, electronic, or mechanical, including photocopying, recording, taping, Web distribution, or information storage and retrieval systems—without the written permission of the publisher.
For permission to use material from the text or product, contact us by
Tel.   (800) 730-2214
Fax   (800) 730-2215
www.thomsonrights.com

Library of Congress Cataloging-in-Publication Data

Wentzell, Timothy H.
  Machine design/Timothy H. Wentzell.
    p. cm
  ISBN 1-4018-0517-5
  1. Machine design. I. Title.

TJ233. W45 2003
621.8'15--dc21   2003055993

NOTICE TO THE READER

Publisher does not warrant or guarantee any of the products described herein or perform any independent analysis in connection with any of the product information contained herein. Publisher does not assume, and expressly disclaims, any obligation to obtain and include information other than that provided to it by the manufacturer.

The reader is expressly warned to consider and adopt all safety precautions that might be indicated by the activities herein and to avoid all potential hazards. By following the instructions contained herein, the reader willingly assumes all risks in connection with such instructions.

The publisher makes no representation or warranties of any kind, including but not limited to, the warranties of fitness for particular purpose or merchantability, nor are any such representations implied with respect to the material set forth herein, and the publisher takes no responsibility with respect to such material. The publisher shall not be liable for any special, consequential, or exemplary damages resulting, in whole or part, from the readers' use of, or reliance upon, this material.

Contents

Preface .... *xv*
Symbols and Abbreviations .... *xix*

CHAPTER 1 What Is Mechanical Design?

   1.1 Objectives 2
   1.2 The Mechanical Design Process 3
   1.3 Factors of Safety 7
   1.4 Properties of Materials 8
   1.5 The Design Calculation 12
   1.6 Preferred Sizes 15
   1.7 Unit Systems 16
   1.8 Codes and Standards 17
   1.9 Summary 18
   1.10 Problems 18

CHAPTER 2 Force, Work, and Power

   2.1 Objectives 22
   2.2 Weight, Force, and Mass Distinctions 22
   2.3 Work and Power 23
   2.4 Torque 25

2.5 Power and Rotational Speed 26
2.6 Pressure, Force, and Area 29
2.7 Moments of Inertia and Section Modulus 31
2.8 Summary 32
2.9 Problems 32
2.10 Cumulative Problems 34

CHAPTER 3 Stress and Deformation

3.1 Objectives 38
3.2 Categorization of Stress Types 39
3.3 Tensile and Compressive Axial Loads 39
3.4 Stresses and Deflection Due to Bending 42
3.5 Shear Stresses 44
3.6 Torsional Shear Stresses 46
3.7 Columns 48
3.8 Design Factors and Factors of Safety 56
3.9 Summary 57
3.10 Problems 58
3.11 Cumulative Problems 62

CHAPTER 4 Combined Stress and Failure Theories

4.1 Objectives 64
4.2 Grouping of Similar and Different Types of Stresses 65
4.3 Combined Axial and Bending Stresses 65
4.4 Coplanar Shear Stresses in More Than One Direction ...67
4.5 Combined Shear and Torsional Stress 68
4.6 Mohr's Circle 70

4.7 Combined Normal and Shear Stresses 70
4.8 Combined Maximum Shear Stress 74
4.9 Distortion Energy Theory 74
4.10 Summary 75
4.11 Problems 76
4.12 Cumulative Problems 80

CHAPTER 5 Repeated Loading

5.1 Objectives 82
5.2 Mechanisms of Fatigue 83
5.3 Endurance Limit and Endurance Strength 85
5.4 Modifying Factors for Endurance Limits 86
5.5 Variation of Stresses 89
5.6 Theories of Failure: Soderberg Equation 90
5.7 Stress Concentration Factors 99
5.8 Limited Life 101
5.9 Surface and Other Effects 103
5.10 Mitigating Stress Concentrations 103
5.11 Summary 104
5.12 Problems 105

CHAPTER 6 Fasteners and Fastening Methods

6.1 Objectives 112
6.2 Types of Threads and Terms 113
6.3 Materials and Designations 115
6.4 Tightening Methods and Initial Tension 119
6.5 Elastic Analysis of Bolted Connections 122

6.6	Gasketed Connections	126
6.7	Other Fastening Methods	127
6.8	Strength of Welded Connections	128
6.9	Summary	128
6.10	Problems	128
6.11	Cumulative Problems	132

CHAPTER 7 Impact and Energy Analysis

7.1	Objectives	134
7.2	Impact Energy	134
7.3	Velocity and Impact	138
7.4	Impact on Beams	141
7.5	Designing for Impact	142
7.6	Summary	145
7.7	Problems	145
7.8	Cumulative Problems	148

CHAPTER 8 Spring Design

8.1	Objectives	150
8.2	Types of Springs	150
8.3	Helical Compression Spring Design	154
8.4	Spring Rate	155
8.5	Spring Index	156
8.6	Number of Coils	156
8.7	Pitch	156
8.8	Types of Materials	157

Contents

8.9	Spring Stresses	158
8.10	Solid Heights and Solid Stresses	161
8.11	Deflection of Springs and Spring Scale	161
8.12	Spring Buckling	164
8.13	Design Process	165
8.14	Flat Springs	165
8.15	Energy Absorption	166
8.16	Summary	167
8.17	Problems	168
8.18	Cumulative Problems	170

CHAPTER 9 Electric Motors

9.1	Objectives	173
9.2	AC and DC Power	174
9.3	Motor Mounting and Frame Types	181
9.4	Motor Enclosures	184
9.5	Motor Control	188
9.6	Other Types of Motors	188
9.7	Summary	189
9.8	Problems	190

CHAPTER 10 Pneumatic and Hydraulic Drives

10.1	Objectives	193
10.2	Principles of Operation	193
10.3	Types of Hydraulic and Pneumatic Motion Systems	197

10.4 Miscellaneous Pneumatic and Hydraulic Motion Systems 202
10.5 Summary 204
10.6 Problems 204
10.7 Cumulative Problems 206

CHAPTER 11 Gear Design

11.1 Objectives 208
11.2 Types of Gears 209
11.3 Categories of Gears 215
11.4 Velocity Ratios and Gear Trains 216
11.5 Spur Gear Styles 220
11.6 Spur Gear Geometries 223
11.7 Helical Gear Geometries 232
11.8 Bevel Gear Geometries 233
11.9 Worm Gear Geometries 237
11.10 Gear Train Configurations 239
11.11 Summary 245
11.12 Problems 245
11.13 Cumulative Problems 249

CHAPTER 12 Spur Gear Design and Selection

12.1 Objectives 252
12.2 Forces on Spur Gear Teeth 253
12.3 Strength of Gear Teeth 257
12.4 Classes of Gears and Manufacturing Methods 260

12.5	Force Transmitted and Dynamic Loads	260
12.6	Design Methods	261
12.7	Buckingham's Method of Gear Design and Expected Error	268
12.8	Wear of Gears	272
12.9	Summary	273
12.10	Problems	274
12.11	Cumulative Problems	276

CHAPTER 13 Helical, Bevel, and Worm Gears

13.1	Objectives	278
13.2	Helical Gears	279
13.3	Helical Gear Stresses	282
13.4	Bevel Gear Forces	285
13.5	Worm Gear Forces and Geometries	287
13.6	Worm Gear Thermal and Mechanical Ratings	295
13.7	Summary	296
13.8	Problems	298
13.9	Cumulative Problems	302

CHAPTER 14 Belt and Chain Drives

14.1	Objectives	304
14.2	Belt Drives	305
14.3	V-Belts	311
14.4	Flat-Belt Systems	315
14.5	Timing Belts	316

14.6 Chain Drives 316
14.7 Cable Chains and Other Specialty Chain Drives 323
14.8 Summary 323
14.9 Problems 323
14.10 Cumulative Problems 326

CHAPTER 15 Keys and Couplings

15.1 Objectives 328
15.2 Types of Keys 329
15.3 Design of Keys 331
15.4 Splines 334
15.5 Couplings 337
15.6 Universal Joints 339
15.7 Other Shaft Attachment Methods 341
15.8 Summary 343
15.9 Problems 344
15.10 Cumulative Problems 345

CHAPTER 16 Clutches and Brakes

16.1 Objectives 349
16.2 Types of Clutches and Brakes 349
16.3 Friction Materials and Coefficients of Friction 354
16.4 Torques and Forces on Clutches and Brakes 357
16.5 Rotational Inertia and Brake Power 363
16.6 Design of Brake and Clutch Systems 366
16.7 Summary 367

16.8	Problems	.367
16.9	Cumulative Problems	.369

CHAPTER 17 Shaft Design

17.1	Objectives	.371
17.2	Sources of Loads on Shafts	.373
17.3	Design Stresses in Shafts	.379
17.4	Combined Stresses in Shafts	.382
17.5	Comparison of Stresses to Allowable Values and Endurance Limits	.384
17.6	Standard Shaft Sizes	.386
17.7	Critical Speed	.386
17.8	Summary	.389
17.9	Problems	.389
17.10	Cumulative Problem	.391

CHAPTER 18 Power Screws and Ball Screws

18.1	Objectives	.394
18.2	Types of Power Screws	.394
18.3	Torque, Power, and Efficiency in Power Screws	.397
18.4	Torque and Ball Screws	.403
18.5	Summary	.405
18.6	Problems	.406
18.7	Design Problem	.407
18.8	Cumulative Problem	.408

CHAPTER 19 Plain Surface Bearings

- 19.1 Objectives 411
- 19.2 Types of Plain Surface Bearings 412
- 19.3 Selection Criteria and Use of the Pressure-Velocity Factor 415
- 19.4 Shaft Considerations 416
- 19.5 Shaft Clearances 418
- 19.6 Wear 418
- 19.7 Bearing Materials 421
- 19.8 Hydrostatic Bearings 423
- 19.9 Thrust Bearings 423
- 19.10 Summary 423
- 19.11 Problems 424
- 19.12 Cumulative Problems 424

CHAPTER 20 Ball and Roller Bearings

- 20.1 Objectives 427
- 20.2 Life Expectancy of Ball Bearings 428
- 20.3 Types of Rolling Contact Bearings 437
- 20.4 Mounted Bearings 440
- 20.5 Lubrication and Bearing Sealing Methods 444
- 20.6 Summary 445
- 20.7 Problems 445
- 20.8 Cumulative Problems 446

CHAPTER 21 The Design Process and Design Projects

- 21.1 Objectives 449
- 21.2 Review of the Machine Design Process 449
- 21.3 Stimulating the Creative Process 452
- 21.4 Patents, Copyrights, and Protection of the Creative Process 458
- 21.5 Summary 462
- 21.6 Problems (Design Projects) 462

APPENDIXES ILLUSTRATIONS AND TABLES

- 1 Categorization of Stress Types (Basic Stress Theory) 466
- 2 Beam Moment and Deflections 467
- 3 Properties of Common Shapes 469
- 4 Properties of Common Steels 470
- 5.1 Properties of American Standard Steel Channels: C-Shapes 473
- 5.2 Properties of Steel Angles, Equal Legs, and Unequal Legs: L-Shapes 474
- 5.3 Properties of Steel Wide-Flange Shapes: W-Shapes 475
- 5.4 Properties of American Standard Steel Beams: S-Shapes 476
- 5.5 Properties of Steel Structural Tubing: Square and Rectangular 477
- 5.6 Properties of American National Standard Schedule 40 Welded and Seamless Wrought Steel Pipe 478

5.7	Properties of Aluminum Association Standard Channels	479
5.8	Properties of Aluminum Association Standard I-Beams	480
6	Stress Concentration Factors	481
7	Typical Properties of Some Non-Ferrous Metals	485
8	Typical Properties of Some Stainless Steels	486
9	Properties of Cast Iron	487
10	Wire Gages and Diameters for Springs	488
11	Mechanical Properties of Wire for Coil Springs	489
12	Stock Compression Springs: Music Wire and Stainless Steel	490
13	Values of Limiting Wear-Load Factor K_g	493
14	Conversion Factors (as a Product of Unity)	494

Selected Formulas 497
Answers to Selected Problems 503
Glossary 509
Bibliography 519
Index 521

Preface

The field of machine design can be thought of as consisting of both art and science. The creative part of the field can be among the most interesting and rewarding aspects of mechanical design. However, without an understanding of how machines work, how to determine appropriate sizes for parts so that the machine can function properly, and last through the desired lifetime, the creative part, in and of itself, cannot result in an acceptable design.

Having worked for many years developing, testing, and building remote robotic systems for use in the nuclear field, I developed an appreciation for both the art and science of design. I later developed numerous industrial and commercial products, and I hold numerous patents on many of these designs. This experience created and nurtured my interest in the field of machine design, both in the analytical and the creative aspects of the design process.

In a course of this type, students will probably start with a basic understanding of how many machines work, whether it is an automobile or other consumer products. What students probably lack is knowledge and experience in the analytical aspects of these designs. This includes calculating appropriate factors for loading on a design, appropriate stresses in order to result in a safe product, and such aspects as factors of safety and how they vary in different industries or even between products within an industry. For example, when we compare an economy car to a luxury car, the same manufacturer may use different factors of safety in different parts of each design. It is important to understand that the field of design, which includes machine design, involves the design of nearly everything from the very simple to the very complex. This includes many items that might not properly be thought of as machines.

As we think about the whole process of designing a machine, it can be helpful to imagine an ideal machine. For example, if our task were to design an automobile, we would first need to decide what we want from that automobile. Let us think for a moment about the perfect automobile. It would need to be fun to drive. Also, when seen driving it, our ideal automobile would make us appear to be more attractive to the opposite sex. It would likely go forever on the initial tank of fuel, require no maintenance during its lifetime, and would turn to dust or, preferably, recyclable dust once it had reached the end of its desired life so that it could be reincarnated into the next generation of automobile. Obviously no such automobile exists, but certainly this image would be a worthy goal for our design process.

In the design of a complex machine like an automobile, there are many factors we need to consider. Consider something that might appear to be simple: the feeling of stiffness of an automobile to the driver and passengers. In order to obtain an appropriate stiffness we need to factor in a whole assortment of components, including the tires, the suspension system, the stiffness of the frame, and the stiffness of the seats, as well as how the automobile feels when empty and also when fully loaded. These all can have an effect on something that seems as straightforward as how comfortable a ride we receive from this assembly of many machine parts. Also factoring into the stiffness of this car is the safety component. We need to insure that there is a rigid enough cage around the occupants to minimize injury in an accident. We need appropriate methods of cushioning the interior of the automobile, which includes the use of bumpers and the crushing of either the engine or trunk compartment in a crash. All of these are factors that would go into the design of the stiffness of an automobile. Therefore, as we think about complex designs in this course, it is important to keep in mind that we are first going to look at many small pieces of the design process. Eventually, as students progress in the field of mechanical design, they will feel comfortable adding all these multiple pieces together.

As students progress through this book, we will be trying to add to their understanding of this evolving field of knowledge while trying not to lose sight of the artistic aspects of machine design, since these aspects are what drew many of us to the field. This is why we keep a pad of paper and a pencil beside our beds at night to record our ideas and thoughts when we dream about the solution to the problem at hand. Hopefully, the students' enjoyment of this field of knowledge will grow as their study of this field progresses.

This textbook will try to incorporate the technical aspects of analyzing the strengths of many machine products such as gears, springs, shafts, belts, pulleys, and fasteners, and it will show how to design for long life. Students will also learn how to select standard products, where suitable, for the design. This textbook will encourage them to pursue actual designs through the use of homework problems and through lab exercises if a lab is incorporated with this course.

In addition to the analytical aspects of machine design, a chapter is included on stimulating creativity in the design process and how patents and copyrights are issued. The instructor may want to move this material, which is covered in Chapter 21, to earlier in the course, especially if design projects are part of this course.

Students should have had courses in structural mechanics (statics), strength of materials, college algebra, and trigonometry. It would be helpful, but not required, to have studied material science and manufacturing processes.

This text is intended to be used at the associate or bachelor's degree level in an engineering technology program. It attempts to relate the technical aspects of machine design to the more practical aspects inherent in a course such as this. The idea for writing this text came from the fact that, as an instructor of this course, it always appeared that the texts available were either at far too low a level or included complexities that were either above the level of the student or simply not applicable

to the practitioner in the field. It is hoped that this book will be a reference material for engineers or engineering technicians doing machine design, as well as for a course of this type. Included in the instructor's package for this text is a solution manual showing the solutions for the problems at the end of each chapter, as well as sample test problems and solutions. It also includes a PowerPoint presentation that uses the sketches and drawings from the book where appropriate and many of the example and homework problem solutions in order to expedite the process of explaining these problems.

The test bank and solutions are included in the Instructors Guide and also on the Power Point Disc. These can be used as supplemental problems, as quizzes, or as exams.

Chapters 1 and 3 are reviews of the material that were likely covered in a strength of materials course. They may be omitted or used in an expedited manner, depending on the student group. If it has been many years since the student took a strength of materials course or the student believes a review is necessary, these chapters can be used independent of the instructor. The material covered in Chapter 2 likely is a review of material covered in a Newtonian physics course and, likewise, could be skimmed or used for independent review, depending on the desire of the instructor. Chapter 4 on combined stresses may also be a review of the material covered in a strength of materials course. This chapter offers a straightforward, simplified approach that is used throughout this text. It probably will need to be reviewed even if students have a strong understanding of strength of materials. It may be studied in an accelerated manner, however.

To help the student to develop an understanding of loads and how they are developed in the design process, the chapter on force, work, and power (Chapter 2) has been placed at the beginning of this book and the chapter on motors (Chapter 9) at the middle. Many of the homework problems require the student to decide on the appropriate design loads and the appropriate factors of safety as these are also important aspects in undertaking an actual design.

The homework problems at the end of each chapter are directly related to the material covered in the chapter, and there are fewer problems offered than in many texts. If the instructor feels that the material in the chapter should be covered in significant depth, all of the problems could be assigned. However, if the instructor wants to cover any particular chapter as an overview, only selected problems could be chosen. The answers to some of the problems are located in the back of the text. Also located in the back of the text (Appendix 14) are conversion factors, using the principles of the property of unity and delineating the conversion factors as fractions, as ultimately this is how they are used in equations. I believe this is less confusing than the format used in most conventional conversion factor tables.

There is also a list of common formulas used in the text that may be useful. If the instructor decides to give a closed-book test, this list could be used as a handout for those tests or it may simply be handy for the students to use in later chapters when they need to refer to chapters from earlier in the text.

The end-of-chapter problems often end with additional cumulative problem(s) that build upon each other in order to help students understand that many projects are far more complex than simple homework problems and that many different aspects of design need to be undertaken in order to complete the overall design of complex machines.

Lastly, as this is a new text, although it has been carefully checked, there likely are mistakes, and I would appreciate any of them being forwarded to me for correction in potential later editions. Also, I want to thank my students in my Machine Design class from the Spring 2002 semester who used this book in its draft form for finding mistakes and offering suggestions on the content and layout of this book.

DEDICATION

This book is dedicated to my deceased father, J. Harwood Wentzell, who instilled in me the belief that a task could and should be undertaken, even if how to do the task was not understood. He knew that, at least when finished, one would have gained that knowledge. This has been an inspiration and a driving force over the course of my life.

Also, to my daughter, Kate Wentzell, who is a keen observer of life and my spirited debate partner.

Furthermore, I would like to dedicate this book to the many students who may use it who have spent their lives, either knowingly or unknowingly, with ADD or ADHD. I would like to remind them that, unlike common wisdom, this can also be a gift.

ACKNOWLEDGMENTS

I would like to thank Pat Wentzell and Susan Talbot for their assistance and patience in the hard work of transcribing, clarifying, and editing my often-garbled dictation in order to turn this book into, hopefully, a clear and understandable delineation of the engineering design process. My thanks also to my friend Dianna Roberge, whose constant encouragement made this task seem easier.

The author and Delmar Learning gratefully acknowledge the contributions of the following reviewers, whose valuable comments helped to shape this textbook: Jim Adkins, Danville Community College, Danville, VA; Bonnie Mills, Augusta Technical College, Augusta, GA; Dr. John Steele, Colorado School of Mines, Golden, CO; Dr. Wendy Reffeor, Grand Valley State University, Grand Rapids, MI; and Ken Means, West Virginia University, Morgantown, WV.

Timothy H. Wentzell, P.E.
twentzell@trcc.commnet.edu
860-885-2347

Symbols and Abbreviations

b	width; width of gear tooth
C	centerline distance; circumference; torque coefficient
c	distance from neutral axis to fiber where stress is desired, usually to extreme fiber
D, d	diameter; distance
	D_e equivalent diameter (bolting)
	D_i inside diameter
	D_m mean coil diameter (springs)
	D_o outside diameter
	D_p pitch diameter (belting, bolting, chain, gearing)
	D_w wire diameter (springs)
E	modulus of elasticity in tension (Young's modulus)
e	eccentricity of load; effective error in gear-tooth profiles; efficiency
F	force; load
f	coefficient of friction
G	modulus of elasticity in shear or torsion
g	local acceleration due to gravity
	g_o standard acceleration of gravity (32.2 f/s² or 9.81 m/s²)
hp	horsepower
I	moment of inertia
J	joule; polar moment of inertia (italic)
K	Wahl factor for design
K_f	fatigue strength reduction factor

K_t	theoretical stress concentration factor
k	spring rate; the load-per-unit deflection; stiffness
M	moment
m	mass; meters; metric module (gears)
m_w	velocity ratio
N	newton; design factor or factor of safety (italic); load normal to a surface (italic); number of teeth (gears) (italic)
n	angular velocity; revolutions per minute
P	pressure; load or force; power; pitch of springs, gear teeth, threads
$\quad P_c$	circular pitch; critical load (columns)
$\quad P_d$	diametral pitch
r	radius
S	stress
$\quad S_a$	alternating stress
$\quad S_m$	mean stress
$\quad S_n$	endurance strength
$\quad S_p$	proof strength
$\quad S_s$	shear stress
$\quad S_u$	ultimate stress
$\quad S_{us}$	ultimate stress in shear
$\quad S_y$	yield stress
$\quad S_{ys}$	yield stress in shear
$\quad \sigma$	(sigma) resistant stress in the part (normal stress); combined normal stress
$\quad \tau$	(tau) resultant shear stress; combined shear stress
SI	International System (metric system)
T	torque; temperature
t	time; thickness
U	energy

V, v	velocity; volume
V_r	velocity ratio; gear train ratio
W	total weight or load; force; work
Y	Lewis's factor in gearing
Z	section modulus, I/c
Z'	polar section modulus based on polar moment of inertia, J/c
α	(alpha) coefficient of thermal expansion; an angle; angular acceleration
δ	(delta) total elongation; total deflection of a beam; spring deflection
Δ	(capital delta) any change
ε	(epsilon) strain; unit deflection
γ	(gamma) cone angle bevel gears
λ	(lambda) lead angle of worm or screw threads; pitch angle for springs
μ	(mu) coefficient of friction
ω	(omega) rotational speed
ψ	(psi) an angle
φ	(phi) an angle
π	(pi) 3.1416 . . .
ρ	(rho) density
θ	(theta) an angle
in. or ″	inch
ft or ′	foot
lb	pound force

About the Author

Timothy H. Wentzell is Professor of Mechanical Engineering Technology at Three Rivers College, where he has taught machine design for over twenty years. He has served as Program Coordinator for the Mechanical Engineering department and, for many years, as the Department Chairman for the Engineering Technology department. He has written many articles and technical papers on a variety of topics. Before becoming a faculty member, Professor Wentzell designed industrial, commercial, and consumer products, ranging from developing remote robotic inspection devices in the nuclear industry to automated products in the garment field. Since then as a consultant he has designed many commercial and consumer products.

Mr. Wentzell has received over fifty U.S. and foreign patents on varied and complex devices such as remote mobile robotic systems and automated car washing systems. He is a licensed professional engineer in Connecticut and Vermont. A graduate of the University of Vermont and Rensselaer Polytechnic Institute, he is a member of the American Society of Mechanical Engineers and the American Society of Engineering Education.

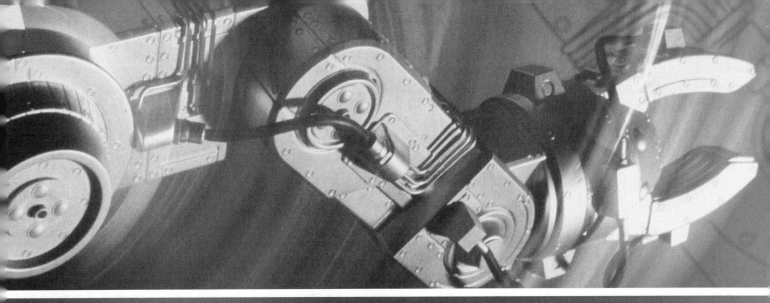

CHAPTER 1

What Is Mechanical Design?

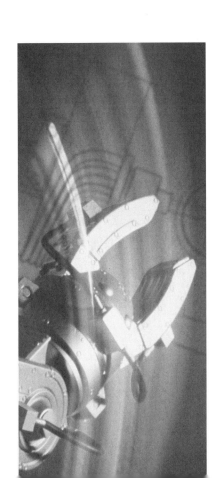

The field of mechanical design is very broad and includes the field of machine design, the topic of this text. Machine design can include the design of anything from a product as simple as a manual can opener to something as complex as an automobile or jet engine. Many of the products that we see around us every day are probably designed by specialists in machine design. Mechanical design includes not only machine design but also process or system design. In a power plant, for instance, systems must be designed for fuel handling; coal pulverizing, if coal is the fuel source; burner systems to regulate both the feeding of the fuel and the air for combustion; and steam turbines and generators or gas-scrubber systems for cleaning the effluent from combustion by-products. Another example could be in a refinery where pumps, valves, and heating systems must be included in the design in order to effectively break the crude oil into different products.

It may be useful as we think of the field of machine design to envision the perfect automobile that was discussed in the preface. There we determined that the requirements for this product included: it had to be extremely fun to drive; it had to make us feel more attractive to members of the opposite sex; it had to run for a lifetime on one tank of fuel; it would never need maintenance during this period; and when we had decided an appropriate life was fulfilled, it would turn to dust or, more specifically, recyclable dust so that we did not have problems with disposal and had the raw materials to create the next generation of automobile.

Obviously, we don't presently have such a product available to us. But if we understand the goal of the design process and consider what we can actually attain, we can get a feel for how demanding and exciting this field can be. Many players are involved in the field of machine design because creating a complex product often requires a team of people. This design team could include a mechanical engineer; possibly an electrical engineer if electrical drives are used; a computer engineer to design the control systems; industrial designers who may focus on the cosmetic features of the product, such as shape or dashboard layout in the case of a car; and many other specialists such as chemists, physicists, or other experts. Generally, when we talk about the field of machine design, we assume it does not include **basic research,** but often it may involve a fair amount of investigation into the problem and into possible solutions, and it often includes extensive testing to prove feasibility and, later, durability.

basic research
The scientific investigation of basic principles that typically does not include the application of those principles.

1.1 OBJECTIVES

This chapter will discuss many of the steps involved in the design process. In concert with the material in Chapter 21 that discusses some of the

creative and artistic parts of the machine design process, this chapter will help you to understand what is involved in actually creating a product for use by consumers, industry, or government. The second part of the chapter will review the properties of materials, which most students will have touched on previously in a materials science course, and will include a brief discussion of the appropriate format for calculations, sizes, and unit systems.

After completing this chapter, you should be able to:

- Recognize examples of mechanical design to which the principles of this book apply.
- Understand the steps involved in the mechanical design process and the importance of creating functional requirements.
- Describe the importance of factors of safety in the design and how they vary between products.
- Distinguish between yield strength, ultimate strength, and the relationship to different materials.
- Understand how the stiffness of different materials relates to their modulus of elasticity and ductility.
- Calculate the change in size of materials due to changing temperature conditions.
- Prepare an engineering calculation.

1.2 THE MECHANICAL DESIGN PROCESS

To better understand the design process, you should learn some of the basic steps that are usually involved. The following sections describe some of the common steps in an effective design and implementation process.

Problem Definition

In order to effectively approach a problem to be solved or to create a new design, we must first define the task at hand. Often we begin by trying to define the problem to be solved. In listing the requirements for the proposed design, we try to think of it in terms of **attributes,** not solutions. There are cases where the design needs to solve a specific problem or the company where we work wants to create a new product for sale, such as a new model of automobile or a new aircraft size. The task may be to broaden the product line or replace a dated product in order to enhance the position of our company in the marketplace. Often, however, we have a specific problem that needs to be solved. In order to come up with an effective solution, we must first understand the problem, the known constraints, and even the potential market for this product if it has not been predetermined.

> **attributes**
> The basic properties of a solution that would not include specifics about solutions.

problem statement
A description of the problem to be solved, often detailing only attributes of the specific problem.

This is why the **problem statement** is often developed as the first step in the design process. Although we must try to correctly define this problem, we must also be very careful not to try to solve the problem at this stage. We merely try to define it by listing the constraints and requirements for the system. By trying to list the attributes of the problem and not the potential solutions themselves, we are far more likely to come up with a creative approach to the problem and avoid getting stuck with solutions that eliminate the possibility of unique and exciting new design approaches. Finally, some level of basic research may be needed into either the problem or the technology available. In many cases, as a designer, you may be asked to solve a problem that may not be clearly defined.

Example Problem 1.1: Functional Requirements for a Grass-Cutting System

The goal of this design is to create a system for shortening the length of grass on a typical residential lawn to an appropriate height. The **functional requirements** are summarized as follows:

functional requirements
A listing of the requirements that the product needs to accomplish, not how it would be accomplished.

1. The system should result in a uniform grass length between two and four inches in height.
2. The system should be relatively easy to operate.
3. The system should be relatively safe to the operator and any animals or people in the vicinity.
4. The system should be fairly flexible for use on somewhat uneven ground and around obstacles.
5. The system should not damage the grass extensively such that immediate regrowth would not occur.
6. The system should be significantly different in principle from other products on the market in order to create a new market niche for an existing lawn-care equipment company.

Note that these requirements only describe the results of the process and some basic functional requirements with regard to performance, leaving the designer to conceptualize all different ways to cut the grass. This process obviously could include lasers, ultrasound, mechanical means, string inertia methods, or a whole array of different potential methods to cut grass.

Concept

brainstorming
Free-flow thought for the creation of many ideas in order to solve a problem.

Creating conceptual designs typically includes a fair amount of what is called **brainstorming.** This is a process in which designers toss around possible ideas, whether they are practical, crazy, or, in some cases, simply silly,

as a way of broadening the scope of the problem and opening up the possibility of creative problem-solving methods. Typically, we list the possible solutions as they are suggested. This may include making basic sketches. It is important to note that, at this point in the process, a good designer tries to be as noncritical of the ideas as possible. Later in the process we will try to narrow down the ideas and focus on one or more concepts that should be pursued. As a way of thinking about how we need to keep an open mind in the brainstorming process, remember that someone at some point in time first had the idea to eat a lobster. Stop and think about this for a minute! This example is offered to help you to understand that ideas do not always need to be practical, logical, or apparently good solutions.

Once we have done some of this brainstorming, either individually or often in groups frequently called **creativity sessions,** we then try to define the solution in a more manageable and applicable way. At this point, we start evaluating the ideas, picking the best parts of different ideas, if appropriate, and putting together one or more concepts to pursue. We may also create some sketches or some more detailed descriptions of the problem and possibly even start a review of whether the idea may be **patentable** if we are hoping to protect it from competitors. At this point, even sleeping on the problem may be useful in further refining the ideas inherent in our new design.

creativity sessions
A more formal process of brainstorming that involves a group of people brainstorming about an idea or problem in order to come up with possible remedies or solutions.

patentable
A process in which the right to use an idea is protected by law, discouraging others from using the idea.

Design

At this point, the sketches and ideas are likely going to become drawings with dimensions. An analysis will probably be done to determine the appropriate sizes for the applied loads. Forces acting on our design, including the potential weight of the design, may require a **deflection analysis** along with a calculation of stresses and other necessary analyses in order to help make the preliminary design feasible. Depending on the application, at this point it may be necessary to incorporate some **corporate knowledge** that has been developed over the years for similar products in this field. Aspects such as weight may be more critical, for example, in an airplane than in construction equipment. We should understand at this point that this analysis is not final and that this is going to be an ongoing process as our design is created, tested, and likely redesigned.

Much of this text is going to focus on the analysis part of the design. The ability to analyze our design effectively will make us much more competent designers.

deflection analysis
An analysis of the movement of a part or assembly under load.

corporate knowledge
Specialized technical knowledge that a company may have obtained over time that is applicable to their products.

Prototype

The next step in the design process often includes the creation of **prototypes.** These prototypes may be scale models, in the case of some large

prototype
A partial or full-scale model of a product that is usually used for testing.

3-D modeling
A computer generated image in 3 dimensions showing what a product or machine might look like.

feasibility testing
The testing of a prototype or product to determine if the principle of operation is feasible.

conceptual testing
Similar to feasibility testing except that it may involve only a specific concept; can be very limited in scope.

functionality testing
Testing to determine whether a prototype or model actually performs in the manner for which it was conceived.

durability testing
Testing to determine whether an appropriate lifetime for a machine or part of a machine can be achieved.

consumer testing
Testing of the product by potential customers; typically includes compiling their opinions.

products, or models made of other more easily formed materials. The automobile industry, for example, has for years used clay models for developing the visual appearance of new models of cars. We may, with **3-D modeling** and CAD programs, be able to duplicate some of these kinds of activities without creating actual models, but typically some form of prototype is almost always created. Sometimes these prototypes may include only certain features of our design. They may be created so that we can perform some feasibility testing on our concept if it is a relatively new and unique idea. Often in the development of prototypes it is not practical to create these prototypes out of the same materials that will be used for the final product. For instance, machined steel may be used instead of a casting, cast plastics as opposed to injection-molded pieces, or other methods that lend themselves to lower-cost prototyping. The development of prototypes in many industries is even a specialized field for technicians because prototype cost can be significant in the overall development process.

Redesign

The chances are that, as we go through the design process and create prototypes, we discover certain features that need to be changed or improved. We may now need to go back, after creating our prototype and possibly doing some **feasibility testing,** and make modifications. We may have discovered, during this process that some features can be simplified while others are no longer feasible. At this point, we can make these modifications before we go into the next phase, which is typically testing.

Testing

Testing can involve any of the following: **conceptual testing,** where we try to test whether the concept appears to be feasible; **functionality testing,** where we try to see if our prototype can function in the manner we assume; **durability testing,** where we try to determine whether or not our product is going to last what we believe is an appropriate lifetime; and **consumer testing,** which could be as simple as asking the public about the concept and whether they like the idea. For instance, auto manufacturers often include the public in evaluating new designs because the cost of introducing a new automobile can be significant. If consumers do not find the car to be pleasing, it can be both financially costly and damaging to the company's image. Most of us can think of a few automobiles that have been introduced that the public did not like. The Ford Edsel and the American Motors Pacer are examples of cars that did not receive broad acceptance in the marketplace.

Design for Manufacture

An important part of the process of creating a new design is making it cost-effective to manufacture. Certainly, for products that are made in large volume, the cost of manufacturing can be critical to the product's success. The public is not likely not to buy a $100 can opener or a $50,000 economy car, so the process of making the design cost-effective can be critical. Often at this point, or preferably throughout the whole design process, we seek comments and ideas from engineers who specialize in the process of manufacturing what we as mechanical designers have created. By working with the manufacturing team, we can often simplify aspects of our design to make it easier to manufacture or consider substituting alternate materials that may be more durable and less expensive to create. The modern automobile, for example, now incorporates a great deal more types of materials than it did twenty or thirty years ago when cars were built primarily of steel, glass, and rubber. A modern car may use many different materials such as magnesium, zinc, multiple types of plastic, and a great deal more galvanized steel or aluminum, along with the typical glass and rubber materials of yesteryear. An automobile is an excellent example of where minor cost savings can be significant because cars are typically created in very large volumes. The competition between manufacturers is intense. Furthermore, an automobile is an excellent example of the importance of weight savings. When we save a few pounds on a car, other components can often become smaller as well. Hence, the whole process of economic manufacturing can be critical.

1.3 FACTORS OF SAFETY

In any design we almost always include some margin in the relationship between the projected actual stresses on the product and the stress capability (stress allowable) of the materials from which the product is made. This is typically done for at least two major reasons. The first and most obvious reason is that the allowable stress for a material is given as a range of values. This can be caused by variations in the manufacturing process—such as in heat treating, actual chemistry, or the presence of unknown defects—or by other variations in the material. The other reason we often include a factor has to do with the level and sophistication of our design analysis. We often use the term **back-of-envelope calculations** to describe cases where our analysis may not include all necessary factors. For instance, our design may have some minor impact loads that we have not calculated, some sharp corners or other **stress-concentration factors** that can affect the

back-of-envelope calculation
A common term for an abbreviated or simplified engineering analysis.

stress concentration
A location or point on a part at which stresses may be higher than average due to changes in shape or other factors.

factor of safety
A margin built into a design to allow for unusual situations or factors not considered in the analysis.

expected lifetime of our product, or just some unknowns in how the product may be used. A good example of this may be a toy truck that we assume, if used in a logical manner, a child is likely to push around, fill with materials, and so on. However, the child may also decide it is appropriate to sit on the truck. If we are building a product that we believe should resist that kind of load, we need to factor that into the design, either in the maximum load or in the **factor of safety.** We should be careful how we combine both of these types of factors in our design. The factor for variability in materials, sizes, and so on is known and expected. The factor for the limitation in our analysis should be clearly documented in the design process. By the time we finish our design, we should have done an analysis appropriate for the situation at hand and therefore will not need to include large factors for the rudimentary nature of the analysis. There are exceptions to this: you may be designing a one-time solution to a problem where a complicated analysis may cost more than simply over-designing for the task at hand. But, as good designers, we need to be aware of the distinction between a factor of safety to account for a variation in properties and conditions and factors that we are including because of the lack of sophistication in our analysis.

It is always interesting to note, and it is often surprising to many young engineers, that the factor of safety used in a typical aircraft is lower than in almost any other field. For example, the factors used in designing construction equipment may be two or three times those used in aircraft design. There are, however, some obvious reasons for this. If we use the same kind of margin in an airplane as we do in designing a bulldozer, the airplane likely would never fly. Therefore, we must do an analysis with a greater level of sophistication in our aircraft design, both incorporating lessons learned from past designs and also using some rather high-level analysis such as **finite element analysis,** or using other sophisticated methods to accurately model and evaluate both the loading and the reaction to the loading. Lastly, you will note as we go through this textbook that many of the example problems and homework problems will gradually start to use lower factors of safety as we do more complete and accurate analyses.

finite element analysis
A structural analysis that identifies loads and stresses on individual sections of a part.

1.4 PROPERTIES OF MATERIALS

For mechanical designers to create effective designs, they must understand the relationship between the stresses that are placed on a part and the ability of the materials to resist those stresses. Obviously, the stresses in the part must always be below the breaking strength of the materials from which they are made. The designer needs to understand how the strength

and other properties of the materials from which we typically build machines are determined. In general, the following relationship must always be true:

$$S_{allowable} \geq S_{actual}$$

As discussed in Section 1.3, a factor of safety would typically be incorporated into this formula. With a stress-versus-strain graph for a typical material, which is usually created by means of a simple tensile test for a standard-size sample, the deflection is measured as the sample is stretched. Then the associated force is used to calculate the stress in the part as follows (see Figure 1.1):

$$S = \frac{F}{A}$$

It is important to recognize how different material properties are determined. For instance, for many materials, a series of tensile test specimens ½ inch in diameter are stretched in a tensile test machine and the elongation-versus-load is measured up to the point of failure. The mean (average) of this test data is often the value used in material property tables, as there is a range of values of tensile test strength. This is another reason why safety factors are used in design: the specific material from which your part is made may have slightly lower or higher value than the tabular value.

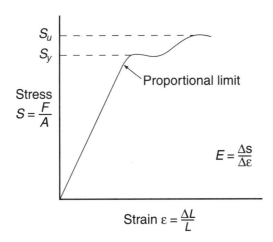

FIGURE 1.1 Typical stress-strain diagram for steel.

Yield Strength (S_y)

yield strength
A stress level at which permanent deformation of the part occurs.

The **yield strength** is the value at which additional deflection occurs in the material with little or no increase in the force. It is also the point at which, when the load is removed, the material returns close to its original shape. (It is actually the point at which .2 percent permanent deformation exists after the load is removed.) However, for practical purposes we usually use the value shown for the **proportional limit** in our design. This may be more relevant because this is the point at which the relationship between stresses and deflection is linear such that we can determine deflections correctly below this value. Because of the predictability of the performance of the material below the proportional limit, we would always try to make sure our designs are below this value. Not only is the material predictable in this range, it will also return to its original shape. Thus, because typically this is slightly less than the yield strength, we gain a level of protection against failure. In practice, the yield strength is typically available but the proportional limit is usually only slightly less.

proportional limit
A point similar to yield strength but often slightly below where the stress-versus-deflection of a material is linear.

Ultimate Strength (S_u)

ultimate strength
The stress level at which failure occurs.

The **ultimate strength** is the point at which the material actually fails. There are some changes in the material before that point. Typically, the ultimate strength is really only important if we are trying to determine what happened to a failed part so we can predict what loads cause it to fail. An exception to that is made with the use of very brittle materials, such as cast iron, where the relationship between the ultimate strength and the yield strength is quite similar. Also, in rare cases, we may be unconcerned about yielding of the part.

Modulus of Elasticity (E)

modulus of elasticity
The relationship between stress or load and the deflection of a material.

The **modulus of elasticity** is the relationship between the stress and the strain in the part or the relationship between deflection and load, depending on the size and the material from which the part is constructed. We use modulus of elasticity in predicting deflection, and typically we select the value of modulus of elasticity from tables for the material. It is interesting to note that most materials of similar composition, such as steels, have a very similar modulus of elasticity, despite the fact that their yield and ultimate strength can vary significantly. From Figure 1.1, it can be seen that the slope of the curve below the proportional limit is the modulus of elasticity.

The modulus of elasticity would therefore be found from the relationship of the stress divided by strain as the strain could be considered to be the unit deflection or the deflection-per-unit length.

$$E = \frac{\text{stress}}{\text{strain}} = \frac{S}{\varepsilon}$$

Ductility

Ductility is simply the measure of how much the material would deflect or stretch before failure. It gives us insight into the ability of the material to accept certain kinds of loading. A material with a high ductility would deflect a greater amount before failure and also would tend to be able to absorb **impact loads** and repeated loadings better than a more brittle material. We measure ductility in terms of a percent elongation, which is a relationship between its original length and how much it stretches before failure. Generally, a ductile material will have a percent elongation of 5 percent or greater. If we are subjecting our part to significant impact or repeated loads, we might try to select a material with a ductility of greater than 12 percent. To compile data on the ductilities of different materials, we typically measure ductility by using a tensile testing system, and we measure impact resistance by means of a Charpy or Izor test (a test that measures the amount of energy that is absorbed under impact loading).

> **ductility**
> The level of deflection before failure; often measured by the amount of energy that can be absorbed before failure.

> **impact load**
> A sharp load that is typically created from a clearance or a gap, such as a space in a machine part or an automobile hitting a guardrail.

Thermal Expansion

The thermal expansion that occurs in a material upon heating or contraction upon cooling is measured in terms of a **coefficient of thermal expansion,** which is the relationship between the change in length and the original length:

$$\delta = \alpha L \Delta T \tag{1.1}$$

where δ = change in length
α = coefficient of thermal expansion
L = original length
ΔT = change in temperature

> **coefficient of thermal expansion**
> The ratio of the expansion of a part to the change in temperature.

Thermal expansion (1.1)

$\delta = \alpha L \Delta T$

We are concerned with the coefficient of thermal expansion when we design products that are subject to significant changes in temperature or

when we are using different materials in this environment or parts that may change in relationship to each other. A good example of this is steel sleeves in an aluminum engine block or the **bimetallic spring** that may exist in your typical home thermostat where the differences in thermal expansion cause the spring to bend with changing temperature. Thermal expansion properties may also be used to join parts together: one part is heated while the other is cooled and, when they become the same temperature, they are permanently attached.

> **bimetallic**
> Created from two or more materials with different coefficients of thermal expansion, such that a curvature or change of shape occurs with temperature changes.

Changes in Properties

Some materials tend to change over time due to a variety of properties. For instance, some aluminums can become harder and stronger over time; some plastics can become brittle due to the effect of ultraviolet radiation; some materials such as steel can have property changes such as a reduction in ductility in the presence of high levels of radiation. When we use materials in unusual environments such as these, we need to be aware of these properties and how they can change the material. For example, for steels we often measure the change in the ductility by means of a Charpy impact test and then measure how this changes in the presence of high levels of radiation.

1.5 THE DESIGN CALCULATION

A good design calculation should be straightforward, easy to follow by someone knowledgeable in the field, and appropriately referenced so it is easy for another technician or engineer with an appropriate academic background to check it. The results should be understandable to someone who may not be an expert knowledgeable in the field. The purpose of a design calculation is often both to document the basis for the design and allow checking to insure that mistakes have not been made in the analysis.

Some factors important in the design calculation are: the part being analyzed should be clearly identified either by description or by reference to a drawing or part number; appropriate sketches should be included, showing the placement of the loads and reactions; free body diagrams should be included as appropriate, and possibly shear and bending diagrams, depending on the type of analysis being undertaken. Any assumptions about the analysis should be clearly stated, especially the derivation

of the loads and any assumptions that went into determining the appropriate safety factors. As the design calculation is intended to be read by others knowledgeable in the field, appropriate standard formulas do not need references. However, if there is an unusual formula or a formula that is not standard practice in the field, a reference should be noted for its source. Allowable stress levels should reference the source unless they are based on a standard material specification; then that notation may be all that is necessary.

In addition to making the design calculation easy to follow, it is important to show the appropriate checks, if any, that were used. The **unit analysis** should be clearly detailed, either as part of the analysis or, in some cases, off to the side if a separate unit analysis was performed. Lastly, the results should be stated, especially when you may be specifying a standard-sized part as opposed to the direct results of your calculations. The following example problem shows a typical design calculation.

unit analysis
The analytical process of checking the units in a calculation. It can also serve to verify that the inputs to the formula are correct.

Example Problem 1.2: Sample Engineering Calculation

Select a round bar to be constructed from an AISI 1020 cold-drawn steel to support the 300-pound ball, as shown in Figure 1.2. Assume a safety factor (SF) of 2, and determine the deflection for the size selected.

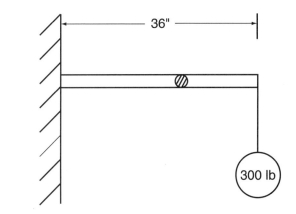

FIGURE 1.2 Cantilever arm for Example Problem 1.2.

Section modulus for a round shape:

$$Z = \frac{\pi D^3}{32} \qquad \textbf{(Appendix 3)}$$

$$M = -FL \quad \text{(Appendix 2)}$$
$$M = -300 \text{ lb } 36 \text{ in.} = -10,800 \text{ in.-lb}$$
$$S = \frac{M}{Z} = \frac{10,800 \text{ in.-lb}}{\frac{\pi D^3}{32}}$$

$$S_y = 61 \text{ ksi} \quad \text{(Appendix 4)}$$

Shear diagram

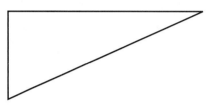
Bending diagram

FIGURE 1.3 Shear and moment diagrams.

Factor of safety of 2:

$$S_{allowable} = \frac{S_y}{N} = S_{actual} = \frac{M}{Z}$$
$$\frac{S_y}{N} = \frac{M}{Z}$$
$$\frac{61,000 \text{ lb/in.}^2}{2} = \frac{10,800 \text{ in.-lb}}{\frac{\pi D^3}{32}}$$
$$D^3 = \frac{10,800 \text{ in.-lb } (2)\,(32)}{61,000 \text{ lb/in.}^2 \, \pi}$$
$$D^3 = 3.607 \text{ in.}^3$$
$$D = 1.533 \text{ in.}$$

Use a 1⅝ in. bar:

$$D = 1.625 \text{ in.}$$

Checking deflection:

$$\delta = -\frac{FL^3}{3EI} \quad \textbf{(Appendix 2)}$$

$$I = \frac{\pi D^4}{64} \quad \textbf{(Appendix 3)}$$

$$I = \frac{\pi(1.625 \text{ in.})^4}{64} = .342 \text{ in.}^4$$

$$\delta = \frac{300 \text{ lb }(36 \text{ in.})^3}{3(30 \times 10^6 \text{ lb/in.}^2).342 \text{ in.}^4}$$

$$\delta = .455 \text{ in.}$$

1.6 PREFERRED SIZES

Depending on the type of machine being designed and its usage, we often make most parts of a machine out of standard-sized products. We do that so that standard mating products such as bearings, and gears can be utilized depending on the application. For standard products that do not need machining, it is typically far cheaper to use standard English or metric sizes where suitable than custom sizes. Certainly, in some cases, such as for aircraft parts where weight may be far more important than the savings in machining costs, parts may be made in whatever size fits our design calculations. However, in general, for most types of products we would round up to the next standard fractional size or standard material size in the case of standard structural items. For example, in Example Problem 1.2, a standard size round bar was specified based on the results of the structural analysis. For many products, such as screw products that would otherwise require special dies, standard sizes are almost always used, and it would be a rare case where a nonstandard item would be specified.

1.7 UNIT SYSTEMS

There are two standard unit systems generally in use today: the U.S. customary unit system, often called the English system, which uses inches, pounds, and seconds as its base units; and the international (SI) system, commonly referred to as the metric system, which uses meters, kilograms, and seconds as its basic units. Most of the work in this text will be in the U.S. customary unit system because most engineering in the United States is still done using this system. However, where appropriate, both will be utilized as most industries are increasingly targeting world markets. Many of the materials in this text will give examples using both types, and homework problems will use both as well. The initial presentation will typically be in the U.S. customary unit system. Table 1.1 shows typical units in both systems.

TABLE 1.1
TYPICAL UNITS

Quantity	U.S. Customary Units	SI Units
Length or distance	inch (in.)	meter (m)
	foot (ft)	millimeter (mm)
Area	square inch (in.^{2})	square meter (m^2) or square millimeter (mm^2)
Force	pound (lb)	newton (N)
Mass	pound mass (lb-sec^2/ft)	kilogram (kg)
Time	second (sec)	second (sec)
Angle	degree (°)	radian (rad) or degree (°)
Temperature	degrees Fahrenheit (°F)	degrees Celsius (°C)
Torque or moment	inch-pound (in.-lb) or foot-pound (ft-lb)	newton-meter (N-m)
Energy or work	inch-pound (in.-lb) or foot-pounds (ft-lb)	joule (J) (1 J = 1 N-m)
Power	horsepower (hp) (1 hp = 550 ft-lb/sec)	watt (W) or kilowatt (kW) (1 W = 1 J/sec = 1 N-m/sec)
Stress, pressure, modulus of elasticity	pounds per square inch (psi) (lb/in^2 or psi)	pascal (Pa) (1 Pa = 1 N/m^2) kilopascal (kPa) (1 kPa = 10^3 Pa)
Section modulus	inches cubed (in.^{3})	meters cubed (m^3) or millimeters cubed (mm^3)
Moment of inertia	inches to the fourth power (in.^{4})	meters to the fourth power (m^4) or millimeters to the fourth power (mm^4)
Rotational speed	revolutions per min (rpm)	radians per second (rad/sec) or revolutions per min (rpm)

1.8 CODES AND STANDARDS

Codes and standards have been developed to create uniformity for parts made by different manufacturers and for different usages. Codes are also created by government agencies for such things as automobile safety, fuel economy, or emission standards. Hence, many different requirements may need to be met in the machine design process. Codes for different products may be similar in a sense to building codes, which may specify snow loading requirements for a roofing system or require a building to resist certain wind loads to minimize or prevent damage from natural disasters such as hurricanes or tornadoes.

A standard is a set of specifications for parts, materials, or processes intended to create uniformity from different sources. One of the important attributes of a standard is that the designer can call out the material, thread, gear shape, or other property in a standard manner for use in a design or product without having to develop the individual standard.

In contrast, a code, which is also a set of specifications, typically is established by some professional society or government agency and creates a minimum requirement in order to insure that the part works as it should. The wind or snow load discussed previously would be part of the national building code, whereas standards are usually voluntarily adopted.

Many organizations and societies have created specifications for standards in many different areas. The following is a partial list of organizations that create standards for voluntary codes which, in many cases, become required by government action.

Aluminum Association (AA)
American Gear Manufacturers Association (AGMA)
American Institute of Steel Construction (AISC)
American Iron and Steel Institute (AISI)
American National Standards Institute (ANSI)
American Society of Mechanical Engineers (ASME)
American Society for Metals (ASM)
American Society of Testing and Materials (ASTM)
American Welding Society (AWS)
Anti-Friction Bearing Manufacturers Association (AFBMA)
British Standards Institution (BSI)
Industrial Fasteners Institute (IFI)
Institution of Mechanical Engineers (I. Mech. E.)
International Bureau of Weights and Measures (BIPM)
International Standards Organization (ISO)
National Institute for Standards and Technology (NIST)
Society of Automotive Engineers (SAE)

1.9 SUMMARY

After reading this chapter, you should be familiar with the design process, including the problem definition, the conceptual process, the actual design process, and the creation of prototypes, testing, and redesign as needed. In addition, you should now understand the use of factors of safety. As we progress through the book, your knowledge and experience with appropriate factors of safety will be further refined.

The basics of the properties of materials should now be better understood, especially if you have had a course in material science. It should be becoming clearer to you how these properties are used in the design process, especially how the properties of the materials relate to the actual effect of stress on a part. You should understand the relationship between the stress allowable for the material being used and the actual stress in the part.

After reading the section of this chapter on design calculations, try to do your homework problems and other exercises using these techniques. The benefit of clear calculations will be increasingly evident as problems become more and more difficult.

The brief discussion on preferred sizes and unit systems should have been material already basically understood. If not, you may need to seek additional resources.

1.10 PROBLEMS

1. Your task is to design a system to stop trucks that have experienced brake failure on a steep hill.
   a. Write the problem statement for this task, focusing on the attributes of what is needed.
   b. Describe at least three concepts for this task.
2. A boat manufacturer would like to develop a propulsion system that is less susceptible to both weeds and shallow water.
   a. Write a problem statement for this task.
   b. Describe at least three concepts for this task.
3. The superhighways of the future may not require automobile drivers to steer or control auto speed.
   a. Describe the attributes of a system of this type.
   b. List as many concepts as possible, both for directional guidance and speed control.
   c. List safety features that may be needed.
   d. Evaluate these ideas and choose one or more to discuss with your class.

4. Explain why an overhead crane used in a factory may be designed with a higher factor of safety than a trolley track on the floor of the same factory.
5. Find an elastic band and carefully stretch it to different lengths while measuring the original length, yield length, and ultimate length before failure. Was there visible necking before failure? Prepare a sketch of deflection versus your estimate of force. Was this a ductile or brittle material?
6. Cut ½-inch-wide strips of paper both from the top and sides of a sheet of paper. Pull on them until failure. Is this a ductile material? Is this material isotropic (same strength in both directions)?
7. List the functional requirements and describe possible concepts for the following applications:
   a. An automobile jack.
   b. A machine to crush juice and soda cans in a kitchen.
   c. A device to open and close garage doors.
   d. A machine to place twelve glass bottles into a box.

CHAPTER 2

Force, Work, and Power

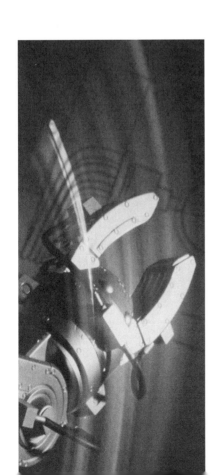

The student taking a course in machine design should have had a basic course in Newtonian physics as well as statics and strength of materials. This chapter is intended to review the principles of weight versus mass, work, torque, power, fluid pressures and forces, and moments of inertia, as well as section modulus. If students are still uncomfortable with this material after completing this chapter, they should seek additional assistance in this area. This chapter is intended to be only a brief review of these principles.

2.1 OBJECTIVES

This chapter is intended to enhance your understanding of how the forces are developed on machine parts, especially as they relate to the forces that are generated by a motive means such as an engine or a motor where the power is transmitted to a mechanism or part of a machine. After completing this chapter, you should be able to:

- Understand the difference between force, work, and power.
- Recognize and be able to convert between mass and weight.
- Convert between English and metric units for force, work, and power.
- Understand the basic principles of fluid mechanics as they apply to hydraulic and air cylinders or similar products.
- Look up and/or calculate moments of inertia and section modulus for different shape parts.
- Apply the principles of work, force, and power to moving machines.

2.2 WEIGHT, FORCE, AND MASS DISTINCTIONS

In the U.S. or English system, we historically have used weight as the basis of analysis, whereas in the metric system, mass has always been used. This is often confusing; in the English system, we factor in a standard gravitational pull of 32.2 feet per second squared or 9.81 meters per second squared to convert from mass to weight. In the English system, we are therefore using a unit that assumes gravity is acting upon it, whereas in the metric system we factor out the effect of gravity. In practice, however, we could go into a store in Europe and buy a kilogram of flour (a mass unit) or into a store in the United States and buy a pound (a weight unit), which causes further confusion. The relationship between these two is

$$W = mg \quad \text{or} \quad m = \frac{W}{g} \tag{2.1}$$

where W = weight
m = mass
g = acceleration due to gravity

When we factor in the gravitational pull of the earth, we can simply relate the mass and weight. It is important to understand this distinction because, if we are accelerating a mass horizontally, the weight of the object is irrelevant, whereas, if we are accelerating that same mass vertically, gravity obviously affects the forces necessary to undertake this change in velocity. Therefore, the equivalent type of unit to the pound would be a Newton in the metric system, and the equivalent of a kilogram in the English system would be a pound-mass. Therefore, the relationship from the standpoint of physics becomes

$$F = ma \quad \text{or} \quad F = \frac{W}{g} a \tag{2.2}$$

where F = force
a = acceleration

Generally, in most analyses except for very high elevations or in aerospace, we assume

$$g = 32.2 \text{ ft/sec}^2 \quad \text{or} \quad g = 9.81 \text{ m/s}^2$$

2.3 WORK AND POWER

Work is often misunderstood because, as lay people, we assume it is related to the process of getting tired. Ironically, we could push all day on a wall, become exhausted, and have done no work as the wall did not move. Work is defined simply as force times distance:

$$W = Fd \tag{2.3}$$

work
The result of a force moving an object a certain distance.

power
The rate at which the work is accomplished.

Note in this formula that there is no time element. We could think of an equivalent amount of work as follows: We could either carry ten boxes at a time up a set of stairs or make ten trips carrying one box at a time, and in both scenarios we would have done exactly the same amount of work. Work would be measured in units of foot-pounds in the English system or Newton-meters in the metric system.

Power, conversely, is the rate of doing work that leads to the following relationship:

$$P = \frac{W}{t} = \frac{Fd}{t} \tag{2.4}$$

This would give us units such as foot-pounds per second in the English system or, in the metric system, units such as Newton-meters per second. Note from our previous example of carrying the ten boxes up the stairs that if in the second case of carrying a single box at a time it took us ten times as long, it would then take only one-tenth of the power to accomplish the same task, as the time element was correspondingly ten times as long. In the English system we often use units of horsepower:

$$\text{hp} = 550 \text{ foot-pounds/second} = \frac{550 \text{ ft-lb}}{\text{sec}}$$

In the metric system we often refer to a Newton-meter per second as a watt; hence, the following relationship also applies:

$$1 \text{ hp} = 746 \text{ Newton-meters/second}$$
$$= 746 \frac{\text{Nm}}{\text{s}}$$
$$= 746 \text{ watts} = .746 \text{ kilowatts}$$

In the metric system we determine power as torque times rotational speed in terms of radians per second in order to obtain watts. Hence, the following relationship still applies:

$$P = T\omega$$

where T, Newton-meters (Nm)

ω, radians per second (rad/sec)

$P, \frac{\text{Nm}}{\text{s}}$ or $\frac{\text{J}}{\text{s}}$ or watts

Note that there are 2π radians per revolution. Hence, if we have a speed in revolutions per second, we would multiply it times 2π, and rotational speed in revolutions per minute would need also to be converted to revolutions per second as follows:

$$n = \frac{\omega}{2\pi} 60$$

where n = revolutions per minute

2.4 TORQUE

The principles of **torque** are quite simple, as torque is nothing more than a force times the distance perpendicular to the force. Examples include using a wrench to create a torque to tighten a nut or the torque in the axle of an automobile rotating the tire. Figure 2.1 illustrates a torque being placed on a nut. Torque is expressed in units of force times distance. In the English system, the units would typically be foot-pounds or inch-pounds and, in the metric system, they would typically be Newton-meters.

$$T = Fd \qquad (2.5)$$

Example Problem 2.1 shows the torque transmitted from an automobile engine through a transmission and differential to the tires and determines the force that the tire can exert against the road to move the automobile forward.

torque
A rotational force created by a force applied at some distance from the centerline that tries to cause rotation.

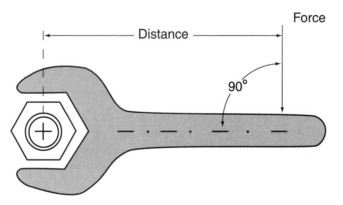

FIGURE 2.1 Torque applied to a nut.

FIGURE 2.2 Force acts between the tire and the road.

Example Problem 2.1: Torque

An automobile engine creates a torque of 300 foot-pounds at a rotational speed of 3,500 rpm. In fourth gear, the transmission has a one-to-one ratio. The differential, however, has a ratio of 3.6 to 1, which creates a torque in the axle of 1,080 foot-pounds. What is the force being exerted on the road by the automobile tire if the diameter of the tire is 30 inches?

$$T = Fd \qquad (2.5)$$

$$F = \frac{T}{d}$$

$$F = \frac{1{,}080 \text{ ft-lb}}{\dfrac{30 \text{ in.}}{2} \dfrac{(\text{ft})}{(12 \text{ in.})}}$$

$$F = 864 \text{ lb}$$

The tire exerts an 864-pound force on the road surface (see Figure 2.2).

2.5   POWER AND ROTATIONAL SPEED

Similar to what we saw in the section on work and power, rotational power is the same for a rotating part. Since work is defined as the product of a force and a distance through which the force moves, rotational work is the rate of work developed by a force at some distance from the center of rotation. If we factor in speed in terms of the number of revolutions per minute, for example, then the force times the distance traveled, which would be equal to two times the radius times π or the circumference of this path on which the force was applied, would then give us the work done as

$$W = Fd \tag{2.3}$$

$$d = (2\pi r) \text{ (the circumference is the distance)}$$

$$W = F(2\pi r)$$

If we were to divide that by the speed, we would then get the corresponding power. Therefore, we can express this relationship as being equal to

$$P = \frac{W}{t} = \frac{F(2\pi r)}{t} \tag{2.4}$$

Factoring in that 1 hp = 550 $\frac{\text{ft-lb}}{\text{sec}}$ and converting rotational speed (ω) to revolutions per minute (rpm):

$$n = \frac{2\pi r}{t} \frac{60 \text{ sec}}{\text{min}} \frac{12 \text{ in.}}{\text{ft}}$$

or

$$hp = \frac{Fn(2\pi r)}{\frac{550 \text{ ft-lb}}{\text{hp}}} \frac{\text{sec}}{60 \text{ sec}} \frac{\text{min}}{12 \text{ in.}} \frac{\text{ft}}{}$$

then

$$hp = \frac{Fn(2\pi r)}{396,000}$$

The formula then becomes

$$hp = \frac{Tn}{63,000} \tag{2.6}$$

where T, in.-lb

n, rpm

Power
(T, in.-lb; n, rpm)   (2.6)

$$hp = \frac{Tn}{63,000}$$

Hence, it can be seen that the same kind of relationship as in our earlier example of carrying boxes up the stairs can be observed in the relationship between torque, speed, and horsepower. Example Problem 2.2 takes the results of Example Problem 2.1, incorporates the speed of rotation of the

automobile engine and the gear ratios, and determines the actual work and power the automobile must generate to move it forward.

Example Problem 2.2: Work and Power

If the automobile in Example Problem 2.1 was going up a long, steep hill at this speed, how much work was done in the period of a second?

First, as work is force times distance, we need to calculate the distance traveled in 1 second:

$$d = \frac{3{,}500 \text{ rpm} \cdot \pi \cdot \frac{30 \text{ in.}}{\text{rev}} \cdot \frac{\text{ft}}{12 \text{ in.}} \cdot \frac{\text{min}}{60 \text{ sec}}}{\frac{3.6}{1}} \quad (2.3)$$

$d = 127$ ft

$W = Fd$

$W = 864 \text{ lb} \times 127 \text{ ft}$

$W = 109{,}728$ ft-lb of work

The power to do this would then be

$$P = \frac{Fd}{t} = \frac{W}{t} \quad (2.4)$$

$$P = \frac{109{,}728 \text{ ft-lb}}{1 \text{ sec}} = 109{,}728 \frac{\text{ft-lb}}{\text{sec}}$$

or if

$$550 \frac{\text{ft-lb}}{\text{sec}} = 1 \text{ hp}$$

$$\text{hp} = \frac{109{,}728 \text{ ft-lb/sec}}{550 \text{ ft-lb/sec/hp}} = 200 \text{ hp}$$

Comparing it to motor output:

$$T = 63{,}000 \frac{P}{n}$$

where n, rpm

T, in.-lb

$$P = \frac{Tn}{63,000} \quad (2.6)$$

$$P = \frac{300 \text{ ft-lb}}{} \cdot \frac{12 \text{ in.}}{\text{ft}} \cdot \frac{3,500 \text{ rpm}}{63,000}$$

$$P = 200 \text{ hp}$$

Note that the power output equaled the power at the tire/road surface.

2.6 PRESSURE, FORCE, AND AREA

In the machine design process, we often use hydraulic cylinders, air cylinders, or similar products in the design of machines. Because this is not a course in fluid mechanics, we will limit our discussion to the forces created by **fluid pressures** and the associated power that may be created without dwelling on frictional losses, flow energy, and so on. We can calculate the force from a fluid as being equal to the pressure in the fluid times the area that it is acting upon. For a basic **pneumatic** or **hydraulic** cylinder, the area of the cylinder would be equal to $\pi D^2/4$, and we multiply that by the pressure in the fluid to determine the force. This principle is the same whether it is a hydraulic or a pneumatic cylinder and whether it is powered by oil, water, air, or even a mixture of air and explosive gases as in an automobile engine. The distance that the cylinder moves multiplied by the force would be the work done in this application and, if we were to factor in the rate at which the cylinder is moving, it gives us the associated power. Example Problem 2.3 demonstrates this for a hypothetical automobile piston that we power by igniting a mixture of gasoline and air, which then creates a corresponding pressure on the piston, pushing it through some distance. This then acts on a lever arm to rotate the shaft of the engine (crankshaft) in order to create the torque found in Example Problem 2.2, where we calculated the work being done by the automobile wheels against the road surface.

fluid pressure
When a force acting on a fluid creates internal pressure within the body of fluid.

pneumatic
The use of a gas; for example, air or compressed air.

hydraulic
The use of a fluid; for example, water or oil.

Example Problem 2.3: Force-Pressure Relationship

For the automobile engine from the two previous example problems, calculate the pressure required in the cylinder under the following assumptions:

1. The cylinder is 2 inches in diameter.
2. The crankshaft has an effective radius of 4 inches.

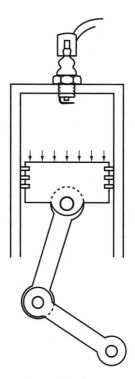

FIGURE 2.3 Automotive cylinder cross-sectional view.

3. Maximum power is achieved when the piston is perpendicular to the crankshaft.
4. At this point, all the power comes from this piston alone. (See Figure 2.3.)

If the T is 300 ft-lb and the effective length of the crankshaft is 4 inches, the force exerted by the piston would be

$$T = Fd \qquad (2.5)$$

$$F = \frac{T}{d} = \frac{300 \text{ ft-lb}}{4 \text{ in.} \frac{\text{ft}}{12 \text{ in.}}} = 900 \text{ lb}$$

Now, analyzing the cylinder:

$$F = PA$$

$$P = \frac{F}{A}$$

$$P = \frac{900 \text{ lb}}{\frac{\pi (2 \text{ in.})^2}{4}}$$

$P = 286 \text{ lb/in.}^2 = 286 \text{ psi}$

(Discuss with your instructor what may be incorrect in our assumptions and why this value is likely to be much higher in a typical engine.)

2.7 MOMENTS OF INERTIA AND SECTION MODULUS

We know from strength of materials that when we have a beam in bending, we have compressive stresses on one side of the beam and tensile stress on the opposite side. Because the material in the beam needs to resist these stresses, we can assume that the larger the part is, the greater its ability to resist stresses as the material is placed farther from the centerline of the part, or the neutral axis. Hence, this greater ability to resist the load on the part by placing the stress elements farther from the neutral axis has the net result of making the beam stronger. This property is what we commonly refer to as the **moment of inertia.** If we envision a fairly typical I-beam, we can see that most of the material in the beam is placed on the outermost surfaces. If we compare this to a solid block of material of equal weight, the stress-resisting elements of the I-beam are farther from the neutral axis; hence, for the same weight, this I-beam would have a significantly larger moment of inertia than would the solid bar. Appendix 3 gives values for some common shapes for the moment of inertia, and it can be seen from this table that certain shapes have a better ability to resist loads from bending stresses and thus have correspondingly larger moments of inertia.

Frequently, when a beam or load-resisting member is a symmetric shape, we can simplify the analysis by using a property called **section modulus,** which is the moment of inertia divided by the distance from the neutral axis to the outermost fiber. Appendix 3 also shows values for the section modulus for symmetric shapes, which can simplify the analysis when this condition is present. As we do increasingly more beam problems, we can see the advantages and disadvantages of different shapes with regard to both their moment of inertia and their section modulus, which will aid us in selecting an appropriate shape. Some shapes, such as round or hollow bars, are ideal when the direction of loading is not known, as the

moment of inertia
Property of shape relating to the distance that the material resisting bending is placed from the neutral axis.

section modulus
Similar to moment of inertia, but for a uniform concentric shape. It is the moment of inertia divided by the distance from the neutral axis to the outermost fiber.

section modulus and moment of inertia are the same in all directions. Shapes such as a rectangle, a wide flange, or an I-beam have different moments of inertia and section modulus in different directions. Therefore, we would select an appropriate shape depending not only on the type of loading but also on whether or not the direction of loading is known.

2.8 SUMMARY

After reading this chapter, you should be familiar with the distinction between mass, weight, and force in both the English and the metric systems; the relationship between work and power and the units in which they are expressed; and the conversion between work and power in both English and metric systems.

The basics of torque, power, and rotational speed should be understood in preparation for study in later chapters of shaft analysis, gears, clutches, motors, and so on, which are often integral to many types of machine design problems. The brief review of pressure, force, and area was intended to help students who have not yet taken a fluid mechanics course to understand later chapters on fluid power. Finally, the discussion of moments of inertia and section modulus, which should be a repeat of information learned in a prior strength of materials course, is intended only as a review.

2.9 PROBLEMS

1. The elevator system shown in Figure 2.4 has a combined weight for the car and occupants of 1,000 pounds (ignore forces to accelerate the car and occupants):
   a. What is the force in the cable?
   b. To raise the car 150 feet, how much work is done?
   c. If it took 10 seconds to travel this distance, what power was required in ft-lb/sec, hp, and kilowatts?
   d. If the effective diameter of the cable drum is 18 inches, determine the torque in the input shaft.
   e. What is the rotational speed of this shaft?
   f. Calculate the power to turn this shaft, and compare this power to that determined in part c.

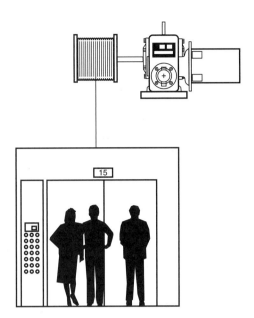

FIGURE 2.4 Elevator for Problems 1 through 4.

2. Redoing problem 1, change the mass of the elevator car and occupants to 500 kg, the distance to 50 meters, and the cable drum diameter to ½ meter.
3. If we now include in problem 1 an acceleration of 16 ft/sec² for the first two seconds of travel for the elevator car, determine:
   a. The force in the cable during the first two seconds.
   b. The distance traveled during that period.
   c. The power required during the acceleration period.
   d. Compare this result to the power calculated in problem 1.
4. Redoing problem 3, change the acceleration to 3 m/sec². (Use the values from problem 2.)
5. If a 3,000-pound automobile has a combined wind resistance and frictional losses of 200 pounds at a speed of 50 mph:
   a. Calculate the power required to maintain this speed on a level road.
   b. If it takes three seconds to accelerate to 60 mph from 50 mph, calculate the power to do this (ignore frictional loss).
   c. Assuming the frictional losses stay the same during this period, what is the total power?
6. A large truck has an overall range of gear ratios that includes the transmission and differential of 12 to 1 through 3 to 1. If the engine torque is 350 ft-lb and the wheels have an effective diameter of 36 in.:

a. What force can be applied to the road at these two ratios?
b. If the truck weighs 40,000 pounds, ignoring any friction, what is the maximum rate of acceleration on a flat road?

7. A hydraulic elevator has a weight of 4,500 pounds for the moving parts, including passengers, and the hydraulic cylinder has an effective piston diameter of eight inches.
   a. Calculate the pressure needed in the fluid when the elevator is not moving.
   b. If the elevator goes up 20 feet in 10 seconds (ignore acceleration), calculate the work done.
   c. Determine the average power required.
   d. If, in the first second, the elevator accelerates to a speed of 2 ft/sec, calculate the force required and the power required during this first second.
   e. In part d, what was the pressure in the fluid during this acceleration?
   f. If, during the last second, the elevator decelerates at the same rate as it accelerated, what was the total work done?

8. The moving parts of a hydraulic elevator have a mass of 1,000 kg and the hydraulic cylinder has an effective piston diameter of 25 cm.
   a. Calculate the pressure needed in the fluid when the elevator is not moving.
   b. If the elevator goes up 6 meters in 10 seconds (ignore acceleration), calculate the work done.
   c. Determine the power required.
   d. If, in the first second, the elevator accelerates to a speed of 1 m/sec, calculate the force required and the power required during this first second.
   e. In part d, what was the pressure in the fluid during this acceleration?

9. Determine the moment of inertia and section modulus for the shapes shown in Figure 2.5.

10. For the 2 × 6 rectangle in Figure 2.5, how much would the 6-inch dimension need to change to double its section modulus?

2.10 CUMULATIVE PROBLEMS

11. A locomotive is driven by a steam piston system with an effective pressure of 300 psig. The steam piston has a diameter of 12 inches (see Figure 2.6).

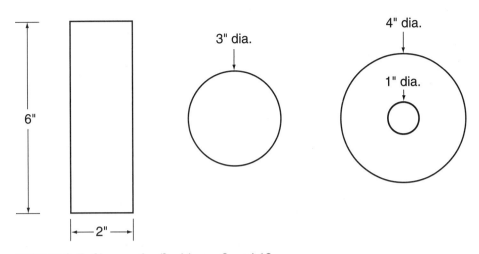

FIGURE 2.5 Shapes for Problems 9 and 10.

    a. Determine the torque created on the locomotive wheel system and the maximum tractive force if the locomotive weighs 80,000 pounds and the coefficient of friction between the wheels and the track is .3.

    b. Determine the power of the locomotive if this situation exists at 50 mph.

12. A truck weighs 60,000 pounds and has 18 wheels, each of which is 8 inches wide. If the tire pressure is 80 psi and assuming no load is carried by the sidewalls, how long is the flat spot on each tire?

13. In problem 12, if the tire pressure is increased to 100 psi, find the new flat spot length.

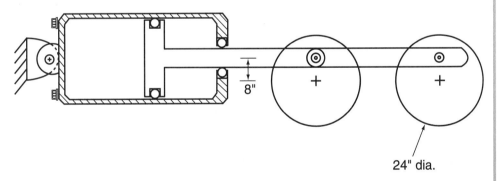

FIGURE 2.6 Locomotive drive for Problem 11.

14. If the truck in problem 12 needs to be able to stop in 500 feet from a speed of 60 mph, what is the average stopping force?
15. If this truck has ten sets of wheels, each of which is 42 inches in diameter, what is the required torque on each set of wheels to accomplish the stop described in problem 14?
16. For this truck to stop in the distance described in problem 14, what is the necessary coefficient of friction between the tires and the road?

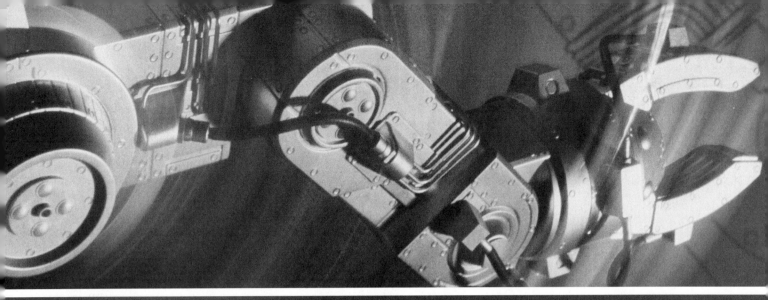

CHAPTER 3

Stress and Deformation

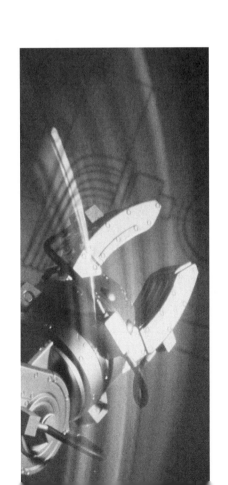

Chapter 3

An important part of any design is insuring that suitable stress levels are maintained in the part being designed. There must be an acceptable relationship between actual stresses and the allowable stress for the materials from which the part is made. You should already have studied the principles of strength of materials. This chapter is a review of the topics covered in a strength of materials course with a special emphasis on reviewing types of stresses and the use of beam formulas for simplifying stress analysis and deflection for common beam types, as opposed to the detailed method of creating shear and bending diagrams that was utilized in the student's previous strength of materials course.

This chapter also discusses how appropriate factors of safety are determined for a design. Most of the topics covered in this chapter will be expanded upon as you progress through this book, especially the review of stress types and their combinations for use in **fatigue analysis** and combined stress problems and the selection of appropriate safety factors.

> **fatigue analysis**
> The study of the effect of repeated loading on a part.

Calvin and Hobbes   by Bill Watterson

FIGURE 3.1 Courtesy of Universal Press Syndicate.

3.1 OBJECTIVES

After completing this chapter, you should be able to:

- Describe the types of stresses caused by axial, bending, shear, and torsion loading.
- Describe the relationship between stresses in the part and the strength or stress-carrying ability of the part, and appreciate the relationship between the two.

- Distinguish between the ability of a material to carry loads in shear versus axial loading, and understand the relationship between these types of stresses.
- Understand the principles of deformation and whether those levels of deformation are acceptable to the design being analyzed.
- Understand beam deflection formulas and their use in design problems.
- Analyze columns using the Euler and Johnson methods.
- Describe the meaning of *factor of safety* and its use in determining safe stress levels and/or allowable load limits and their relationship to the unknowns in the design process.

3.2 CATEGORIZATION OF STRESS TYPES

Often in the structural analysis of machine parts or other mechanical items, we divide the stresses into two types: **normal stress,** which is caused by axial and bending loads, and **shear stress,** which is from shear and **torsional loading.** The reason for this distinction is that most materials have differing abilities to resist stress from these loading types. Most machine part materials, such as steel and aluminum, are much less resistant to shear stress than they are to normal stress. A material's ability to resist a shear or torsion stress can range from 50 to 60 percent of its ability to resist axial or bending stress. To illustrate this, take a sheet of paper, roll it into a tube, and try to shear it between your two hands, as opposed to pulling on the two ends of the tube. You can easily see that pulling on the ends of the tube requires significantly greater force to break the paper. Another way to envision this difference in the strength of materials to resist these different types of loads is to envision the molecules being pulled apart as opposed to being slid along each other. In the sliding analogy, as the molecules slide past adjacent molecules, the molecular attractions are transferred to the next adjacent molecules, and so on, much like sliding two large magnets apart as opposed to pulling them apart. You can slide these two large magnets apart fairly easily, whereas pulling them apart can take a tremendous force. The intermolecular attraction can be thought of in a similar manner.

3.3 TENSILE AND COMPRESSIVE AXIAL LOADS

In typical **axial** tensile and compressive loading as shown in Figure 3.2a and Figure 3.2b, **isotropic** materials such as steel have very similar load-carrying capacities in both tension and compression. In these cases, the material or

> **normal stress**
> Stress caused by an axial or bending load that tends to pull the molecules apart.
>
> **shear stress**
> Stress caused by a torsional or shearing-type load that tends to make the molecules slide by each other.
>
> **torsional loading**
> A load that is created by a torque, such as in an axle, due to a force being applied at some distance from the centerline that causes a shearing-type, twisting stress.
>
> **axial**
> In the same direction as the centerline of a part, generally caused by a tensile "pulling" or compressive "pushing" force.
>
> **isotropic**
> Having similar properties in all directions. For example, steel is generally considered to be an isotropic material because it has the same stress allowables regardless of direction of stress.

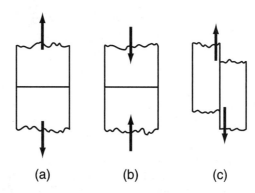

FIGURE 3.2 (a) Tension. (b) Compression. (c) Shear.

stress elements are separated from the adjacent element or pushed closer together, which results in failure. Stresses in this type of loading condition can be summarized by

$$S = \frac{F}{A} \tag{3.1}$$

where $S =$ stress

$F =$ force

$A =$ area

The associated deflection of a part loaded in this manner would be

$$\delta = \frac{FL}{AE} \tag{3.2}$$

where $\delta =$ deflection

$L =$ length

$E =$ modulus of elasticity

The convention for labeling tensile stresses is to call a tensile stress positive and a compressive stress negative. The units that are used to indicate stresses are force per unit area. In the U.S. customary unit system, typical units would be pounds divided by square inches, or psi, and in the SI system, **Newtons** divided by square meters or a **Pascal** (Pa). Since a Pascal is a very small unit, we often substitute a Newton divided by a square millimeter, which is a megapascal (MPa).

Stress pressure (3.1)

$$S = \frac{F}{A}$$

Axial deflection (3.2)

$$\delta = \frac{FL}{AE}$$

Newton
The unit of force in the metric SI system. This is a similar quantity to a pound in the U.S. customary unit system.

Pascal
A unit of stress or pressure in the metric SI system similar to a psi or pound-per-square-inch stress or pressure unit in the U.S. customary system.

Example Problem 3.1

A 20,000-pound load is applied to a one-inch diameter steel bar that is made from AISI 1020 hot-rolled steel with $S_u = 55{,}000$ psi and $S_y = 30{,}000$ psi. (See Figure 3.3.)

a. What is the stress in the bar?

$$S = \frac{F}{A} = \frac{20{,}000 \text{ lb}}{\frac{\pi(1 \text{ in.})^2}{4}} = 25{,}465 \text{ lb/in.}^2 \qquad (3.1)$$

b. By how much did the bar shorten?

$$\delta = \frac{-FL}{AE} = \frac{-20{,}000 \text{ lb } (6 \text{ in.})}{\frac{\pi(1 \text{ in.})^2}{4} \; 30 \times 10^6 \text{ lb/in.}^2} = -.005 \text{ in.} \qquad (3.2)$$

c. Will this bar return to its original length when the load is removed? The stress of 25,465 psi is less than S_y of 30,000 psi, so it will return to its original size because the yield limit was not exceeded.

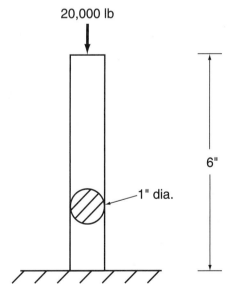

FIGURE 3.3 Example Problem 3.1.

3.4 STRESSES AND DEFLECTION DUE TO BENDING

Bending stresses can be seen in a diving board under the weight of a diver, floor joists in a building, a leaf spring in an automobile, or a flagpole bending in the wind. A beam in bending has simultaneous axial stresses both in tension and compression on opposite sides of the beam. These are the same types of axial stresses exhibited in axial tensile or compressive loads discussed in Section 3.3. Since the maximum tensile and compressive loads are on opposite sides of the beam, there is always a section, typically in the middle of the beam, in which there is no stress, and we refer to that section as the **neutral axis.** A beam's ability to minimize bending is related to its moment of inertia, which is an index of how far the material under stress is located from the neutral axis and the material's properties. This is why I-beams or wide-flange beams are often used—as they have a greater amount of material at a greater distance from the neutral axis than, for example, a solid beam would.

Once we understand the principles of bending in a strength of materials course, we frequently use tables (like the one in Appendix 2) to determine both bending moments and deflection.

Example Problem 3.2 reflects a beam in bending, along with the shear and bending diagram and an analysis using the tables of Appendix 2 for the deflection and stress formulas. Appendix 3 can be used to determine the moment of inertia. Often when using this technique we simplify the analysis by using a quantity called Z, which is equal to the moment of inertia (*I*) divided by *c*. This value (*c*) is the distance from the outermost fiber to the neutral axis, which is at the center for a concentric beam. This quantity (Z) is referred to as the *section modulus*. Hence the stresses from bending are

$$S = \frac{Mc}{I} = \frac{M}{Z} \qquad (3.3)$$

This technique is based on the following conditions:

1. The beam must be primarily in pure bending. Shearing stresses must be negligible, and separate axial loads must not exist.
2. The beam must not twist or be subject to torsional loading.
3. The material from which the beam is made must have the same modulus of elasticity in both tension and compression.
4. The loads must be kept low enough that the beam is not subject to localized buckling.

---

neutral axis

An axis at which, under bending or torsional stress, no stress exists. For a symmetric part, this is the centerline.

Bending stress (3.3)

$$S = \frac{Mc}{I} = \frac{M}{Z}$$

For typical beams, the deflection can also be analyzed using the formulas from Appendix 2, and the acceptability of these deflections is often based on the designer's experience or, in many cases, on engineering codes. The following example summarizes a beam on which the stresses are calculated along with the level of deflection.

Example Problem 3.2

For the 2 × 2–inch, simply supported beam in Figure 3.4, made from the same 1020 steel as the beam in Example Problem 3.1, assume a safety factor of 2 based on the ultimate stress allowable:

  a. What is the maximum stress in the beam?

$$I = \frac{bh^3}{12} = \frac{2 \text{ in. } (2 \text{ in.})^3}{12} = 1.33 \text{ in.}^4 \qquad \textbf{(Appendix 3)}$$

$$M_m = \frac{FL}{4} \qquad \textbf{(Appendix 2)}$$

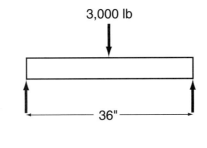

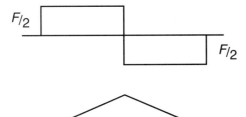

$$M_m = \frac{F}{2} \cdot \frac{L}{2} = \frac{FL}{4}$$

FIGURE 3.4 Simply supported beam for Example Problem 3.2 and shear and bending moment diagrams.

$$S = \frac{Mc}{I} = \frac{(FL)c}{(4)\,I} \qquad (3.3)$$

$$S = 3{,}000 \text{ lb } \frac{36 \text{ in. }(1 \text{ in.})}{4(1.33 \text{ in.}^4)} = 20{,}300 \text{ lb/in.}^2$$

b. What is the maximum deflection in this beam?

$$\delta = -\frac{FL^3}{48\,EI} \qquad \textbf{(Appendix 2)}$$

$$\delta = -\frac{3{,}000 \text{ lb }(36 \text{ in.})^3}{48\,(30 \times 10^6 \text{ lb/in.}^2)(1.33 \text{ in.}^4)} = -.073 \text{ in.}$$

c. Will this bar return to its original shape when this load is removed?
Yes, the stress of 20,300 lb/in² is less than the yield of $S_y = 30{,}000$ lb/in.²

d. For a safety factor of 2, based on ultimate, is this an acceptable design?

$$\frac{S_u}{N} = \frac{55{,}000 \text{ lb/in.}^2}{2} = 27{,}500 \text{ lb/in.}^2$$

It is acceptable because this is less than the actual stress of 20,300 lb/in.²

3.5 SHEAR STRESSES

Shear stresses are created when a force is applied perpendicular to an element, creating a slicing action in the material, as scissors slice through paper, a keyway is sheared off the surface of a shaft, or an untensioned bolt connecting two pieces of material. The method for calculating shear stress appears to be very similar to that for calculating tensile stress. When we assume the applied force is uniform across the area of the part, shear stress would then be equal to

$$S_s = \frac{F}{A_s} \qquad (3.4)$$

Note, however, as discussed earlier in this chapter, the strength of the material to resist shear is less than its axial strength, and a factor of 50 to 60 percent will need to be applied to that value.

Example Problem 3.3

The yoke shown in Figure 3.5 has a ½-inch diameter pin made from AISI 1040 cold-drawn steel. For a load of 20,000 pounds, will this fail?

$$S_u = 80 \text{ ksi} \qquad S_y = 71 \text{ ksi} \qquad \textbf{(Appendix 4)}$$

$$S_s = \frac{F}{A_s} \qquad \textbf{(3.4)}$$

Note double shear.

$$S_s = \frac{F}{\frac{(2)\pi D^2}{4}}$$

$$S_s = \frac{20,000 \text{ lb}}{\frac{(2)\pi(.5 \text{ in.})^2}{4}} = 50,930 \text{ lb/in.}^2$$

Comparing to .5 or .6 of S_u, this stress is too high and the pin would fail.

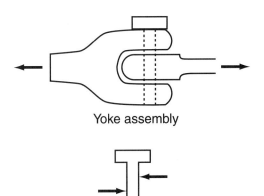

Yoke assembly

Free body diagram of pin

FIGURE 3.5 Example Problem 3.3.

3.6 TORSIONAL SHEAR STRESSES

Torsional shear is very similar to direct shear in its effect on material and the ability of materials to resist this shear. However, it is also related to the distance the material is from the centerline of the part making it very similar to bending. Torsional shear stresses most commonly occur in cylindrical shafts that transmit power. Torsional shear is calculated as

$$S_s = \frac{Tc}{J} \quad (3.5)$$

where c = radius of the shaft

J = **polar moment of inertia**

Appendix 3 lists formulas for J. We often use a property similar to section modulus called the **polar section modulus** (Z'). In this case, the formula becomes

$$S_s = \frac{T}{Z'} \quad (3.6)$$

When a shaft is subject to a torque, it undergoes twisting. The angle of twist is computed as follows:

$$\theta = \frac{TL}{JG} \quad (3.7)$$

where θ = angle of twist in radians

L = length of shaft over which the angle of twist is acting

G = **shear modulus of elasticity** of the shaft

Example Problem 3.4

A cylindrical shaft transmits power delivered by a ten horsepower electric motor turning at 1,750 rpm. If the shaft is made from AISI 1117 hot-rolled steel, ½ inch in diameter and 12 in. in length:

a. What is the stress in this shaft?

$$S_u = 62 \text{ ksi} \qquad S_y = 34 \text{ ksi} \qquad \textbf{(Appendix 4)}$$

Sidebar notes:

Torsional stress (3.5)

$$S_s = \frac{Tc}{J}$$

Torsional stress (3.6)

$$S_s = \frac{T}{Z'}$$

polar moment of inertia

This is a similar property to the moment of inertia. However, it is the moment of inertia that would resist a torsional load.

polar section modulus

This is a similar property to section modulus. However, the polar section modulus would be the ability of a material to resist a torsional load.

Angular deflection (3.7)

$$\theta = \frac{TL}{JG}$$

shear modulus of elasticity

This is a similar property to the modulus of elasticity except that it is the relationship between force and deflection from a torsional load.

$$T = \frac{63,000 \text{ hp}}{\text{rpm}} \qquad (2.6)$$

$$T = \frac{63,000 \ (10)}{1,750} = 360 \text{ in.-lb}$$

$$S_s = \frac{T}{Z'} \qquad (3.6)$$

$$Z' = \frac{\pi D^3}{16} \qquad \text{(Appendix 3)}$$

$$Z' = \frac{\pi (.5 \text{ in.})^3}{16} = .025 \text{ in.}^3$$

$$S_s = \frac{T}{Z'} = \frac{360 \text{ in.-lb}}{.025 \text{ in.}^3} = 14,400 \text{ lb/in.}^2$$

b. What would the angular deflection of this shaft be?

$$\theta = \frac{TL}{JG} \qquad (3.7)$$

$$J = \frac{\pi D^4}{32} = \frac{\pi (.5 \text{ in.})^4}{32} \qquad \text{(Appendix 3)}$$

$$J = .0061 \text{ in.}^4$$

$$\theta = \frac{TL}{JG} = \frac{360 \text{ in.-lb} \ (12 \text{ in.})}{.0061 \text{ in.}^4 \ (11.5 \times 10^6 \text{ lb/in.}^2)}$$

$$\theta = .062 \text{ radians}$$

c. If a safety factor of 3 is desired, based on ultimate strength, how large should this shaft be?

$$\frac{(.5)S_u}{(N)} = \frac{T}{Z'} = \frac{T}{\frac{\pi D^3}{16}}$$

$$D^3 = \frac{(N) \ 16T}{.5 S_u \pi}$$

$$D = \left(\frac{(N) \ 16T}{.5 S_u \pi} \right)^{1/3}$$

$$D = \left(\frac{3(16)\ 360\ \text{in.-lb}}{(.5)62,000\ \text{lb/in.}^2 \pi}\right)^{1/3}$$

$$D = .562\ \text{in.}$$

Note that the stresses caused by torsion have a similar effect on the material as those from shear and generally the same .5 to .6 S_u factor would be used for shear and torsion stress problems. A summary of the types of stresses is shown in Figure 3.6.

			Basic stress theory		
Method of molecule separation	Type	Example	Stress	Deflection	Stress allowable
Push together or pull apart	Axial		$S = \dfrac{F}{A}$	$\delta = \dfrac{FL}{AE}$	S_y or S_u
	Bending		$S = \dfrac{MC}{I}$ or $\dfrac{M}{Z}$ I or Z from App. 3 Calculate M from App. 2	δ from App. 2	S_y or S_u
Slide by	Torsion		$S = \dfrac{TC}{J}$ or $\dfrac{T}{Z'}$ J or Z' from App. 3	$\theta = \dfrac{TL}{JG}$	S_s if known; otherwise use .5 or .6 S_y or S_u
	Shear		$S = \dfrac{F}{A}$ $S = \dfrac{VQ}{IB}$	Usually don't care!	S_s if known; otherwise use .5 or .6 S_y or S_u

FIGURE 3.6 Types of stress.

3.7 COLUMNS

A specialized type of stress analysis is the analysis of long, slender columns. In members of this type, failure can occur by buckling as opposed to failure due to high stresses. Columns are often used as machine parts (for instance, for long connecting rods or slender structural members), so it is important to be able to recognize when column action might occur and to make the appropriate analysis. Figure 3.7 reflects buckling of a thin rectan-

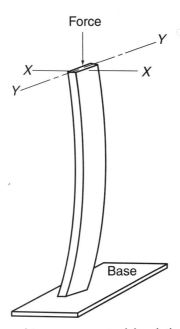

FIGURE 3.7 Column subject to a central load that causes buckling.

gular column where the center portion bows to an unstable geometry and the continuation of this load would cause column failure.

In order to analyze columns, we need to identify the type of column. For a very long column, an analysis developed by Euler appears to duplicate the effect when the column is long and slender. Intermediate-length columns, though long in comparison to typical axial load problems where a similar effect occurs, can best be analyzed by a method developed by Johnson. Very short columns are not considered columns; they are analyzed as axial load problems, as discussed previously.

To identify column type, we need to know how the column is supported and its size and shape. This is done by calculating the **radius of gyration,** which is a product of the ratio of the moment of inertia and the cross-sectional area as follows:

$$r = \sqrt{\frac{I}{A}} \qquad (3.8)$$

where r = radius of gyration
  I = moment of inertia
  A = column cross-sectional area

radius of gyration
A measure of the stability of a column found from the square root of the moment of inertia to the cross-sectional area.

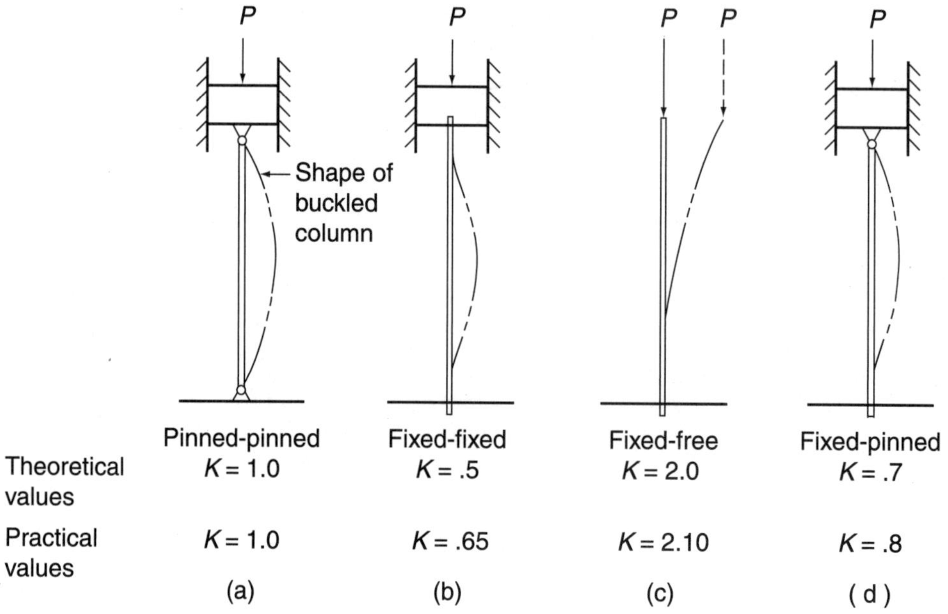

FIGURE 3.8
Types of end support for columns.

	Pinned-pinned	Fixed-fixed	Fixed-free	Fixed-pinned
Theoretical values	$K = 1.0$	$K = .5$	$K = 2.0$	$K = .7$
Practical values	$K = 1.0$	$K = .65$	$K = 2.10$	$K = .8$
	(a)	(b)	(c)	(d)

Note that for a rectangular column or a column that has a smaller moment of inertia in different planes, the smallest value of I would typically be used for the column analysis.

We also need to understand how the ends of the columns are supported, as a pinned-end connection would offer less lateral stability than a fixed end. Figure 3.8 reflects different end conditions for typical types of columns. Note in this figure that, if the theoretical value and a practical value are given for different types of columns due to the less-than-absolute end conditions that may exist, the practical values should generally be used in the analysis.

The factors from Figure 3.8 are then used to determine an "effective" length of the column. Based on this effective length and the radius of gyration, we can determine a value called the "slenderness ratio," which is the effective length of the column in the plane of its lowest radius of gyration. The slenderness ratio is used to determine the method of analysis for a straight, centrally loaded column.

$$L_e = KL \tag{3.9}$$

where L_e = effective length of the column
  K = end support condition factor
  L = actual column length between supports

The slenderness ratio can be found from any of the following ratios:

$$(SR) = \frac{L_e}{r} = \frac{KL}{r} = \frac{L_e}{\sqrt{\frac{I}{A}}} \qquad (3.10)$$

where r = radius of gyration

This slenderness ratio would then be compared to a value called the column constant, which is a property of the modulus of elasticity of the column and the yield strength of the material from which it is constructed. The column constant can be found as follows:

$$C_c = \sqrt{\frac{2\pi^2 E}{S_y}} \qquad (3.11)$$

The relationship between the slenderness ratio and the column constant can be used to determine the type of column to be analyzed as follows:

$$(SR) \text{ or } \frac{L_e}{r} > C_c \quad \text{(Euler)}$$

or

$$(SR) \text{ or } \frac{L_e}{r} < C_c \quad \text{(Johnson)}$$

Once we have determined the type of column, we can make an appropriate analysis using the Euler or Johnson formulas that follow. In determining the slenderness ratio, we must make assumptions about the dimensions of the column and the material from which it is made.

The Euler formula is as follows:

$$P_c = \frac{2\pi EA}{\left(\frac{L_e}{r}\right)^2} = \frac{\pi^2 EA(r^2)}{L_e^2}$$

where P_c = critical load that would cause failure by buckling

or, substituting

$$r = \sqrt{\frac{I}{A}} \quad \text{or} \quad r^2 = \frac{I}{A}$$

then

$$P_c = \frac{\pi^2 EAI}{L_e^2 A}$$

or

$$P_c = \frac{\pi^2 EI}{L_e^2} \tag{3.12}$$

If it is necessary to incorporate a safety factor, the Euler column would be modified as follows:

$$F_{allowable} = \frac{P_c}{N} = \frac{\pi^2 EI}{N(L_e^2)} \tag{3.13}$$

For a Johnson column, the slenderness ratio is less than the column constant; the **force critical** can be found as follows:

$$P_c = AS_y \left[1 - \frac{S_y\left(\frac{L_e}{r}\right)^2}{4\pi^2 E} \right] \tag{3.14}$$

Or, if a safety factor is to be incorporated, the force critical can be modified as follows:

$$F_{allowable} = \frac{P_c}{N} = \frac{AS_y}{N} \left[1 - \frac{S_y\left(\frac{L_e}{r}\right)^2}{4\pi^2 E} \right] \tag{3.15}$$

In either of these formulas, for any shape of column, it is important to note that the direction of failure would be along the plane of the lowest moment of inertia. In many cases, where columns are obviously either long or short and are both round and solid, the Johnson and Euler formulas can be modified as follows:

Euler column (3.13)

$$F_{allowable} = \frac{P_c}{N} = \frac{\pi^2 EI}{N(L_e^2)}$$

force critical

Often referred to as the critical force, this is the force that could act on a column to the point of causing instability, at which point a lower force could continue to cause the column to fail.

Johnson column (3.15)

$$F_{allowable} = \frac{P_c}{N} = \frac{AS_y}{N} \left[1 - \frac{S_y\left(\frac{L_e}{r}\right)^2}{4\pi^2 E} \right]$$

For a round, solid Euler column:

$$D = \left(\frac{64 N P_{act} (L_e)^2}{\pi^3 E}\right)^{1/4} \quad (3.16)$$

Or, correspondingly, for a Johnson column:

$$D = \left(\frac{4 N P_{act}}{\pi S_y} + \frac{4 S_y (L_e)^2}{\pi^2 E}\right)^{1/2} \quad (3.17)$$

The following two example problems demonstrate the use of the Johnson and Euler formulas.

Example Problem 3.5

A connecting link is 30 inches long and ⅝ inch in diameter as shown in Figure 3.9. The link is made from 1040 HR steel. Determine the critical load.

Find the radius of gyration:

$$r = \sqrt{\frac{I}{A}} \quad (3.8)$$

$$r = \sqrt{\frac{\pi D^4}{64} \frac{4}{\pi D^2}} = \sqrt{\frac{D^2}{16}} = \frac{D}{4} \quad \text{(Appendix 3)}$$

$$r = \frac{.625 \text{ in.}}{4} = .156 \text{ inch}$$

Note: $r = \frac{D}{4}$ can be found directly from Appendix 3 for some shapes.

$$K = 1.0 \quad \text{(from Figure 3.8)}$$

Find the effective length:

$$L_e = KL \quad (3.9)$$
$$L_e = (1.0)\, 30 = 30$$

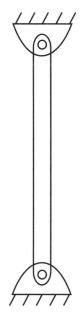

FIGURE 3.9 Pinned end column for Example Problem 3.5.

Find the slenderness ratio:

$$(SR) = \frac{L_e}{r} = \frac{30}{.156} \tag{3.10}$$

$$(SR) = 192$$

Find the column constant:

$$C_c = \sqrt{\frac{2\pi^2 E}{S_y}} \tag{3.11}$$

$$S_y = 42 \text{ ksi} \qquad \textbf{(Appendix 4)}$$

$$C_c = \sqrt{\frac{2\pi^2 \; 30 \times 10^6 \text{ lb/in.}^2}{42,000 \text{ lb/in.}^2}}$$

$$C_c = 118$$

Compare the slenderness ratio to the column constant:

$$\text{Compare } \frac{L_e}{r} \text{ to } C_c$$

$$192 > 118 \qquad \text{Use the Euler formula.}$$

Find the moment of inertia:

$$I = \frac{\pi D^4}{64} \qquad \textbf{(Appendix 3)}$$

$$I = \frac{\pi (.625 \text{ in.})^4}{64}$$

$$I = .0075 \text{ in.}^4$$

Find the critical force:

$$P_c = \frac{\pi^2 E I}{L_e^2} \tag{3.12}$$

$$P_c = \frac{\pi^2 (30 \times 10^6 \text{ lb/in.}^2) \; .0075 \text{ in.}^4}{(30 \text{ in.})^2}$$

$$P_c = 2,470 \text{ lb}$$

Example Problem 3.6

A built-in column with the top free to slide, but not rotate, is made from an A-36 structural steel S-shape beam. S4 × 7.7, 60 inches long, and S_y = 36 ksi. (See Figure 3.10.) Find the acceptable load for a safety factor of 2.

For S4 × 7.7: $\quad A = 2.26$ in.^{2} $\quad\quad I = .764$ in.^{4} $\quad$ **(Appendix 5.4)**

$$r = \sqrt{\frac{I}{A}} \quad (3.8)$$

$$r = \sqrt{\frac{.764 \text{ in.}^4}{2.26 \text{ in.}^2}}$$

$$r = .581 \text{ inch}$$

Find the effective length:

$$L_e = KL \quad (3.9)$$

$$K = .65 \quad \text{(from Figure 3.8)}$$

$$L_e = .65 \ (60 \text{ in.})$$

$$L_e = 39 \text{ inches}$$

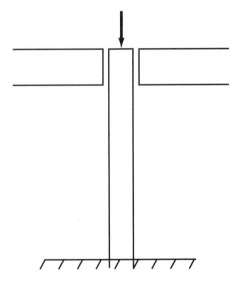

FIGURE 3.10 Column with fixed base and laterally constrained top for Example Problem 3.6.

Find the slenderness ratio:

$$(SR) = \frac{L_e}{r} = \frac{39 \text{ in.}}{.581 \text{ in.}} \quad (3.10)$$

$$(SR) = 67$$

Find the column constant:

$$C_c = \sqrt{\frac{2\pi^2 E}{S_y}} \quad (3.11)$$

$$C_c = \sqrt{\frac{2\pi^2 30 \times 10^6 \text{ lb/in.}^2}{36,000 \text{ lb/in.}^2}}$$

$$C_c = 128$$

Compare the slenderness ratio to the column constant:

Compare (SR) to C_c

$67 < 128$ Use the Johnson formula.

$$F = \frac{AS_y}{N}\left(1 - \frac{S_y\left(\frac{L_e}{r}\right)^2}{4\pi^2 E}\right) \quad (3.15)$$

$$F = \frac{2.26\,(36,000)}{2}\left(1 - \frac{36,000\,(67)^2}{4\pi^2\,30 \times 10^6}\right)$$

$$F = 35,100 \text{ lb}$$

3.8 DESIGN FACTORS AND FACTORS OF SAFETY

When a designer analyzes a part, there are often conditions that are not known, such as the magnitude of the load, the possibility of defects in the material, and conditions that may change over time, such as corrosion and changes in use, that need to be factored into the size of the parts to be used. It is customary to add a margin either by adding a factor to the load acting on the part or by reducing the allowable stresses in our design. For example, designing a part for a load of 1,000 lb if 500 lb is what is expected, or allowing a design stress of 30,000 if a material has yield strength of 60,000. Both of these result in what may be called a factor of safety of 2.

In practice, however, we often group a factor of safety for unknowns into our analysis with a factor for the sophistication of the level of the analysis. As engineers we often do an abbreviated analysis for sizing purposes, at least in the preliminary part of a design. We often refer to this simplified analysis as a "back-of-an-envelope" analysis where all factors may not be taken into account, such as repeated loading, impact, or other factors that a good analysis would include. The designer should be very careful when combining these factors, and simplified analysis should be used only in preliminary sizing calculations when a more detailed analysis is still to be done. Otherwise, the distinction between actual factor of safety and perceived factor of safety can be easily lost.

It is interesting to note the range of factors of safety used in different industries. For instance, the factor of safety used in an aircraft frame design may be much lower than that used in designing farm or construction equipment. When we consider the potential for tragedy in an aircraft design, we might expect the opposite. However, if the same factors were used for airplanes as were used for tractors, typical aircraft would never leave the ground. Obviously when small factors of safety are used, the analysis must include all pertinent conditions.

The appropriate factor of safety for a given product is often developed over time in different industries or different companies. This factor may even vary between different products in a company's product line, such as a luxury car as opposed to an economy car or an expensive toy as opposed to a more "disposable" toy. As you gain more experience in design, you will become more comfortable selecting the appropriate safety factors.

3.9 SUMMARY

After completing this chapter, you should have gained an understanding of the relationship between axial, bending, shear, and torsional loading, and their related stresses. The grouping of stresses into the categories of axial/bending and shear/torsion and the effects of these two categories on the ability of a part to carry loads and stresses should now be better understood. You should better understand how to view the standards for stress resistance as opposed to the presence of actual stresses. You should also understand how a factor of safety determines the relationship between actual stress and stress-allowable levels. The principles of deflection or deformation and how to determine these deflections for typical parts, along with the use of beam deflection formulas, should now be understood, and you should be able to undertake these types of analyses. Lastly, you should be gaining an understanding of what would be appropriate factors of safety and what factors are applicable in determining appropriate levels.

3.10 PROBLEMS

1. Four round rods, 48 inches long, support the weight of an overhead space heater from the ceiling that weighs 2,400 lb. If these rods are made of AISI cold-drawn 1020 steel, how large should they be? Select a factor of safety you think is appropriate for your design. How much are these rods going to stretch under this load?
2. A rock climber's rope is laid out on a flat road and measures 200 feet long. If this rope has an effective diameter of ½ inch and a modulus of elasticity of 5×10^5 lb/in.², how long does the rope become when a 200-lb man is hanging on the end? What is the stress in this rope?
3. If the horizontal beam in Figure 3.11 is to remain horizontal after the load is applied, where should the load be applied? Assume the horizontal beam is rigid and no column action occurs.

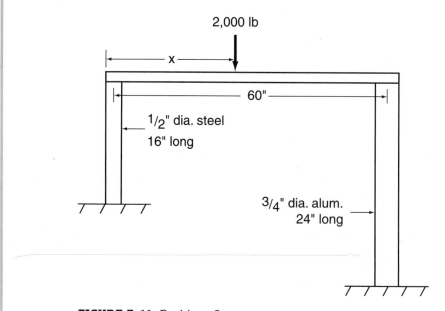

FIGURE 3.11 Problem 3.

4. The cantilevered beam shown in Figure 3.12 is made from steel. What are the maximum stress and the maximum deflection?

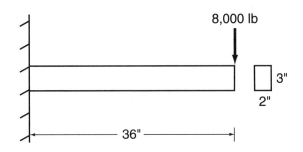

FIGURE 3.12 Problem 4.

5. a. A simply supported aluminum beam, as shown in Figure 3.13, is 30 inches long and carries no weight. Calculate the stress and deflection for its own weight. Assume a density of .07 lb/in.^{3}

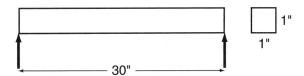

FIGURE 3.13 Problem 5a.

b. A force of 200 pounds is now applied to the center (see Figure 3.14). Calculate the stress and deflection from this load. Ignore the weight of the beam.

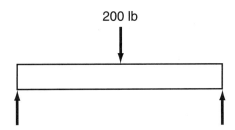

FIGURE 3.14 Problem 5b.

c. What would you now expect the total stress and deflection to be?
d. In your opinion, is it necessary to consider the weight of the beam in this problem?

6. a. In the steel frame shown in Figure 3.15, calculate the maximum stress in the cantilevered beam, the vertical bar, and the ½-inch diameter pin.

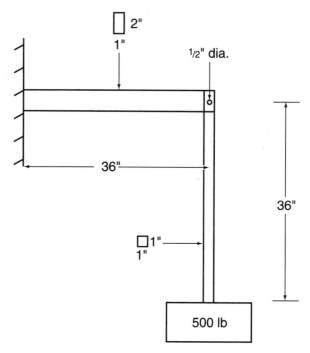

FIGURE 3.15 Problem 6.

b. Calculate the deflection of both the beam and the vertical bar.
c. Ignoring pin deflection, what is the total vertical deflection of the 500-pound weight?

7. For a 2 × 8 inch rectangular cross section of wooden floor beams, the deflection under a given load must be reduced by one-half. Leaving the width the same, by what amount should the height be increased? Assume load in center and beams are simply supported.

8. The same result for problem 7 could have been obtained by reducing the distance between floor beams. By what percentage should this distance be reduced (assume uniform floor load)?

9. A 38-inch-long torsion bar is made from a .25-in.-diameter C1050 steel bar, as rolled. What is the maximum twisting angle in degrees to which the bar can be subjected before failure?

10. A hollow shaft has an outside diameter of $D_o = 1.250$ in. and inside diameter of $D_i = 1.000$ in. The steel it is made from has an $S_{ys} = 24,000$ psi and $G = 11.5 \times 10^6$ psi.
    a. What hp can this shaft transmit at 600 rpm when $N = 2$?
    b. What is the torsional deflection under this load when $L = 40$ in.?
11. A solid shaft has a diameter of 1.25 inches. It is replaced by a hollow shaft of 1.5 in. outside diameter (OD). For the same value of outer fiber stress (torque and rpm remaining the same), what should the inside diameter (ID) be? What savings in weight-per-foot length will be realized?
12. In a car with front-wheel drive, the wheel drive shafts are 10 and 17 inches long, respectively. In order to avoid pulling to one side during acceleration, these shafts should have the same torsional deflection angle for a given torque. What should be the ratio of their respective diameters? (Hint: Write the results for the twisting angle for each shaft and equate.)
13. If, in Example Problem 3.5, the column is 50 inches long, made from AISI 1040 HR steel, is round, and is required to carry a load of 8,000 pounds, determine the required diameter for a safety factor of 3.
14. For a sling as shown in Figure 3.16, select a beam. Assume it is made of steel, a safety factor of 2 is desired, and the length between pins is 96 inches.

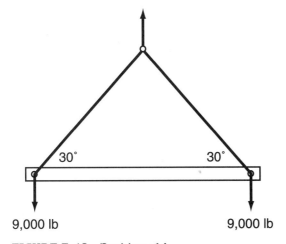

FIGURE 3.16 Problem 14.

15. For the assembly shown in Figure 3.17:
    a. Select an appropriate size solid pipe, assuming $S_y = 36,000$ lb/in.^{2} and a safety factor of 3.
    b. Select a 4 × 4 hollow square tube for this application.

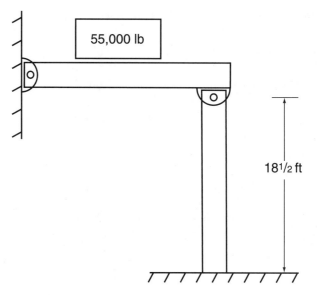

FIGURE 3.17 Problem 15.

3.11 CUMULATIVE PROBLEMS

16. If the truck shown in Figure 3.18 carries a uniformly distributed load of 40,000 pounds and the frame of this truck is made from four Aluminum Association 12 × 7 (14.292 lb/ft) standard I-beams, find the maximum stress in these beams. Compare this value to the yield strength of 6061-T6 aluminum. Considering the unknowns to which the truck may be subjected, do you think this safety factor is suitable?

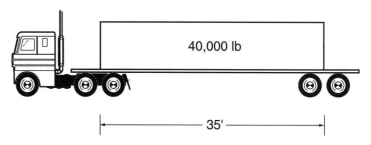

FIGURE 3.18 Problems 16 and 17.

17. Find the deflection in problem 16. Do you think this is suitable? If not, suggest a possible change to this design (see Figure 3.18).

CHAPTER 4

Combined Stress and Failure Theories

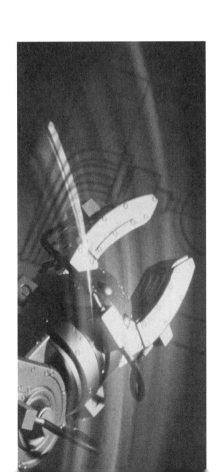

Combined stresses occur in a part when we have multiple types of loading or more than one type of stress from a single load. An example of combined stresses is the case of a shaft that is subject to a torque that creates a torsional stress and also a bending load. This could be caused by a cable winding on the shaft, which then causes both torsional and bending stresses. Another example might be a conventional C-clamp on which there is bending in the frame of the clamp as well as tension from the lead screw assembly. Because multiple stresses can occur in the same location on a part, we need to be able to combine these stresses in order to evaluate their compounding effect on the ability of the material to resist this load.

Because of the multiple ways in which loads can be combined, different failure theories have been developed to predict when a material will fail under combined stresses. This chapter will review three of the most common failure theories.

4.1 OBJECTIVES

As you learned in your strength of materials course, multiple types of stresses can occur in a part. This chapter is going to take a slightly simplified but practical approach. The stresses we will calculate with this approach will generally be slightly higher than the actual, so this method will give conservative results. This simplified approach, however, may result in a better understanding of how to combine different types of stresses. If you are going to specialize in an area such as shaft design or high-level structural analysis, further investigation should be done into a more complete analysis of the factors involved in combined stresses. After completing this chapter, you should be able to:

- Group stresses by type, separating the stresses into bending and axial versus shear and torsional stresses.
- Combine like types of stresses in an appropriate manner.
- Combine different types of stresses, utilizing appropriate combined stress theories.
- Gain further understanding into how these combined stresses should be compared to the stress allowables for the materials being used in the design.

4.2 GROUPING OF SIMILAR AND DIFFERENT TYPES OF STRESSES

As reflected in Appendix A, there are two different types of stresses that must be considered in structural analysis. The materials that we usually use, such as steel and aluminum, have differing abilities to resist these two types of stresses. Generally, axial and bending stresses can be compared to yield or ultimate strength directly, whereas when a part is subject to torsion or shear stresses, the ability of the material to resist these types of stresses is usually between 50 and 60 percent of the typical values for axial and bending (normal stresses). In order to determine their combined effect on a part, therefore, we would generally divide the kinds of loads into those that either create axial or bending stress or those that create torsion and shear stresses. Also, in almost all cases we will ignore the beam **transverse shearing stress** as it is typically insignificant.

4.3 COMBINED AXIAL AND BENDING STRESSES

Frequently we encounter applications where a part is subject to both a bending and axial stress. Envision a short column loaded eccentrically such that the **eccentricity of the load** causes a bending in the column along with the axial load along its main axis. We can see from this example that both of these stresses would have an effect on the part and, as they may be either tensile or compressive normal stresses, they can be easily combined. We take into account whether they are tensile or compressive as follows:

$$S = \pm S_{axial} \pm S_{bending} \quad (4.1)$$

or

$$S = \pm \frac{F}{A} \pm \frac{M}{Z} \quad (4.2)$$

(Note that we will arbitrarily call tensile stresses positive and compressive stresses negative.)

---

transverse shearing stress
A stress in a beam parallel to the neutral axis caused by the difference in deflection longitudinally between the tensile and compressive stresses acting upon a beam.

eccentricity of the load
A load that is applied to a column but not on the column's centerline. This could cause a bending stress along with the column loading.

Combined normal stress (4.2)

$$S = \pm \frac{F}{A} \pm \frac{M}{Z}$$

Example Problem 4.1: Design of Short Column with Eccentric Load

Determine the stress in the two-inch-diameter vertical column shown in Figure 4.1.

First, determine stresses.

Axial stress:

$$S = \frac{F}{A} = \frac{10,000 \text{ lb}}{\frac{\pi(2 \text{ in.})^2}{4}} = 3,183 \frac{\text{lb}}{\text{in.}^2}$$

Bending stress:

$$S = \frac{M}{Z} = \frac{Fe}{\frac{\pi D^4}{32}} = \frac{10,000 \text{ lb }(2 \text{ in.})}{\frac{\pi(2 \text{ in.})^4}{32}} = 25,460 \text{ lb/in.}^2$$

$$S = \pm S_{axial} \pm S_{bending} \quad \quad (4.1)$$

$$S = -3,183 \text{ lb/in.}^2 - 25,460 \text{ lb/in.}^2$$

$$S = -28,647 \text{ lb/in.}^2$$

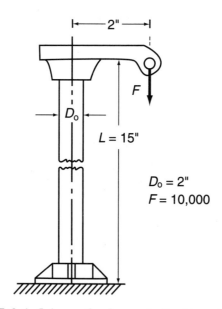

FIGURE 4.1 Column for Example Problem 4.1.

Note that the bending stress and axial stress are added on the inner side of the column. The stresses are subtracted on the outer side, so we are primarily concerned about the inner surface.

4.4 COPLANAR SHEAR STRESSES IN MORE THAN ONE DIRECTION

Shear stresses created by forces acting in different directions at the same point in a part may be added vectorially, or the resultant force could be determined first and then the shear calculated. Either approach typically results in the same stress level.

It should be remembered that this stress is a shear stress and, as such, should be compared to the yield or ultimate shear strength of the material.

Example Problem 4.2: Coplanar Shear

Determine stress in the ½-inch-diameter pin in Figure 4.2.

As the bell crank is free to rotate, both forces create shear stresses in the pin.

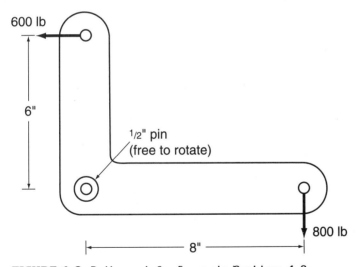

FIGURE 4.2 Bell crank for Example Problem 4.2.

Adding forces vectorially:

$$F_t = \sqrt{(F_1^2 + F_2^2)} \quad \text{(perpendicular forces)}$$

$$F_t = \sqrt{(600 \text{ lb})^2 + (800 \text{ lb})^2}$$

$$F_t = 1,000 \text{ lb}$$

$$S = \frac{F}{A} = \frac{1,000 \text{ lb}}{\frac{\pi(1/2 \text{ in.})^2}{4}} = 5,093 \text{ lb/in.}^2$$

Note that if the forces were not perpendicular, they would be added vectorially, or the individual shear stresses could be determined and added in a like manner.

4.5 COMBINED SHEAR AND TORSIONAL STRESS

Another example may be a case where a part is subjected to a combined shear and torsional stress. This may be a rotating shaft for which the torque creates a torsional stress in the shaft and the load location creates a shear stress. Because these stresses are of the same general type with regard to the load capacity of materials to resist them, they can simply be added together.

This, however, is not exactly correct because there are some angular differences at which the plane of the stress occurs. However, adding the stresses will give us a reasonable, though conservative, answer. As the shaft is also typically rotating, we do not need to be concerned with direction and, hence, we would add the stress from shear and torsion directly in accordance with the following formula. Note that these stresses again should be compared to the shear stress allowable which, as discussed previously, is typically 50 to 60 percent of the yield or ultimate stress allowable.

$$S = S_{torsion} + S_{shear} \quad (4.3)$$

Example Problem 4.3: Combined Torsion and Shear

A roller chain system transmits 50 hp at a speed of 300 rpm. If the chain sprocket has an effective **pitch diameter** of ten inches, calculate the combined stress in the 1-inch-diameter shaft. (See Figure 4.3.)

> **pitch diameter**
> The point on a sprocket or gear at which the load is assumed to be applied. This is sometimes referred to as the "running circle."

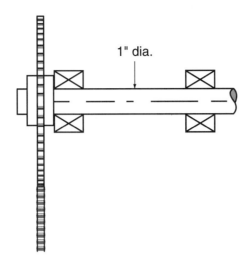

FIGURE 4.3 Shaft subject to both shear and torsional stress.

$$T = \frac{63{,}000 \text{ (hp)}}{n} \qquad (2.6)$$

$$T = \frac{63{,}000 \text{ (50 hp)}}{300 \text{ rpm}}$$

$$T = 10{,}500 \text{ in.-lb}$$

Driving force in chain:

$$F = \frac{T}{r} = \frac{10{,}500 \text{ in.-lb}}{\frac{10 \text{ in.}}{2}} \quad \text{as } T = Fr \qquad (2.5)$$

$$F = 2{,}100 \text{ lb}$$

Calculate shear and torsional stresses.

Shear in shaft:

$$S_s = \frac{F}{A} = \frac{2{,}100 \text{ lb}}{\frac{\pi(1 \text{ in.})^2}{4}}$$

$$S_s = 2{,}674 \text{ lb/in.}^2$$

Torsional stress:

$$S_s = \frac{T}{Z'}$$

$$Z' = \frac{\pi D^3}{16} \qquad \text{(Appendix 3)}$$

$$Z' = \frac{\pi (1 \text{ in.})^3}{16} = .196 \text{ in.}^3$$

$$S_s = \frac{10,500 \text{ in.-lb}}{.196 \text{ in.}^3}$$

$$S_s = 53,570 \text{ lb/in.}^2$$

$$S_{total} = S_{torsion} + S_{shear} \qquad (4.3)$$

$$S_{total} = 53,570 \text{ lb/in.}^2 + 2,674 \text{ lb/in.}^2$$

$$S_{total} = 56,255 \text{ lb/in.}^2$$

This value would be compared to shear stress allowable for the shaft material.

4.6 MOHR'S CIRCLE

Mohr's Circle
A graphical method for combining stresses. In Mohr's Circle, the stress values are added based on their directions graphically, and then the combined stresses are determined from this graphical combination.

You probably studied **Mohr's Circle** in your strength of materials course; it is an excellent method for combining stresses. Mohr's Circle will not be reviewed again in this text. However, you should refresh your understanding of this method if you will be doing significant amounts of combined stress analysis because, though slightly more complicated than the techniques given in this text, Mohr's Circle has some analytical benefits, including the ability to calculate angles of maximum stress.

4.7 COMBINED NORMAL AND SHEAR STRESSES

In this case, we have what is typically called a principal or normal stress that would result from a bending or axial loading condition, in combination with either a shear or a torsional stress. Examples of this could be a rotating shaft where a bending load is also present or a rotating shaft subject to torque and an axial load. These can be combined as follows:

$$\sigma = \frac{S}{2} \pm \left[S_s^2 + \left(\frac{S}{2}\right)^2 \right]^{1/2} \quad (4.4)$$

where σ = equivalent combined normal stress
 S = normal stress from bending or axial loads
 S_s = shear or torsional stress

Note that the normal stress would be calculated from axial loads such as $\frac{F}{A}$ and bending loads such as $\frac{Mc}{I}$ or $\frac{M}{Z}$. The shearing stress could be from an $\frac{F}{A}$ shear, $\frac{Tc}{J}$ torsion shear, or $\frac{VQ}{IB}$ vertical beam shear stress, as appropriate. As this formula gives us equivalent normal stresses, these values should then be compared to S_y or S_u as appropriate, including factors of safety.

Example Problem 4.4: Combined Normal and Shear Stress

A center-mounted chain drive system transmits 20 hp at a speed of 500 rpm. If the sprocket has a pitch diameter of eight inches, would this be an acceptable design if the shaft is made of hot-rolled AISI 1020 steel and a safety factor of 2 based on yield is desired? (See Figure 4.4.)

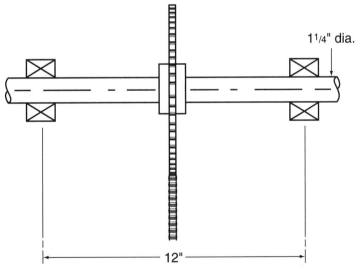

FIGURE 4.4 Shaft subject to torsional shear and bending (normal) stresses.

Principal stress theory (4.4)

$$\sigma = \frac{S}{2} \pm \left[S_s^2 + \left(\frac{S}{2}\right)^2 \right]^{1/2}$$

Torque on shaft:

$$T = \frac{63{,}000 \text{ hp}}{n} \quad \text{(2.6)}$$

$$T = \frac{63{,}000 \ (20 \text{ hp})}{500 \text{ rpm}}$$

$$T = 2{,}520 \text{ in.-lb}$$

Force in chain:

$$F = \frac{T}{r} = \frac{2{,}500 \text{ in.-lb}}{\dfrac{8 \text{ in.}}{2}} = 630 \text{ lb}$$

Bending moment in shaft (see Figure 4.5):

$$M_m = \frac{FL}{4} \quad \text{(Appendix 2)}$$

$$M_m = \frac{630 \text{ lb } 12 \text{ in.}}{4}$$

$$M_m = 1{,}890 \text{ in.-lb}$$

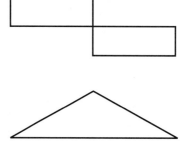

FIGURE 4.5 Shear and bending moment diagram for Example Problem 4.4.

Calculate torsional and bending stresses.

Torsional shear stress:

$$S_s = \frac{T}{Z'}$$

$$S_s = \frac{2{,}520 \text{ in.-lb}}{\frac{\pi D^3}{16}} \quad \text{(Z' from Appendix 3)}$$

$$S_s = \frac{2{,}520 \text{ in.-lb}}{\frac{\pi(1.25 \text{ in.})^3}{16}}$$

$$S_s = 6{,}570 \text{ in.-lb}$$

Bending (normal) stress:

$$S = \frac{M}{Z} = \frac{1{,}890 \text{ in.-lb}}{\frac{\pi D^3}{32}} \quad \text{(Z from Appendix 3)}$$

$$S = \frac{1{,}890 \text{ in.-lb}}{\frac{\pi(1.25 \text{ in.})^3}{32}}$$

$$S = 9{,}860 \text{ lb/in.}^2$$

Combining stresses:

$$\sigma = \frac{S}{2} \pm \left[S_s^2 + \left(\frac{S}{2}\right)^2 \right]^{1/2} \quad (4.4)$$

$$\sigma = \frac{9{,}860 \text{ lb/in.}^2}{2} + \left[\left(6{,}570 \text{ lb/in.}^2\right)^2 + \left(\frac{9{,}860 \text{ lb/in.}^2}{2}\right)^2 \right]^{1/2}$$

$$\sigma = 13{,}150 \text{ lb/in.}^2$$

$$S_y = 30{,}000 \text{ lb/in.}^2 \quad \text{(Appendix 4)}$$

$$\frac{S_y}{2} = 15{,}000 \text{ lb/in.}^2$$

This is greater than the combined stress of 13,150 lb/in², so it is acceptable.

4.8 COMBINED MAXIMUM SHEAR STRESS

In this case, we would also have a combination of shear and normal stresses acting at a point on a part. This slightly different theory of failure gives slightly different results because in this case we would compare the resultant values to the shear stress allowables that, as discussed previously, are typically 50 to 60 percent of yield or ultimate.

$$\tau = \left[S_s^2 + \left(\frac{S}{2} \right)^2 \right]^{1/2} \tag{4.5}$$

where τ = maximum combined shear stress
S_s = shear stress
S = normal stress

In general, ductile materials such as soft steels, which are generally weaker in shear, are often designed using the maximum shear stress theorem, whereas less ductile materials, such as cast iron or hardened steel, are usually designed using the principal stress methods.

Shear stress theory (4.5)

$$\tau = \left[S_s^2 + \left(\frac{S}{2} \right)^2 \right]^{1/2}$$

4.9 DISTORTION ENERGY THEORY

The **distortion energy theory** is often called the von Mises or the Mises-Hencky theory. This theory appears to closely duplicate the failure of ductile materials under static, repeated, and combined stresses. Hence, it is often used in combination with the combined stress methods discussed in this chapter or the fatigue stress methods that will be discussed in Chapter 5. In order to use this method, we need to determine the two principal stresses using the methods described in this chapter or using Mohr's Circle. We then combine them as follows to determine the von Mises stress:

$$\sigma' = \sqrt{\sigma_1^2 + \sigma_2^2 - \sigma_1 \sigma_2} \tag{4.6}$$

where σ' = von Mises stress
σ_1 = principal stress or combined stress
σ_2 = principal stress or combined stress

distortion energy theory
A method of combining principal stresses in ductile materials to predict failure.

From this theory, we could obtain a value at which predicted failure would occur when this stress level exceeds the allowable yield for the material as follows:

$$\sigma' > S_y$$

More complicated cases involving multiple normal or multiple shear stresses will not be covered in this chapter. When those circumstances arise, you should realize that additional training or the possible use of Mohr's Circle would be appropriate.

There are times when we may be concerned about the angle of maximum stress. Examples would include cases where we are using **strain gages** to measure stress in a part, or sometimes, in **forensic engineering**, where we are trying to determine what forces and stresses were acting on a part that had unexpectedly failed. In cases like these, Mohr's Circle could be used or additional research could be done on the appropriate methods for determining the angle of maximum stress.

It is interesting to note that, when a ductile material fails under a simple axial stress, the angle of failure is cone-shaped and is usually about 45 degrees. For a very brittle material, this angle would be close to 90 degrees. In the presence of combined stresses, determining this failure angle is quite difficult.

4.10 SUMMARY

After completing this chapter, you should be able to recognize the different kinds of stresses and, where they are similar, add them together, properly noting the difference between compressive and tensile stresses for normal stresses and adding directly the shear and torsional stresses when they exist in the same location.

You should now understand how to use the two most common combined stress theories and how to compare the results of these theories to the stress allowables. Remember that not all cases were covered in this chapter. When more complicated cases arise, the methods in this chapter should be expanded upon.

strain gages
Used to measure stress in a part. A strain gage is essentially a miniature resistor that changes resistance as it is stretched and can measure strain or unit deflection when epoxied to a part with the appropriate instrumentation.

forensic engineering
The process of investigating what happened to a part, machine, or structure that unexpectedly failed. Typically, this process tries to determine what loads and associated stresses were acting on a part to cause the unexpected failure.

4.11 PROBLEMS

In Problems 1 through 7, using the same hypothetical machine member, calculate individual stresses and then combine them to find the net resultant stresses for each different combination of loading. In all cases, you should determine the maximum stresses.

1. Axial load only. Determine the axial (normal stress) in the part shown in Figure 4.6.

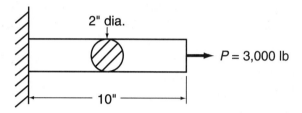

FIGURE 4.6 Problem 1: Axial load only.

2. Determine the maximum bending stress in the part shown in Figure 4.7.

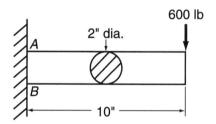

FIGURE 4.7 Problem 2: Bending stress only.

3. Determine the maximum torsional stress in the part shown in Figure 4.8.

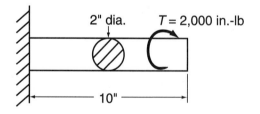

FIGURE 4.8 Problem 3: Torsional stress only.

4. Determine the combined stress in the part from the bending and torsional loads. Determine this stress using both the principal stress method and the maximum shear stress theorems. (See Figure 4.9.)

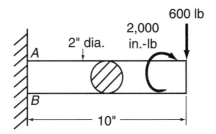

FIGURE 4.9 Problem 4: Combined bending and torsional loads.

5. Determine the maximum stress in the part from the bending and axial loads. (See Figure 4.10.)

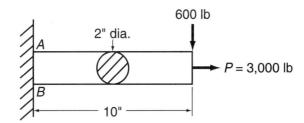

FIGURE 4.10 Problem 5: Combined bending and axial loads.

6. Determine the maximum combined stress from the torsional and axial loads using both the principal stress method and the maximum shear stress theorems. (See Figure 4.11.)

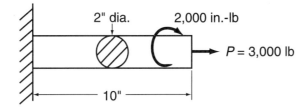

FIGURE 4.11 Problem 6: Combined torsional and axial loads.

7. Determine the stress from the bending, axial, and torsional loads, using both the principal stress method and the maximum shear stress theorem (see Figure 4.12). Note that in this problem the bending and axial load stresses should be combined before entering them into the principal normal stress method or maximum shear stress theorems.

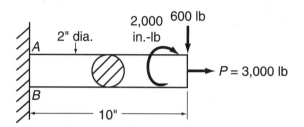

FIGURE 4.12 Problem 7: Combined axial, bending, and torsional loads.

8. From Figure 4.13, in the cantilevered beam assembly that supports a pulley and cable, determine the maximum combined stress. Select the appropriate theorem based on the fact that this assembly is made from a low-strength ductile steel material.

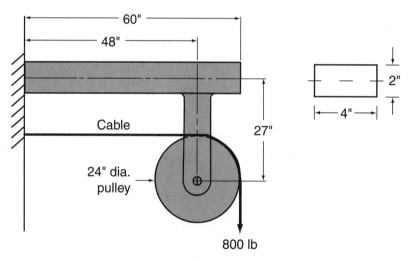

FIGURE 4.13 Problem 8: Combined stresses on beam.

9. In Figure 4.14, calculate the maximum combined stress at section A-A. Select the appropriate method of combining these stresses, based on the fact that this part is made of a fairly brittle cast iron material.

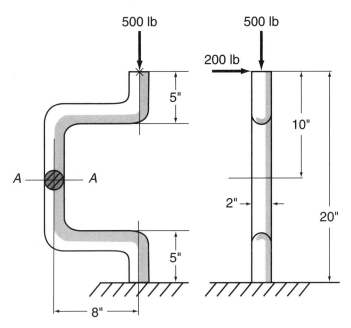

FIGURE 4.14 Problem 9: Combined stresses on cast iron part.

10. In Figure 4.15, a crank arm assembly is loaded as shown. Using the maximum shear stress theorem, determine the combined stress at points *A* and *B*.

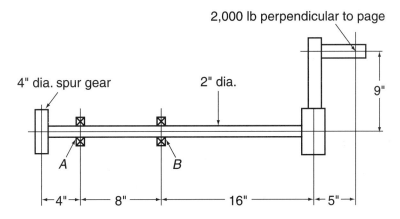

FIGURE 4.15 Problem 10: Crank arm assembly.

11. In Figure 4.16, an elephant with a mass of 120 kg stands on the edge of a table that is fixed rigidly in a concrete base. Calculate the combined stress in the vertical tube. If this column was made of AISI 1020 HR steel, would this be an acceptable design if a safety factor (SF) of 3 based on yield was desired?

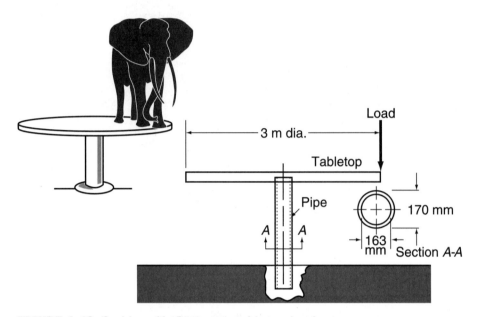

FIGURE 4.16 Problem 11: Eccentric tabletop load.

4.12 CUMULATIVE PROBLEMS

12. For the locomotive engine in Problem 11 in Chapter 2, determine the combined stress at the maximum point for a 2½-inch-diameter axle, assuming that the weight of the engine on flat ground is equally distributed and results in a vertical load on the axle of 40,000 pounds, which results in a force per wheel of 20,000 pounds. Assume that the bearings supporting this wheel are immediately adjacent such that no bending loads are exerted in the shaft and one-half of the total torque is transmitted by each axle.
13. For Problem 12, if a safety factor of 2 based on ultimate is desired, select a material.

CHAPTER 5

Repeated Loading

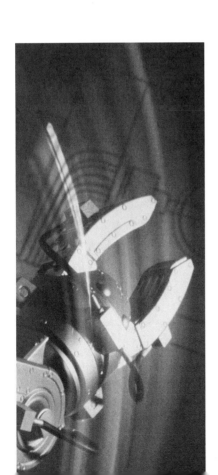

When a part is subjected to a load that is repeatedly applied and removed, we are concerned about a phenomenon called fatigue failure. If we assume that any material from which we would construct a part inherently has **microscopic flaws** either because of the material, the manufacturing methods, or the shape of the part where sharp changes in geometry lend themselves to stresses being concentrated under load, we can expect that each time the load is removed and reapplied, the concentrated stresses in these areas could cause a progressive microscopic failure in the intermolecular attraction within the material. This phenomenon is frequently referred to as **crack propagation,** whereby a microscopic defect grows very slowly over time due to the application of repeated loads. Often when a load is reapplied, this postulated crack may not close in exactly the same manner, resulting in some prying actions around the tip of the crack as well. Because of this action, we often see something that resembles the receding lines on a beach as the tide is going out. We refer to these lines on a failed part as **beach marks,** that tell us that the part failed because of fatigue as opposed to a one-time load.

Because of this phenomenon, we need to design parts that may be subject to repeated loading so that the stresses at the critical locations are low enough that a hypothetical defect will not propagate. In this chapter, we will learn how to analyze the effects of these stress concentrations and how to reduce the localized stress to the level at which crack propagation will not occur.

microscopic flaws
Defects in a part typically between grain boundaries, usually visible with a microscope but not with the naked eye.

crack propagation
A condition in which the microscopic flaw would increase in size during each flexure of a repeated load.

beach marks
The result of crack propagation where, during each flexure cycle as the part recloses in a slightly different position, some wearing occurs on the part, resulting in an appearance that looks like the tide going out on a beach.

5.1 OBJECTIVES

After completing this chapter, you should be able to:

- Identify the various kinds of loading encountered on a part and learn to combine them as appropriate.
- Determine the level of stress in a material at which a hypothesized defect would not propagate.
- Recognize what types of factors affect this endurance limit for the material.
- Factor in the effect of shapes and discontinuities as they affect the stress concentration factors.
- Gain experience in using fatigue equations when designing parts subject to repeated loads.

5.2 MECHANISMS OF FATIGUE

Picture the stresses on a hypothetical defect as being similar to the flow of water in a pipe as it goes around a burr on the pipe's surface or changes velocity where the pipe changes shape. We can envision the stresses on a part, which could be best thought of as velocity profiles, undergoing a similar phenomenon. Figures 5.1, 5.2, and 5.3 show examples of the flow being concentrated as it flows through a pipe. This is similar to what would happen at the tip of our postulated defect that is caused by a change in shape. If our part is subject to repeated loading, this concentration of stresses causes the defect to grow over time until failure could occur solely by means of a one-time application of the stress. This is illustrated in Figure 5.4. The beach marks would be present up until the point at which a conventional failure occured due to a one-time application of the load.

Envision for a moment or, better yet, get a stick of chewing gum, hold it between your fingers, and flex it back and forth. After a few cycles, stop and look at the gum carefully and you will see microscopic cracks starting to develop. As you continue to flex the gum back and forth, you will begin

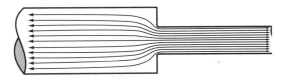

FIGURE 5.1 Stress concentration from change in cross section.

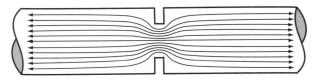

FIGURE 5.2 Stress concentration caused by a snap ring or O-ring groove.

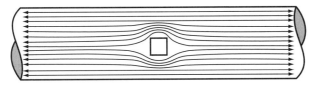

FIGURE 5.3 Stress concentration caused by central hole.

FIGURE 5.4 Failure caused by repeated loading.

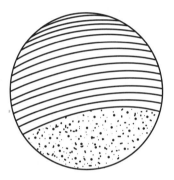

FIGURE 5.5 Note the smooth curves resembling those left by the receding tide up until the point of abrupt failure.

to see larger cracks propagating, likely from the corners, until such point that a single flexure will cause the gum to fail. This is simply an acceleration of the phenomenon that would likely occur in a part subject to fatigue loading. Note in our gum example that the growth of the cracks occurs on the side of the gum in tension and that the beach marking would be occurring on the side in compression, which is reversed during each cycle (see Figure 5.5).

5.3 ENDURANCE LIMIT AND ENDURANCE STRENGTH

We know from strength of materials and from the first part of this text that if a part is stressed past its ultimate strength it will fail under a single loading. Sometimes we also need to know how much stress a part could be subject to for an infinite number of cycles without failing. Figure 5.6 shows this relationship between the failure stress and number of cycles. You can see from this figure that a limited number of cycles would not result in failure if the stress was below the ultimate stress allowable and that there is a value (usually around a million cycles) at which a material could be expected to last forever if the stresses are below this level. It is this stress level that we call the **endurance limit** or endurance strength (S_n). If the stresses around our **postulated defect** are below this level, they would not propagate. This is the value that we would normally refer to as the endurance limit. Later in this chapter we will learn about the factors that need to be incorporated to modify this limit, such as changes in shapes, but for a smooth, polished part, we can assume that this value of S_n is the stress level at which our part would last forever. If our design had a known number of cycles under a million, we could use Figure 5.6 to find the allowable stress for this number of cycles.

endurance limit
If a load were applied repeatedly to a part, this is the stress level that would not cause a postulated defect to propagate.

postulated defect
An assumed microscope flaw that could propagate when subjected to repeated loads.

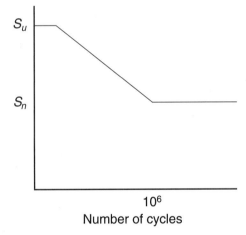

FIGURE 5.6 Endurance limit versus number of cycles.

For many materials, we can find a published value for endurance strength. Often, however, this information is not available. For a typical steel, when we have no better source of information, we will use one-half of the ultimate strength as an acceptable conservative value.

$$S_n = .5\, S_u$$

5.4 MODIFYING FACTORS FOR ENDURANCE LIMITS

There are many factors that modify the endurance limit and affect the ability of a material to survive under repeated loading conditions. Among these factors are the type of stress, the size of the part, and the surface finish conditions as follows:

$$S_n = C_{type}\, C_{size}\, C_{surface}\, S_n' \tag{5.1}$$

S_n' = endurance limit from literature

Type of Stress

Tests have shown that the ability of materials to survive under repeated loading varies depending on the type of stresses. Bending, which is the method typically used to determine the endurance strength, generally requires no factor. However, axial and shear loads require factors of .8 and .6, respectively, and these factors would be used to modify the endurance limit.

$$C_{type} = .8 \quad \text{axial}$$
$$C_{type} = .6 \quad \text{shear or torsion}$$
$$C_{type} = 1 \quad \text{bending}$$

Size

The size of our part also has an effect on its ability to resist repeated loads. There are a couple of major reasons for this. First, as parts become larger, their material properties are more likely to vary throughout their thickness because of the processes used in their manufacture. This is especially true when heat-treating or other processes such as hard drawing are used because the resulting internal condition of the part can be significantly different than the surface condition. Second, in bending and torsional conditions, as well as somewhat in shear, the loads typically are concentrated on the outer surface. Therefore, we will modify the endurance limit for size as follows:

$$C_{size} = 1 \quad \text{for} < \tfrac{1}{2} \text{ inch}$$
$$C_{size} = .85 \quad \text{for } \tfrac{1}{2} \text{ to 2 inches}$$
$$C_{size} = .75 \quad \text{for 2 inches and over}$$

Surface

Surface conditions can have a significant effect on a material's ability to resist repeated loading. Very rough surfaces, such as as-forged or cast surfaces, have inherent **stress risers** due to the surface roughness and the hot or cold working that may have been done as part of the manufacturing process. Even such things as machining marks on the surface can become potential sources of stress risers that would then propagate under certain loading conditions. Figure 5.7 shows modifying factors for these surface conditions. Note that Figure 5.7a gives S_n directly, whereas 5.7b results in a percentage reduction in values. These two techniques yield similar yet different results. We can summarize these factors as follows:

$$S_n = C_{type} \, C_{size} \, C_{surface} \, S_n'$$

S_n' from literature

or

$$S_n = C_{type} \, C_{size} \, C_{surface} \, .5 \, S_u$$

S_u if values for S_n' are not known

> **stress risers**
> Often referred to as discontinuities or stress concentrations at a point where due to shape changes in a part where stresses are concentrated when subject to a load.

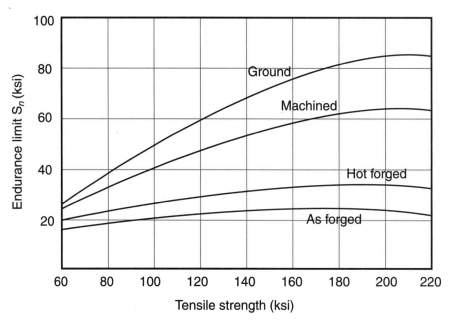

FIGURE 5.7(a) Surface effect: Note that this graph incorporates the surface factor and $.5S_u$ to obtain S_n.

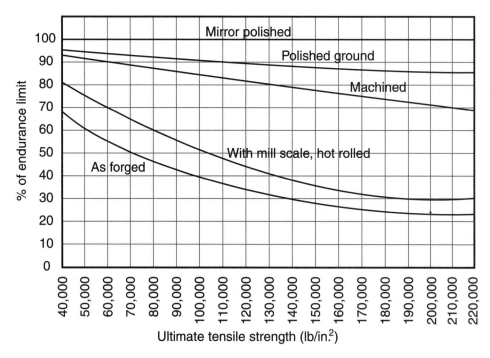

FIGURE 5.7(b) This graph modifies the endurance limit: This will yield results similar to those of Figure 5.7a.

5.5 VARIATION OF STRESSES

The effect of repeated loading on a part is also related to whether the loading and the associated stress is a **repeated and reversed stress,** a **repeated stress,** or if there is a steady stress in addition to the fluctuating stresses.

If we go back to the chewing gum example earlier in this chapter and, instead of flexing the gum back and forth, we flex it so that the load is applied repeatedly to one side but never the other. The amount of time to failure in this new scenario is much longer than in the other. There are many reasons for this, but the most important is that, when there is no alternating between compressive and tensile stress, the prying action about the point of the postulated defect is reduced, and the total level of change in the stress is correspondingly less. Therefore, the effect on a part is significantly less from a repeated load than from a repeated and reversing load. Consistent stress on a part affects the general stress-carrying ability of the material, typically the yield strength (S_y), but does not appear to have the same effect on the alternating stress capability (S_n).

Therefore, to distinguish between the tensile and compressive stress, we will label the tensile stress as a positive stress and the compressive stress as negative. We then calculate our minimum and maximum stress, so that compressive stress is negative and the minimum stress and tensile stress is the maximum positive stress.

For example, envision a rotating shaft with a side load on it. As the shaft rotates, the stresses are reversed due to the difference in orientation. Conversely, envision truck traffic repeatedly crossing the deck of the bridge, resulting in loading that, although repeated, is not reversed. Or, in a third scenario, envision the weight of the concrete deck being placed upon the support beams and then adding the truck traffic, resulting in a situation where there is a consistent stress that is then added to by the load of the trucks crossing the bridge. These different scenarios can be thought of as a repeated and reversed stress, a repeated but not reversed stress, and a repeated but not reversed with a residual consistent, steady stress. In order to analyze these conditions, we must distinguish between the loading that causes a tensile stress and that which causes a compressive stress in the part.

The first step in the process is to calculate the minimum and maximum stress based on the type of loading present. Note that our minimum stress may often then be the negative compressive stress. First, determine the maximum and minimum stress for the different types of loading as follows:

> **repeated and reversed stress**
> A stress that varies from tensile to compressive or reversing direction shear that can exacerbate the effect of a postulated defect on a part.
>
> **repeated stress**
> A stress that is repeated over and over again but does not go through a reversal cycle.

$$S = \frac{F}{A} \qquad S = \frac{M}{Z} \qquad S = \frac{T}{Z'}$$

or

$$S_{min} = \frac{F_{min}}{A} \quad \text{or} \quad S_{min} = \frac{M_{min}}{Z} \quad \text{or} \quad S_{min} = \frac{T_{min}}{Z'}$$

$$S_{max} = \frac{F_{max}}{A} \quad \text{or} \quad S_{max} = \frac{M_{max}}{Z} \quad \text{or} \quad S_{max} = \frac{T_{max}}{Z'}$$

Next calculate the alternating and mean stresses (note in many cases S_{min} may be negative) as follows:

Mean stress (5.2)

$$S_{mean} = \frac{S_{max} + S_{min}}{2} \tag{5.2}$$

Alternating stress (5.3)

$$S_{alt} = \frac{S_{max} - S_{min}}{2} \tag{5.3}$$

Graphically, we can represent these different conditions as in Figure 5.8.

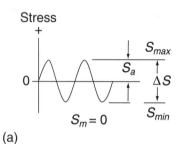
(a)

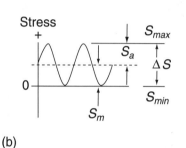
(b)

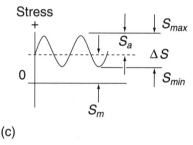
(c)

FIGURE 5.8 Varying stress conditions: (a) fully reversed, (b) repeated, (c) fluctuating with consistent, steady load.

5.6 THEORIES OF FAILURE: SODERBERG EQUATION

Soderberg equation
A method for evaluating the effect of different types of repeated stresses on a part.

There are several theories on how to analyze the effect on a part from repeated loading. The most common theory is the **Soderberg equation**, which evaluates the relationship between the mean stresses, the alternating stresses, and the allowable yield stress and endurance stress limits. The formula for the graph of the Soderberg equation is as follows:

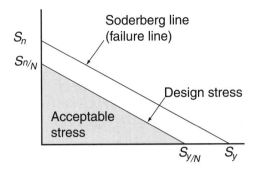

FIGURE 5.9 Graph of the Soderberg equation.

$$1 = \frac{S_m}{S_y} + \frac{S_a}{S_n} \qquad (5.4)$$

You can see from a careful review of this formula and Figure 5.9 that we are, in essence, comparing the mean stress in the part to the yield stress allowable and the alternating stress to the endurance stress limit. This is logical because the mean stress acts much like a one-time application of a load, whereas the alternating stress is the fluctuating component of the stress that we would therefore compare to the endurance stress limit. If we wish to add in a safety factor, the Soderberg equation becomes as follows:

$$\frac{1}{N} = \frac{S_m}{S_y} + \frac{S_a}{S_n} \qquad (5.5)$$

Soderberg equation (5.5)

$$\frac{1}{N} = \frac{S_m}{S_y} + \frac{S_a}{S_n}$$

To use this formula, you should select an unknown such as the part size. For instance, this could be done in terms of area for an axial load problem or perhaps Z for a bending problem or Z' for a torsion problem. If a known size exists, we could solve for the factor of safety and then, if that was inappropriate, adjust the size accordingly.

If the Soderberg equation is used for shear loads, it is modified to incorporate the endurance limit for shear and yield for shear as follows:

$$\frac{1}{N} = \frac{S_{ms}}{S_{ys}} + \frac{S_{as}}{S_{ns}} \qquad (5.6)$$

It should be noted that the Soderberg diagram and the associated formula represent a reasonable approximation of the ability of a part to resist repeated stresses when we are dealing with a relatively ductile material. Other theories, such as the **Gerber theory** or **Goodman criteria,** are more suitable for brittle or other nonductile materials. The Gerber formula follows so that you can understand how it differs from the Soderberg equation. Although slightly less conservative, the results are similar.

$$\frac{1}{N} = \left(\frac{S_m}{S_u}\right)^2 + \frac{S_a}{S_n} \qquad (5.7)$$

> **Gerber theory**
> Similar, albeit slightly different, to the Soderberg equation, this theory of fatigue failure is sometimes used for brittle or nonductile materials.
>
> **Goodman criteria**
> Quite similar to the Gerber theory; used for fatigue failure of brittle or nonductile materials.

As the Soderberg equation is often considered to be very conservative, the Gerber or modified Goodman criteria is often used. Figure 5.10 reflects a plot of these three different curves, you can see that the Soderberg line is the most conservative of the three. It is often useful because the results are below the yield line in all cases without the inclusion of a factor of safety. However, the overly conservative nature of the Soderberg line occasionally offers excess conservatism. Hence, the Gerber or modified Goodman are often used in cases where less conservatism is desired or, in some cases, for less ductile materials. The modified Goodman formula is similar to the Gerber and Soderberg equations.

$$\frac{1}{N} = \frac{S_m}{S_u} + \frac{S_a}{S_n} \qquad (5.8)$$

Gerber theory (5.7)

$$\frac{1}{N} = \left(\frac{S_m}{S_u}\right)^2 + \frac{S_a}{S_n}$$

Modified Goodman (5.8)

$$\frac{1}{N} = \frac{S_m}{S_u} + \frac{S_a}{S_n}$$

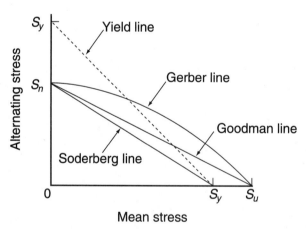

FIGURE 5.10 Failure diagram that shows the relationship between different analysis methods.

Example Problems 5.1 through 5.4 show the use of the Soderberg equation.

Example Problem 5.1

For a smooth, rotating shaft with no sharp corners or change in shape, determine the required diameter under the loading condition shown in Figure 5.11 (ignore any torque). The surface of the shaft is highly polished. The shaft is made from annealed AISI 4140 steel. Use a safety factor of 2.

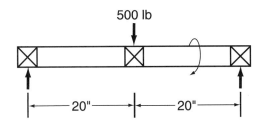

FIGURE 5.11 Rotating shaft for Example Problem 5.1.

For the first trial, assume the resultant diameter will be between ½ and 2 inches.

$$S_u = 95 \text{ ksi} \qquad S_y = 60 \text{ ksi} \qquad \textbf{(Appendix 4)}$$

First calculate S_{max} and S_{min}.

$$M_{max} = \frac{FL}{4} \qquad \textbf{(Appendix 2)}$$

$$M_{max} = \frac{500 \text{ lb } 40 \text{ in.}}{4} = 5,000 \text{ in.-lb}$$

$$M_{min} = \frac{FL}{4} = -5,000 \text{ in.-lb}$$

$$S_{max} = \frac{M_{max}}{Z} = \frac{5,000 \text{ in.-lb}}{Z}$$

$$S_{min} = \frac{M_{min}}{Z} = -\frac{5,000 \text{ in.-lb}}{Z}$$

Next, find S_{mean} and $S_{alternating}$.

$$S_{mean} = \frac{S_{max} + S_{min}}{2} \qquad (5.2)$$

$$S_{mean} = \frac{\dfrac{5,000 \text{ in.-lb}}{Z} - \dfrac{5,000 \text{ in.-lb}}{Z}}{2} = 0$$

$$S_{alt} = \frac{S_{max} - S_{min}}{2} \qquad (5.3)$$

$$S_{alt} = \frac{\dfrac{5,000 \text{ in.-lb}}{Z} - -\dfrac{5,000 \text{ in.-lb}}{Z}}{2} = \frac{5,000 \text{ in.-lb}}{Z}$$

Find the endurance limit modifying factors.

$$S_n = C_{size}\ C_{surface}\ C_{type}\ S_n' \qquad (5.1)$$

As surface is polished, $C_{surface} = 1$. (Figure 5.7b)

As S_n' is unknown, use $.5\ S_u$.

Use $C_{size} = .85$ assuming $.5 < D < 2$
Bending $C_{type} = 1$

$$S_n = .85\ (1)\ (1)\ (.5)\ 95 \text{ ksi}$$
$$S_n = 40.375 \text{ ksi}$$

Substituting into the Soderberg equation:

$$\frac{1}{N} = \frac{S_m}{S_y} + \frac{S_a}{S_n} \qquad (5.5)$$

$$\frac{1}{2} = \frac{\dfrac{5,000 \text{ in.-lb}}{Z}}{40,375 \text{ lb/in.}^2}$$

$$Z = .247 \text{ in.}^3$$

$$Z = \frac{\pi D^3}{32} \quad \text{(Appendix 3)}$$

$$D^3 = \frac{32Z}{\pi}$$

$$D = \sqrt[3]{\frac{32\,(.247)\,\text{in.}^3}{\pi}}$$

$$D = 1.36 \text{ in.}$$

Use $D = 1\tfrac{3}{8}$ or 1.375 in.

Example Problem 5.2

A 12-inch-long round cantilever beam is loaded repeatedly, but no load reversal occurs (see Figure 5.12). If the bar is made from annealed AISI 302 stainless steel and the surface is polished for a safety factor of 1.6, find the required diameter using the Soderberg method.

$$S_u = 90 \text{ ksi} \quad S_y = 37 \text{ ksi} \quad S_n = 34 \text{ ksi} \quad \text{(Appendix 8)}$$

First, find S_{max} and S_{min}.

$$M_{max} = FL \quad \text{(Appendix 2)}$$
$$M_{max} = 1{,}000 \text{ lb } 12 \text{ in.}$$
$$M_{max} = 12{,}000 \text{ in.-lb}$$
$$M_{min} = 0$$
$$S_{max} = \frac{M_{max}}{Z} = \frac{12{,}000 \text{ in.-lb}}{Z}$$
$$S_{min} = 0$$

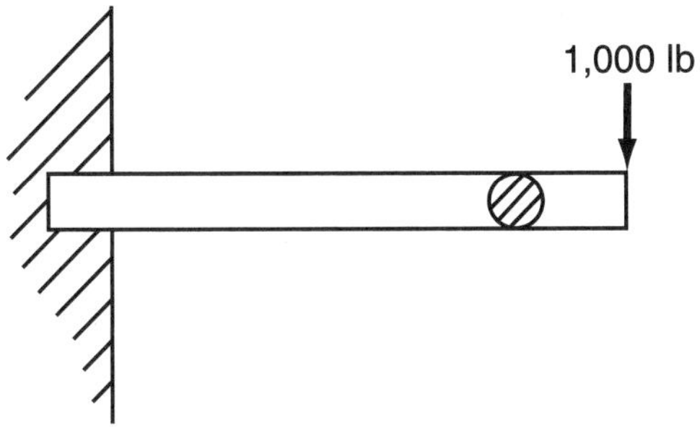

FIGURE 5.12 Cantilever beam for Example Problems 5.2 and 5.3.

Next, find S_{mean} and $S_{alternating}$.

$$S_{mean} = \frac{S_{max} + S_{min}}{2} \quad (5.2)$$

$$S_{mean} = \frac{\frac{12,000 \text{ in.-lb}}{Z} + 0}{2}$$

$$S_{mean} = \frac{6,000 \text{ in.-lb}}{Z}$$

$$S_{alt} = \frac{S_{max} - S_{min}}{2} \quad (5.3)$$

$$S_{alt} = \frac{\frac{12,000 \text{ in.-lb}}{Z} - 0}{2}$$

$$S_{alt} = \frac{6,000 \text{ in.-lb}}{Z}$$

Find the endurance limit modifying factors.

$$C_{size} = .85 \quad (\text{assume } \tfrac{1}{2} < D < 2 \text{ inches})$$
$$C_{type} = 1 \quad (\text{bending})$$
$$C_{surface} = 1 \quad (\text{polished surface})$$
$$S_n = C_{size} \, C_{surface} \, C_{type} \, S_n'$$
$$S_n = .85 \, (1) \, (1) \, 34 \text{ ksi}$$
$$S_n = 28.9 \text{ ksi}$$

Substituting into the Soderberg equation:

$$\frac{1}{N} = \frac{S_m}{S_y} + \frac{S_a}{S_n} \quad (5.5)$$

$$\frac{1}{1.6} = \frac{\frac{6,000 \text{ in.-lb}}{Z}}{37,000 \text{ lb/in.}^2} + \frac{\frac{6,000 \text{ in.-lb}}{Z}}{28,900 \text{ lb/in.}^2}$$

$$Z = .592 \text{ in.}^3$$

$$Z = \frac{\pi D^3}{32} \qquad \text{(Appendix 3)}$$

$$D^3 = \frac{32Z}{\pi}$$

$$D = \sqrt[3]{\frac{32(.592 \text{ in.}^3)}{\pi}}$$

$$D = 1.82 \text{ inches}$$

Example Problem 5.3

Repeat Example Problem 5.2 for the same conditions using the modified Goodman method. Compare these results to the solution found using the Soderberg method. (S_u, S_y, S_{mean}, S_{alt}, and S_n would be the same as in Example Problem 5.2.)

Substitute into the modified Goodman formula.

$$\frac{1}{N} = \frac{S_m}{S_u} + \frac{S_a}{S_n} \qquad (5.8)$$

$$\frac{1}{1.6} = \frac{\frac{6,000 \text{ in.-lb}}{Z}}{90,000 \text{ lb/in.}^2} + \frac{\frac{6,000 \text{ in.-lb}}{Z}}{28,900 \text{ lb/in.}^2}$$

$$Z = .439 \text{ in.}^3$$

$$Z = \frac{\pi D^3}{32} \qquad \text{(Appendix 3)}$$

$$D^3 = \frac{32Z}{\pi}$$

$$D = \sqrt[3]{\frac{32(.439 \text{ in.}^3)}{\pi}}$$

$$D = 1.65 \text{ inches}$$

Note that the modified Goodman results in a less conservative size as would be expected from Figure 5.10.

Example Problem 5.4

If the connecting rod in Figure 5.13 is made from the same AISI 4140 steel material as the shaft in Example Problem 5.1 except with a machined surface, find the required square size in the center section where there are no discontinuities. The piston pump has a tensile load of 1,000 pounds pulling water into the pump and a compressive load of 6,000 pounds pushing water out. Assume SF = 1.5. Use the Soderberg equation.

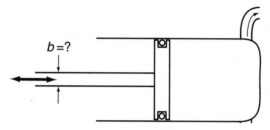

FIGURE 5.13 Water pump connecting rod: Example Problem 5.4.

$$S_u = 95 \text{ ksi} \qquad S_y = 60 \text{ ksi} \qquad \textbf{(Appendix 4)}$$

$$A = b \bullet b = b^2$$

$$S_{max} = \frac{F_{max}}{A} = \frac{1{,}000 \text{ lb}}{b^2}$$

$$S_{min} = \frac{F_{min}}{A} = -\frac{6{,}000 \text{ lb}}{b^2}$$

Finding S_{mean} and $S_{alternating}$:

$$S_{mean} = \frac{S_{max} + S_{min}}{2} = \frac{\dfrac{1{,}000 \text{ lb}}{b^2} - \dfrac{6{,}000 \text{ lb}}{b^2}}{2} \qquad (5.2)$$

$$S_{mean} = \frac{2{,}500 \text{ lb}}{b^2}$$

$$S_{alt} = \frac{S_{max} - S_{min}}{2} = \frac{\dfrac{1{,}000 \text{ lb}}{b^2} - -\dfrac{6{,}000 \text{ lb}}{b^2}}{2} \qquad (5.3)$$

$$S_{alt} = \frac{3{,}500 \text{ lb}}{b^2}$$

Endurance limit modifying factors:

$$S_n = C_{size}\, C_{surface}\, C_{type}\, S_n' \qquad (5.1)$$

Assume: $b < \tfrac{1}{2}$; use $C_{size} = 1$.

$$C_{surface} = .86 \qquad \text{(Figure 5.7b)}$$
$$C_{type} = .8 \text{ (axial load)}$$
$$S_n = C_{size}\, C_{surface}\, C_{type}\, S_n' \qquad (5.1)$$
$$S_n = 1\,(.86)\,(.8)\,.5\,(95 \text{ ksi})$$
$$S_n = 32{,}680 \text{ lb/in.}^2$$

Solving Soderberg equation for b:

$$\frac{1}{N} = \frac{S_m}{S_y} + \frac{S_a}{S_n} \qquad (5.5)$$

$$\frac{1}{1.5} = \frac{\dfrac{2{,}500 \text{ lb}}{b^2}}{60{,}000 \text{ lb/in.}^2} + \frac{\dfrac{3{,}500 \text{ lb}}{b^2}}{32{,}680 \text{ lb/in.}^2}$$

$$b = .47$$

5.7 STRESS CONCENTRATION FACTORS

When we have parts with significant changes in geometry, such as sharp notches, steps in a rotating shaft, or size changes in a flat plate, we need to add a factor as the stresses are concentrated at this point, and therefore this is the likely point of failure. We would modify the Soderberg equation to account for this as follows:

$$\frac{1}{N} = \frac{S_m}{S_y} + \frac{K_t\, S_a}{S_n} \qquad (5.9)$$

Soderberg with concentration factor (5.9)

$$\frac{1}{N} = \frac{S_m}{S_y} + \frac{K_t\, S_a}{S_n}$$

The Gerber and Goodman formulas would have the stress concentration factor applied to the alternating stresses in a similar manner.

To determine the value for this factor (K_t), we would identify the closest approximate shape from Appendix 6. When the Soderberg equation, along with stress concentration factors, is used in design, we often need to make assumptions in order to make an initial trial solution. The process may then need to be repeated to come to a final solution.

Example Problem 5.5

The machined shaft shown in Figure 5.14 is made from AISI C1040 cold-drawn steel. If a repeated but not reversed load of 200 in.-lb is applied, what is the factor of safety for this design?

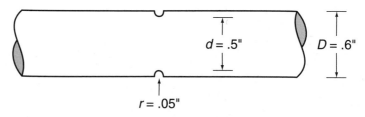

FIGURE 5.14 Machined shaft for Example Problem 5.5.

$$S_u = 80 \text{ ksi} \qquad S_y = 71 \text{ ksi} \qquad \textbf{(Appendix 4)}$$

Find minimum and maximum stress.

$$S_{max} = \frac{T}{Z'} \qquad Z' = \frac{\pi D^3}{16} \qquad \textbf{(Appendix 3)}$$

$$S_{max} = \frac{200 \text{ in.-lb}}{\frac{\pi (.5 \text{ in.})^3}{16}}$$

$$S_{max} = 8{,}150 \text{ lb/in.}^2$$

$$S_{min} = 0$$

$$S_{mean} = \frac{S_{max} + S_{min}}{2} \qquad (5.2)$$

$$S_{mean} = \frac{8{,}150 \text{ lb/in.}^2 + 0}{2} = 4{,}075 \text{ lb/in.}^2$$

$$S_{alt} = \frac{S_{max} - S_{min}}{2} \qquad (5.3)$$

$$S_{alt} = \frac{8{,}150 \text{ lb/in.}^2 - 0}{2} = 4{,}075 \text{ lb/in.}^2$$

Find the endurance strength.

$$S_n = C_{size} \ C_{surface} \ C_{type} \ .5 \ S_u \tag{5.1}$$

$C_{size} = .85$ Assume: $.5 < D < 2$ in.

$C_{surface} = .88$ (machined)   **(Figure 5.7b)**

$C_{type} = .6$ (torsion)

$S_n = C_{size} \ C_{surface} \ C_{type} \ .5 \ S_u$

$S_n = .85 \ (.88) \ (.6) \ .5 \ (80 \text{ ksi}) = 18 \text{ ksi}$

Find the stress concentration factor from Appendix 6c, using the stepped round shaft in torsion.

$$\frac{D}{d} = \frac{.6}{.5} = 1.2$$

$$\frac{r}{d} = \frac{.05}{.5} = .10$$

$$K_t \approx 1.32 \qquad \text{(estimate between the curves)}$$

Solve for the safety factor using the Soderberg criteria.

$$\frac{1}{N} = \frac{S_{ms}}{S_{ys}} + \frac{K_t S_{as}}{S_{ns}} \tag{5.9}$$

$$\frac{1}{N} = \frac{4,075 \text{ lb/in.}^2}{.5 \ (71,000 \text{ lb/in.}^2)} + \frac{1.32 \ (4,075 \text{ lb/in.}^2)}{18,000 \text{ lb/in.}^2}$$

$$N = 2.4$$

5.8 LIMITED LIFE

If a part has a known life of something less than the theoretical endurance stress limit life, we need to adjust the endurance limit to account for this. In this case, we would add a modifying factor to the endurance limit equation. This involves calculating a ratio of the slope of Figure 5.6 to accommodate for this less-than-endurance-limit lifetime. Note that this equation assumes that the material would have an endurance limit of 10^6 or a million cycles. Other materials may have different values. In that case, this formula would be adjusted as follows:

Endurance limit life adjustment (5.10)

$$S_n = S_n' \left(\frac{10^6}{N_c}\right)^{.09}$$

$$S_n = S_n' \left(\frac{N_{c-\text{endurance life}}}{N_{c-\text{cycles needed}}}\right)^{\text{slope}}$$

For most steels, this becomes

$$S_n = S_n' \left(\frac{10^6}{N_c}\right)^{.09} \quad (5.10)$$

Example Problem 5.6

For the shaft in Example Problem 5.5, if the desired life is 100,000 cycles, calculate the new factor of safety for this condition.

$$S_{max}, \; S_{min}, \; S_{mean}, \text{ and } S_{alt} \quad \text{(remain the same)}$$

$$S_n = S_n' \left(\frac{10^6}{N_c}\right)^{.09} \quad (5.10)$$

$$S_n = 18 \text{ ksi} \left(\frac{10^6}{100,000}\right)^{.09}$$

$$S_n = 22.1 \text{ ksi}$$

Solve again using Soderberg.

$$\frac{1}{N} = \frac{S_{ms}}{S_{ys}} + \frac{K_t \, S_{as}}{S_{ns}} \quad (5.9)$$

$$\frac{1}{N} = \frac{4,075 \text{ lb/in.}^2}{.5(71,000) \text{ lb/in.}^2} + \frac{1.32(4,075 \text{ lb/in.}^2)}{22,100 \text{ lb/in.}^2}$$

$$N = 2.8$$

Hence, if the same margin was needed as previously, the shaft could be made smaller.

5.9 SURFACE AND OTHER EFFECTS

If we envision a typical part subject to **reciprocating loads,** it is easy to imagine that in nearly every case the postulated defect that we have discussed throughout this chapter started on the surface. The reason for this is that typically the maximum stresses are at the surface of the part. For example, under bending stresses, our maximum stress is on the outermost fibers, and the same is true for torsional stresses. Therefore, many techniques have been developed to try to minimize the potential for crack propagation from the surface. Almost all of these techniques involve placing a **residual compressive stress** on the surface of the material so that a greater tensile stress is required to reach the point at which crack propagation would occur. If you have had a course in material science, you will remember that many heat treatment processes such as flame hardening, induction hardening, carburizing, nitriting, shot peening, and surface rolling have a net effect of leaving a residual compressive stress in the surface of the part. Hence, all of these processes improve the ability of a part to withstand fatigue loading. These processes are often used together with other techniques where radiuses and other effects can be minimized in order to make parts that can be smaller while having similar resistance to repeated loading.

> **reciprocating loads**
> A term used for either repeated or repeated and reversed loads, implying that the load is applied over and over again in some manner.

> **residual compressive stress**
> A compressive stress that is induced in a part such that it remains when no load is acting on the part.

5.10 MITIGATING STRESS CONCENTRATIONS

One of the more interesting aspects of designing for repeated loading is the way in which changing radiuses, the placement of stress concentration points, and even creating false stress concentrations can sometimes reduce concentrations in certain areas. Figure 5.15 shows some common techniques for doing this. A good way to envision how these configurations can be beneficial to a part would be to envision water flowing through the insides of these parts. You want to prevent turbulence or high-velocity areas from occurring in these water flows by trying to direct them around obstacles. Think for a moment of playing in a stream where there is a sharp obstruction along the bank. By placing something upstream from the obstruction with smoother angles and directing the water flow away from the obstruction, we can actually reduce the turbulence in the water. This analogy applies to stresses in a part. Figure 5.15 shows several methods by which we can accomplish a similar reduction in the effects of sharp changes in geometry. However, these are often difficult to quantify mathematically with regard to their modifying effect on the part in question.

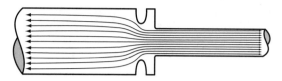

FIGURE 5.15(a) Note that the addition of an "upstream" radius smooths out the flow lines and directs them away from the sharp corners.

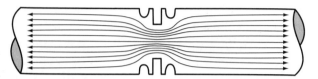

FIGURE 5.15(b) By placing a radius on both sides of the sharp edge groove, the stress lines are "directed" away from the sharp edge.

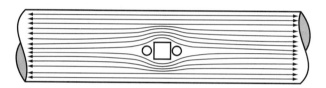

FIGURE 5.15(c) Note how the addition of the smooth hole smooths out the stress lines and "directs" them away from the sharp corners.

5.11 SUMMARY

mitigate defect propagation
A term for different methods used to reduce the effect of a postulated defect on a part. These methods may include leaving residual compressive stresses, changing radiuses, and changing dimensions.

You should now be familiar with the concept of endurance limit and how the allowable stresses need to be reduced in a part in order to **mitigate defect propagation.** The modifying factors for endurance limits and how to calculate them should also now be better understood. You should be able to quantify the different types of modifying factors for material, size, surface finish, loading type, and finite life into solving basic problems in endurance analysis.

Lastly, you should understand how surface irregularities, changes in shape, and other factors affect the ability of a part to withstand repeated loading. You should be able to apply fatigue failure theories such as the Soderberg and modified Goodman equations, and you should be able to apply factors for modifying material endurance limits and considering shape factors in calculating the stresses in parts subject to the varying types of repeated loading.

5.12 PROBLEMS

1. A circular rod with a diameter of 7/16 inch is smoothly polished and is subject to a repeated and reversed axial load of 1,500 pounds. The bar is made of AISI 1144 hot-rolled steel. What is the factor of safety for this design based on the Soderberg criteria?
2. For the rod in Problem 1, solve if the load is repeated but not reversed. Explain why this result is different.
3. For the case described in Problem 2, what is the factor of safety using the Gerber equation?
4. For the case described in Problem 2, what is the factor of safety using the modified Goodman criteria?
5. A strut in a truss-type space frame assembly has a cross section of 15 mm by 25 mm. It has a load that varies from a tensile load of 20 KN to a compressive load of 15 KN. If the material is AISI 1040, the surface is hot-rolled, and a safety factor of 2 is desired, using the Soderberg equation, is this an acceptable design?
6. If the link shown in Figure 5.16 is to be made of AISI 1050 hot-rolled steel and is subject to a repeated and reversed load of 2,200 pounds, and if a safety factor of 1.5 is required, find an appropriate diameter if the surface is as-forged (use the Soderberg equation).
7. For the link shown in Figure 5.16, if the load is 7,500 newtons and repeated but not reversed, and the surface is as-forged, find an appropriate diameter based on the Soderberg criteria. The material and safety factor are the same as in Problem 6.

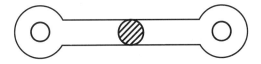

FIGURE 5.16 Link for Problems 6 and 7.

8. In Figure 5.17, if the bar is made from machined cold-drawn AISI 1040 steel, solve for the width of the bar for a safety factor of 1.75 using the Soderberg criteria. The 500-pound load is repeated and reversed.

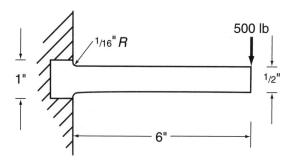

FIGURE 5.17 Bar for Problem 8.

9. In Figure 5.18, the rotating shaft is machined from AISI 3140 annealed steel for the following parameters. Is this an acceptable design based on the Soderberg line?

$$F = 125 \text{ pounds}$$
$$L = 10 \text{ inches}$$
$$D_1 = 1 \text{ inch}$$
$$D_2 = 1.25 \text{ inches}$$
$$r = .05 \text{ inch}$$
$$SF = 1.5$$

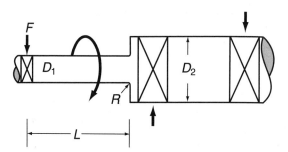

FIGURE 5.18 Rotating shaft for Problems 9 through 13.

10. Use the same conditions as Problem 9, except the radius is increased to .125 inch. Is this now acceptable?
11. If it is necessary in Problem 9 to use the .05 radius, calculate how much larger D_1 would need to be.

12. In Problem 9, if the rotating shaft is mirror polished instead of machined, is this design acceptable?
13. In Figure 5.18, solve for the safety factor using Soderberg under the following conditions (the surface is machined and the material is the same annealed 3140 as in Problem 9):

$$F = 200 \text{ newtons}$$
$$L = 250 \text{ mm}$$
$$D_1 = 20 \text{ mm}$$
$$D_2 = 40 \text{ mm}$$
$$r = 1 \text{ mm}$$

14. In Figure 5.19, the cantilever beam is subject to a repeated and reversed load as shown. If the bar is made from AISI 1040 cold-drawn steel and the surface is machined, what is the factor of safety for this design (based on Soderberg)?

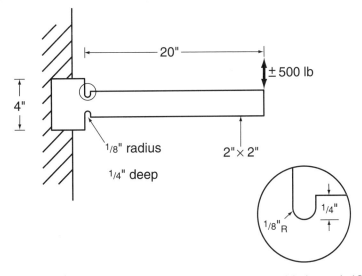

FIGURE 5.19 Cantilever beam for Problems 14 through 16.

15. For Problem 14, which appendix is appropriate to use and what value of h do you think should be used?
16. Prepare a sketch of how the design shown in Figure 5.19 might be modified to improve its resistance to fatigue failure.

17. In Figure 5.20, a stepped rotating shaft transmits 10 hp at 1,750 rpm continuously but has a pulsating torque of 250 in.-lb in addition (repeated but not reversed). The step is needed for a self-aligning coupling. Determine the necessary standard shaft size (D), if a 1/16 radius is allowed. A safety factor of 2 using Soderberg should be used. The shaft would need to be machined and is made of AISI 4140 annealed steel.

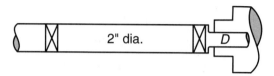

FIGURE 5.20 Stepped rotating shaft for Problems 17 through 20.

18. Solve for the shaft size for the conditions in Problem 17 using the Gerber equation.
19. Solve for the shaft size for the conditions in Problem 17 using the modified Goodman equation.
20. Compare the results of Problems 17 through 19 and discuss the differences.
21. The round beam shown in Figure 5.21 has a repeated load that varies from 1,000 to 3,000 pounds. It is made from AISI 1020 cold-drawn steel. The surface is as-rolled. Based on Soderberg, determine D if $r = .1D$ and a safety factor of 2 is desired (the shaft does not rotate).

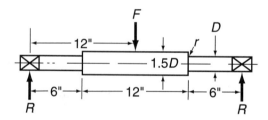

FIGURE 5.21 Round beam for Problems 21 and 22.

22. Using Figure 5.21, solve if the load is steady at 3,000 pounds, the shaft rotates, and the other conditions are the same as in Problem 21.

23. In Figure 5.22, the shoulder screw acts as a cam follower (supports a bearing). The screw is made of AISI 3141 OQT 1000°. The bearing turns at 100 rpm for a 20-hour life and is subjected to a repeated but not reversed load of 900 pounds each revolution. Determine the safety factor for this condition using Soderberg (note that the shoulder screw is subjected to both bending and shear stresses).

$$a = \frac{5}{8} \text{ in.}$$

$$b = \frac{7}{16} \text{ in.}$$

$$c = \frac{3}{4} \text{ in.}$$

The radius at B = .02 in.

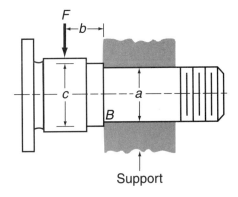

FIGURE 5.22 Assembly for Problems 23 and 24.

24. In Figure 5.22, solve for the repeated and reversed load capacity for this shoulder screw if the material is AISI 3140 OQT 1000°, and if a desired life of 100,000 cycles and a SF of 1.5 based on Soderberg is desired for the following dimensions:

$$a = 10 \text{ mm}$$
$$b = 8 \text{ mm}$$
$$c = 15 \text{ mm}$$

The radius at $B = .5$ mm

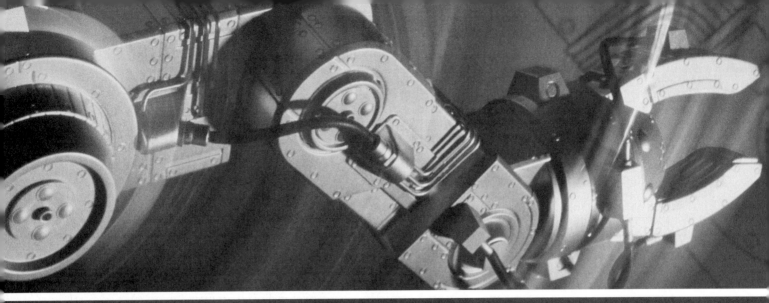

CHAPTER 6

Fasteners and Fastening Methods

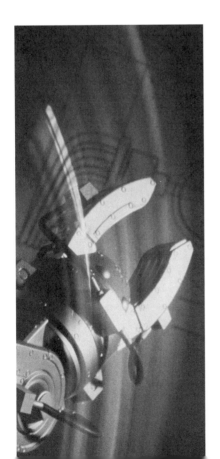

Many types of assembly are used in complex machine design. For example, in an automobile engine and transmission assembly, you can recognize many different items that are manufactured separately and then assembled. In some cases, this may be due to the size of the parts or to the fact that they are made by different companies. Fasteners may be required for those parts that need to be disassembled for maintenance and/or replacement. In some applications, such as the attachment of an automobile engine block to the valve assembly, the need to seal internal pressures may require certain preloads in the bolting assembly in order to resist these pressures during periods of changing temperature. Other applications such as pressure vessels and valves that have internal pressures may require certain levels of preloading as well.

Other applications may require specialized fasteners. These may include items such as sheet metal screws that are self-drilling and tapping and require no hole to be drilled, quick-release fasteners, and set screws for anchoring collars on shafts. Fasteners come in a wide variety of types and forms. This chapter will only cover the basic principles.

In other assemblies, permanent connections may require welding, brazing, or the use of adhesives. In Section 6.8, we will review the basic principles of welded connections.

6.1 OBJECTIVES

After completing this chapter, you should be able to:

- Describe many types of fastening systems and their uses.
- Understand the principles of stress area, pitch diameters, and thread types and forms.
- Describe the common designations of thread forms and grades.
- Understand the principles of the strength of different grades of bolting and the relationship between proof strength, yield, and ultimate strength for bolting materials.
- Understand different types of tensioning systems and how preloaded joints are created, and on appropriate tightening loads.
- Describe the principles of elastic analysis and how the preload affects the ability of joints to resist future loading and/or pressures.

- Describe the principles of gasketing and the different ways in which gaskets are specified and used in pressure-connecting joints.
- Recognize types of fastening systems their basic principles, and where they may be applicable.
- Understand the basic principles of welded joints and how to analyze the strength inherent in a welded joint.

6.2 TYPES OF THREADS AND TERMS

Because the threading on fasteners made by different manufacturers must be compatible, standards have been developed both for specifying and manufacturing thread types. Figure 6.1 shows a pictorial representation of the **major** and **minor diameters,** which are the largest and smallest diameters of a threaded part, along with the **pitch,** the lead of the thread, and other parts of the thread. Note that typically the major diameter is also approximately the nominal size of a bolt and that the stress area of a bolt is approximately equal to the area of the minor diameter. However, if you identify the stress area from a thread table such as Table 6.1, this would vary slightly from the minor diameter because the actual stress area includes the area of part of one thread. The lead or pitch of a threaded part is important because threaded mechanisms are often used for positioning a part. One rotation of a nut around a bolt moves the nut a distance equal to the pitch. When we refer to the lead, the lead is the reciprocal of the number of threads per inch.

major diameter
The larger or outside diameter of a thread, sometimes referred to as nominal diameter.

minor diameter
The smallest diameter of a thread, sometimes referred to as the root diameter.

pitch
The axial centerline spacing of a thread or the reciprocal of the number of threads per unit length.

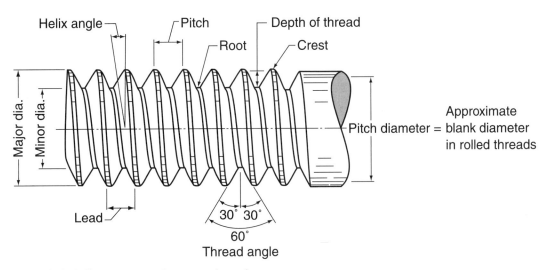

FIGURE 6.1 Properties of screw threads.

TABLE 6.1
AMERICAN STANDARD THREAD DIMENSIONS

Size	Basic Major Diameter (in.)	Coarse Threads: UNC		Fine Threads: UNF	
		Threads per Inch	Tensile Stress Area (in.^{2})	Threads per Inch	Tensile Stress Area (in.^{2})
Numbered Sizes					
0	.060 0			80	.001 80
1	.073 0	64	.002 63	72	.002 78
2	.086 0	56	.003 70	64	.003 94
3	.099 0	48	.004 87	56	.005 23
4	.112 0	40	.006 04	48	.006 61
5	.125 0	40	.007 96	44	.008 30
6	.138 0	32	.009 09	40	.010 15
8	.164 0	32	.014 0	36	.014 74
10	.190 0	24	.017 5	32	.020 0
12	.216 0	24	.024 2	28	.025 8
Fractional Sizes					
¼	.250 0	20	.031 8	28	.036 4
5⁄16	.312 5	18	.052 4	24	.058 0
⅜	.375 0	16	.077 5	24	.087 8
7⁄16	.437 5	14	.106 3	20	.118 7
½	.500 0	13	.141 9	20	.159 9
9⁄16	.562 5	12	.182	18	.203
⅝	.625 0	11	.226	18	.256
¾	.750 0	10	.334	16	.373
⅞	.875 0	9	.462	14	.509
1	1.000	8	.606	12	.663
1⅛	1.125	7	.763	12	.856
1¼	1.250	7	.969	12	1.073
1⅜	1.375	6	1.155	12	1.315
1½	1.500	6	1.405	12	1.581
1¾	1.750	5	1.90		
2	2.000	4½	2.50		

We specify standard bolting as follows:

¼-20UNC-2A

This designation specifies a ¼-inch nominal size having 20 threads per inch. The *UNC* refers to a unified national course thread series that has become the type of thread most commonly used in the United States. The 2 designates the class of the thread, and the *A* designates that it is an internal thread. A *B* signifies an external thread. There are three types of threads in common usage; coarse thread, UNC; fine thread, UNF; and an extra-fine thread, UNEF. As can be seen in Table 6.1, in each of these thread series, the number of threads per inch varies with the nominal size of the bolting. In additional, fixed pitch series, such as an 8- or 12-series, are used in some special applications.

The class of the thread refers to the accuracy of manufacture, with a class 3 being the most accurate and class 1 being the least accurate. Generally, class 2 and class 3 are the most common and would be found in normal practice. If we were manufacturing threads, we would use a reference such as *Machinery's Handbook,* which would give us the accuracy tolerances necessary to meet the requirements for these different classes.

In the metric system, we use a similar designated series starting with an M for metric. The first number is the basic major diameter followed by the pitch in millimeters as follows:

$$M3 \times .5$$

This would be for a basic major diameter of 3 mm and a .5 mm pitch in a standard coarse thread. This system does not specify a class of thread or whether it is an internal or an external thread. Common metric thread dimensions are listed in Table 6.2.

6.3 MATERIALS AND DESIGNATIONS

Bolting materials are typically made from steel; however, most any material with reasonable formability can be used. Different standards agencies such as the Society of Automotive Engineers (SAE) or the American Society for Testing of Materials (ASTM) have specified standards for different grades of materials as shown in Tables 6.3 through 6.5. In addition to the tensile and yield strengths that are used in most structural analyses, bolting introduces another value: **proof strength.** Proof strength is somewhat similar to the elastic limit discussed in Chapter 2 and is generally between 85 and 95 percent of the yield strength. Generally, we think of the proof strength as a minimum strength for bolting, as opposed to yield strength, which is generally an average value for a material. For certain grades of bolting, a stamping may be engraved on the head of a cap screw that would indicate the grade. (See Figure 6.2.) Metric grades are slightly different and use a numeric code for grading. If nongraded fasteners, such as those made from aluminum, brass, or plastic are used, it would be necessary to specify a proof stress. In some cases, materials from specialized manufacturers may have a breaking strength for the individual sizes specified.

proof strength
A stress value below the yield where bolting material would not typically fail; somewhat similar to the elastic limit.

TABLE 6.2
METRIC THREAD DIMENSIONS

Basic Major Diameter (mm)	Coarse Threads		Fine Threads	
	Pitch (mm)	Tensile Stress Area (mm²)	Pitch (mm)	Tensile Stress Area (mm²)
1	.25	.460		
1.6	.35	1.27	.20	1.57
2	.4	2.07	.25	2.45
2.5	.45	3.39	.35	3.70
3	.5	5.03	.35	5.61
4	.7	8.78	.5	9.79
5	.8	14.2	.5	16.1
6	1	20.1	.75	22.0
8	1.25	36.6	1	39.2
10	1.5	58.0	1.25	61.2
12	1.75	84.3	1.25	92.1
16	2	157	1.5	167
20	2.5	245	1.5	272
24	3	353	2	384
30	3.5	561	2	621
36	4	817	3	865
42	4.5	1,121		
48	5	1,473		

TABLE 6.3
SAE GRADES OF STEELS FOR FASTENERS

Grade Number	Bolt Size (in.)	Tensile Strength (ksi)	Yield Strength (ksi)	Proof Strength (ksi)
1	¼–1½	60	36	33
2	¼–¾	74	57	55
	>¾–1½	60	36	33
4	¼–1½	115	100	65
5	¼–1	120	92	85
	>1–1½	105	81	74
7	¼–1½	133	115	105
8	¼–1½	150	130	120

TABLE 6.4
ASTM STANDARDS FOR BOLT STEELS

ASTM Grade	Bolt Size (in.)	Tensile Strength (ksi)	Yield Strength (ksi)	Proof Strength (ksi)
A307	¼–4	60	(not reported)	
A325	½–1	120	92	85
	>1–1½	105	81	74
A354-BC	¼–2½	125	109	105
A354-BD	¼–2½	150	130	120
A449	¼–1	120	92	85
	>1–1½	105	81	74
	>1½–3	90	58	55
A574	.060–½	180		140
	⅝–4	170		135

TABLE 6.5
METRIC GRADES OF STEELS FOR BOLTS

Grade	Bolt Size	Tensile Strength (MPa)	Yield Strength (MPa)	Proof Strength (MPa)
4.6	M5–M36	400	240	225
4.8	M1.6–M16	420	340*	310
5.8	M5–M24	520	415*	380
8.8	M17–M36	830	660	600
9.8	M1.6–M16	900	720*	650
10.9	M6–M36	1,040	940	830
12.9	M1.6–M36	1,220	1,100	970

*Yield strengths are approximate and are not included in the standard.

Trade Marking	Specification	Material
No Mark	SAE—Grade 1 ASTM—A 307 SAE—Grade 2	Low- or medium-carbon steel Low-carbon steel Low- or medium-carbon steel
(3 radial lines)	SAE—Grade 5 ASTM—A 449	Medium-carbon steel, quenched and tempered
(3 radial lines with center)	SAE—Grade 5.2	Low-carbon martensite steel, quenched and tempered
A 325	ASTM-A 325 Type 1	Medium-carbon steel, quenched and tempered (radial dashes optional)
A 325	ASTM-A 325 Type 2	Low-carbon martensite steel, quenched and tempered
A 325	ASTM-A 325 Type 3	Atmospheric corrosion (weathering) steel, quenched and tempered
BC	ASTM-A 354 Grade BC	Alloy steel, quenched and tempered
(6 radial lines)	SAE-Grade 7	Medium-carbon alloy steel, quenched and tempered, roll threaded after heat treatment
(6 radial lines)	SAE-Grade 8 ASTM-A 354 Grade BD	Medium-carbon alloy steel, quenched and tempered Alloy steel, quenched and tempered
(radial lines)	SAE-Grade 8.2	Low-carbon martensite steel, quenched and tempered
A 490	ASTM-A 490 Type 1	Alloy steel, quenched and tempered
A 490	ASTM-A 490 Type 3	Atmospheric corrosion (weathering) steel, quenched and tempered

FIGURE 6.2 ASTM and SAE grade stamps for steel bolts and screws. (Courtesy of *Engineering Drawing and Design*, Madsen, 3rd ed., Delmar Learning)

6.4 TIGHTENING METHODS AND INITIAL TENSION

There are many methods of tightening bolting: the common use of a **torque wrench** to create the required torque for tightening a bolt or **cap screw**; the turn-of-the-nut method in which the necessary elongation is calculated and the nut is turned through a certain angle to reach the desired preload; the use of heating methods such as a **cal-rod heater** where large bolting or studs are made with a hole in the center so that they can be heated to a certain temperature and the appropriate residual force obtained; or even the use of large **hydraulic tensioning** systems in which a bolt or stud is physically stretched through a certain elongation, the nut is then turned down, and the load released. With any of these methods, it may or may not be easy to then measure the physical **elongation** and ascertain the net preload in order to determine whether the desired preload was achieved.

Torquing Methods

With the torque methods for imparting a preload, the following relationship is used:

$$T \approx CDF_i \qquad (6.1)$$

where T = torque
C = torque coefficient
D = nominal diameter of thread
F_i = desired initial preload

In this technique, you must estimate the **torque coefficient.** Generally, values of .15 for lubricated assemblies, .20 for nonlubricated where traces of cutting oil may be present, and up to .34 for dry assemblies are typically used. Obviously this technique loses accuracy when the extent of lubrication is unknown, especially with used bolting material. Depending on the application, we often preload a bolt to a stress value of approximately 90 percent of the proof strength, depending on the stiffness of the clamped part, which will be discussed later.

---

torque wrench
A wrench with a build-in measuring system that measures the torque that is imparted to a bolt or cap screw.

cap screw
The technical term for what is commonly called a bolt.

cal-rod heater
An electrical element often used to heat large bolting or stud materials. It is similar to what would be found in a circular shape on an electric stove.

Fastener torquing (6.1)
$T \approx CDF_i$

hydraulic tensioning
A large machine that uses hydraulic cylinders to stretch a bolt for imparting an initial tension.

elongation
The amount of stretch or change in length of a material when a load is imparted.

torque coefficient
A value similar to the coefficient of friction used for determining the torque required for turning threaded fasteners.

Example Problem 6.1

A ¾-UNC-grade 5 bolt is to be preloaded to 85 percent of its proof strength. The length of engagement is 5 inches. The bolt is new and nonlubricated but likely has traces of cutting oil present. Determine the required torque

$$A_s = .334 \text{ in.}^2 \quad \text{(Table 6.1)}$$

$$S_p = 85 \text{ ksi} \quad \text{(Table 6.3)}$$

$$F = SA$$

$$F_i = .85 S_p A_s$$

$$F_i = .85(85{,}000 \text{ lb/in.}^2)(.334 \text{ in.}^2)$$

$$F_i = 24{,}130 \text{ lb}$$

Using $C = .2$ nonlubricated with traces of oil

$$T = CDF_i \quad (6.1)$$

$$T = .2(\text{¾ in.}) \, 24{,}130 \text{ lb}$$

$$T = 3{,}620 \text{ in.-lb or } 302 \text{ ft-lb}$$

Turn-of-the-Nut Method

In this method, we must first determine the elongation needed to produce the appropriate preload force by using the following relationship:

$$F = \frac{\delta E A}{L} \quad (6.2)$$

Fastener force (6.2)

This equation permits us to find the clamping force necessary. To find the appropriate elongation, we would use

$$\delta = \frac{FL}{AE} \quad \text{or} \quad \delta = \frac{SL}{E} \quad (6.3)$$

To determine the appropriate angle to which we would turn the nut, we would use the following relationship:

$$\delta = \frac{\text{torque angle}}{360°} \text{ (pitch)} \quad (6.4)$$

or the torque angle would then become

$$\text{torque angle}° = \delta \times \frac{360°}{\text{pitch}}$$

Note that in these methods it is usually appropriate to tighten the nut manually, loosen it, and retighten it snug before turning it to the appropriate angle to remove any residual clearances.

Example Problem 6.2

From Example Problem 6.1, determine the angle of rotation needed, using the turn-of-the-nut method:

$$\delta = \frac{FL}{AE} \text{ or } \frac{SL}{E} \tag{6.3}$$

$$\delta = \frac{24,130 \text{ lb } 5 \text{ in.}}{.334 \text{ in.}^2 \; 30 \times 10^6 \text{ lb/in.}^2}$$

$$\delta = .012 \text{ in.}$$

$$\text{torque angle} = \frac{\delta \; 360°}{\text{pitch}} \tag{6.4}$$

Pitch for ¾ UNC is .1 inch. **(Table 6.1)**

$$\text{torque angle} = \frac{.012 \text{ in. } 360°}{.1 \text{ in.}}$$

$$\text{torque angle} = 43.4°$$

Note again that the nut should be tightened, then turned snug, before turning this angle.

Heating Methods

For the heating method, we would determine the elongation necessary, based on the required preload, the same way we did in the turn-of-the-nut method. However, instead of determining the angle we would turn the nut, we would determine the temperature increase over normal operating conditions that we would need to heat the **stud** in order to obtain the appropriate elongation. We would use the relationship

$$\delta = \alpha L \Delta T \tag{1.1}$$

stud
Similar to a bolt, with the exception that there is no head and typically both ends are threaded.

such that

$$\Delta T = \frac{\delta}{\alpha L}$$

In this method, we would simply increase the temperature, turn the nut snug again, and remove the heater.

Example Problem 6.3

In Example Problem 6.1, to obtain the same preload, determine the temperature we would need to heat this bolt above the **service temperature**:

$$\alpha = \frac{6.5 \times 10^{-6} \text{ in.}}{\text{in. }^\circ F} \quad \text{(Appendix 8)}$$

$$\Delta T = \frac{\delta}{\alpha L} \quad (6.5)$$

$$\Delta T = \frac{.012 \text{ in.}}{\frac{6.5 \times 10^{-6} \text{ in. } 5 \text{ in.}}{\text{in. }^\circ F}}$$

$$\Delta T = 370^\circ F$$

Direct Mechanical Tensioning (Stud Tensioner)

This method uses a very large machine for stretching (in most cases a stud) and involves determining the appropriate force, stretching the bolt to the appropriate force level, and turning the nut snug. Note that, in this type of application, it is often necessary to use a special thread for high stress levels; It is often difficult to turn the nut when the bolt is stretched because this changes the lead of the bolt. A modified special pitch is often used to compensate for the change in lead during stretching.

6.5 ELASTIC ANALYSIS OF BOLTED CONNECTIONS

When we tension a bolted connection, not only do we leave a residual tension in the bolt or stud, we also induce a residual compressive stress in the clamped part. Because of this, when we apply an external load to the joint, depending on the relationship between the clamped joint stiffness and the

---

service temperature
The temperature at which a system operates. This needs to be taken into account in order to determine preload temperatures for tensioning bolting or stud systems.

Thermal expansion (6.5)

$$\Delta T = \frac{\delta}{\alpha L}$$

bolt or stud stiffness, only part of this external load is applied to the bolt or stud. In fact, in many clamped connections where the stiffness of the flange is much greater than the stiffness of the bolt, we will find that almost no additional load is added to the bolt or stud. To better envision this phenomenon, imagine a case where we are tightening a fairly large clamped flange of equal strength material to the bolting. This bolting may have a clamped area, for example, of ten times as great as the cross-sectional area of the bolts. Therefore, the implied load on the bolts would cause a stretch in the bolting material approximately ten times greater than that of the flange compression. When we now apply an external load on this clamped joint, the bolt or stud load only increases by approximately 10 percent of the decreased loading in the flange due to the ratio of their relative stiffness. Therefore, the change in the stress in the bolt is relatively small as this load is applied. This preload is typically needed to prevent leakage in a pressure-containing joint such as an automobile head or pressure vessel.

In order to analyze this, we need to determine the stiffness of both the bolt and the flange. The relationship of the deflection of the bolt or the flange is equal to

$$\delta = \frac{FL}{AE} \qquad (6.3)$$

where k is the stiffness of the joint or bolt and is equal to

$$k = \frac{F}{\delta}$$

Then the stiffness of the joint or bolt becomes equal to

$$k_c = \frac{A_c E_c}{L_c} \quad \text{or} \quad k_b = \frac{A_b E_b}{L_b} \qquad (6.6)$$

where E = Young's modulus for the bolting or the flange material
A = the area of either the clamped joint or the bolt
L = the grip length, which would typically be the same for both

The required preload could then be calculated as follows:

$$F_i = QF_e\left(\frac{k_c}{k_b + k_c}\right) \qquad (6.7)$$

where F_i = recommended preload
Q = margin factor (similar to a safety factor)
F_e = expected external or applied load

The new load on the bolting would then become

$$F_t = F_i + \Delta F_b$$

where

$$\Delta F_b = F_e\left(\frac{k_b}{k_b + k_c}\right) \qquad (6.8)$$

or the total force on the bolt becomes

$$F_t = F_i + \left(\frac{k_b}{k_b + k_c}\right)F_e \qquad (6.9)$$

and the new force on the compressed flange would be

$$F_c = F_i - \left(\frac{k_c}{k_b + k_c}\right)F_e \qquad (6.10)$$

Force in bolt (6.9)

$$F_t = F_i + \left(\frac{k_b}{k_b + k_c}\right)F_e$$

Force in flange (6.10)

$$F_c = F_i - \left(\frac{k_c}{k_b + k_c}\right)F_e$$

Example Problem 6.4

As shown in Figure 6.3, a pressure vessel has a sealed head diameter of 20 inches. It uses ten 1¼ UNF studs that are fully threaded and have an effective clamping length of 6 inches. The clamped area of the flange is 50 in.². The studs are made of class 8 material.

1. Find the torque, assuming new lubricated threads necessary to tighten to 90 percent of the proof strength.

$$T \approx CDF_i \qquad (6.1)$$

$$F_i = S_i A$$

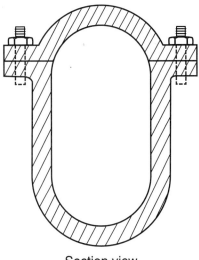

Section view

FIGURE 6.3 Pressure vessel for Example Problem 6.4.

$F = .90(120,000 \text{ lb/in.}^2)(1.073 \text{ in.}^2)$ **(Tables 6.1 and 6.3)**

$F = 115,884 \text{ lb}$

$T = CDF_i$ $\quad C = .15$ lubricating threads

$T = .15(1.25 \text{ in.})(115,884 \text{ lb})$

$T = 21,728$ in.-lb or $1,810$ ft-lb

2. If the pressure vessel is now pressurized to 500 psig, determine the total load on the bolts.

   Find the stiffness of both studs and flanges:

$$k_b = \frac{AE}{L} = \frac{1.073 \text{ in.}^2 \; 30 \times 10^6 \text{ lb/in.}^2}{6 \text{ in.}} \quad (6.6)$$

$k_b = 5.365 \times 10^6$ lb/in. per stud

$$k_f = \frac{AE}{L} = \frac{50 \text{ in.}^2 \; 30 \times 10^6 \text{ lb/in.}^2}{6 \text{ in.}}$$

$k_f = 2.5 \times 10^8$ lb/in.

$F_e = PA \qquad A = \dfrac{\pi D^2}{4} = \dfrac{\pi (20 \text{ in.})^2}{4}$

$F_e = 500 \text{ lb/in.}^2 \; (314 \text{ in.}^2) \qquad A = 314 \text{ in.}^2$

$F_e = 157,000$ lb

$$F_t = F_i + \left(\frac{k_b}{k_b + k_c}\right) F_e \qquad (6.9)$$

$$F_t = 10\,(115{,}884\text{ lb}) +$$

$$\left(\frac{10 \times 5.365 \times 10^6\text{ lb/in.}}{10 \times 5.365 \times 10^6\text{ lb/in.} + 2.5 \times 10^8\text{ lb/in.}}\right) 157{,}000\text{ lb}$$

$$F_t = 1.19 \times 10^6\text{ lb} \qquad \text{or} \qquad 119{,}000\text{ lb/stud}$$

Note the minor increase in bolt load.

Note also that when the stiffness of the clamped joint is much greater than the stiffness of the bolt, we can assume that the total force in the bolting is approximately equal to the initial preload force. On the other hand, if the stiffness of the bolt is much greater than the stiffness of the flange, then the total force is approximately equal to the initial force plus the external load.

We can summarize this as follows:

$$k_c >\!>\!> k_b \qquad F_t \approx F_i$$
$$k_b >\!>\!> k_c \qquad F_t \approx F_i + F_e$$

6.6 GASKETED CONNECTIONS

gaskets
A typically soft material that is used for sealing pressure-retaining parts.

The analysis of **gaskets** is difficult because many gasket materials either do not have a published value for Young's modulus (E) or this modulus changes with compression. Many gasket manufacturers specify a desired pressure versus compression rate and often give a graph for different pressures. In gasket design we usually try not to have the gasket cover the whole compressed surface. We do this by machining a recess or groove for the gasket in the surface so that we can effectively ignore the compressed gasket area or simply determine the necessary additional load required for that area or the amount of deflection. This typically would be the best choice, especially in high-pressure situations. For materials with which we are using metallic gaskets that cover the whole joint, we could determine their equivalent stiffness and simply develop a combined overall stiffness for our compressed flange as follows:

$$\frac{1}{k_c} = \frac{1}{k_1} + \frac{1}{k_2} + \frac{1}{k_3} \ldots$$

where k_1, k_2, and k_3 are the stiffness for the components being clamped, which would be determined from $k = AE/L$.

6.7 OTHER FASTENING METHODS

This chapter emphasized screws and bolts because of their wide application and the known structural analysis associated with them. Obviously many other types of fastening systems are available: rivets that can be installed hot as on old-fashioned bridges and early construction projects; pop rivets that are mechanically expanded; adhesives that use chemical action to bond materials together; or such items as **expansion bolts** and **quarter-turn fasteners,** which are in wide use in many applications. We could even consider in this category such items as Velcro, plastic snap connections, living hinges, or even snaps. Obviously, the joining of many parts can be done using many different methods, and surely the array of products available will continue to grow. Figure 6.4 shows just a few of the many styles of fasteners available.

expansion bolts
A bolting system that has an expanding outer sleeve or ferrule that can be made larger when tightened.

quarter-turn fastener
A type of latching system that, when turned, pulls parts into engagement.

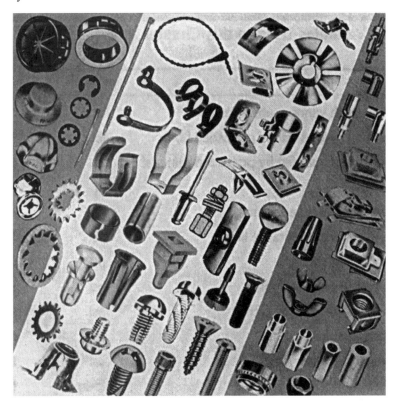

FIGURE 6.4 A selection of commercially available fasteners. (Courtesy of Bolt Products, Inc.)

welding
A process whereby materials are raised to above their recrystalization temperatures and bonded together under high heat, with or without an additional filler element.

soldering
A process for joining parts by use of an alloying element, typically at a significantly lower temperature than that used for welding.

brazing
A process similar to soldering that typically uses a flame and an alloying metal.

residual stress
A stress left over in a part due to such processes as heat treating or welding. This is a stress that exists in the part even in the absence of an external load.

heat-affected zone
A term used when, during such processes as welding, the heat causes changes in the base property of the material or results in a residual stress.

6.8 STRENGTH OF WELDED CONNECTIONS

Many processes could be included under the general heading of **welding**, including such similar processes as **soldering** and **brazing**. However, in general, when we think of welding, we think of creating a material of equal strength to the base material. Typically we assume that the deposited material from the welding process, whether from gas welding, electric arc, or another type, is equal in strength to the base material if a full-penetration weld for the entire joint thickness is completely bonded. Some designers assume that a value of about 70 percent is appropriate for the strength of a welded joint. In any case, a designer should be extremely careful as the welding process creates considerable heat which can result in significant **residual stresses**. In fact, when welded joints fail, they often fail at the edge of the **heat-affected zone** and not in the welded joint. This is due to thermal expansion and contraction during cooling or material phase transformations, which may result in residual tensile stresses at this area, especially in brittle materials such as cast iron where extreme care should be used in the welding process.

The design of welded joints is a specialized engineering topic that this text will not cover in sufficient detail, and additional expertise should be sought in designing welded joints, especially for other than lower-strength, nonheated treated steels.

6.9 SUMMARY

After completing this chapter, you should be familiar with basic types of fasteners, the principles of thread sizes and definitions, and how to determine the strength of different bolting grades and sizes in both the English and metric systems.

The elastic analysis of simple bolted connections should be understood. You should be able to analyze joints of this type and to understand how the bolting or studs can be initially tensioned.

6.10 PROBLEMS

1. A lifting eye on a shipping container is made from grade 5 bolting material and has ¾-UNC threads. If the container weighs 5,000 pounds, is this acceptable if a safety factor of 3 is desired?

2. A ⅝-UNF steel bolt has an effective length of 8 inches, which is also the approximate length of the unthreaded shank area. If the elongation after tightening is .015 inch:
   a. What is the stress in the shank area?
   b. What is the stress in the threaded area?
   c. What grade material should be specified if a SF of 1.5 is desired based on proof strength?
   d. If the threads were lubricated, what torque would be needed to tighten this bolt?
   e. Using the turn-of-nut method, what angle of tightening would be required?
3. In Problem 2, change the bolt to an M16 × 2 metric bolt with an approximate unthreaded length of 250 mm and an elongation of .50 mm and answer questions *a* through *e*.
4. The cylinder head of a 10-inch-diameter compressor is attached by 12 studs made from SAE grade 5 material. The cylinder has a pressure of 300 psi. For a safety factor of 2:
   a. Select a stud size.
   b. To preload these studs with lubrication to 80 percent of proof strength, calculate the required torque.
5. In Problem 4, if we were to substitute ½-inch studs for the size you selected, the flange has an outside diameter of 14 inches, and the grip length is 2 inches:
   a. Find the initial force in the studs if they are preloaded to 85 percent of proof strength.
   b. Find the force in the studs under pressure.
   c. Do you think this load changed enough from Problem 4 that we should analyze for fatigue?
6. In the frame shown in Figure 6.5, which is made from aluminum ($E = 10 \times 10^6$ lb/in.²), the cap screw is tightened to a torque of 350 ft-lb. The threads are not lubricated, but new, with residual cutting oil. Assume a compressed area of 4 in.². Find the following:
   a. The initial preload.
   b. The elongation of the cap screw.
   c. The compressed deflection of the clamped frame.
   d. If an external load of 5,000 pounds is now applied, find the total load in the bolt.
   e. Find the new stress in the bolt under this load.
   f. What load could be applied to just reach the point of the joint opening?

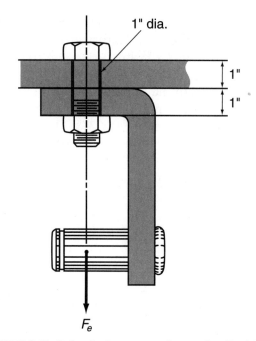

FIGURE 6.5 Bolted aluminum frame for Problem 6.

7. The nut on the milling machine clamp in Figure 6.6 is turned 30° after snug. Assuming the clamp is stiff and the effective length of the stud is 2 inches, find:
   a. The elongation in the stud.
   b. The stress in the stud.
   c. The force at point C.

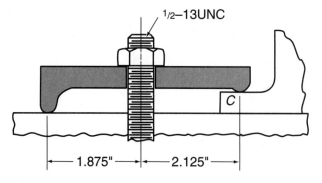

FIGURE 6.6 Milling machine clamp for Problem 7.

8. In the air/hydraulic cylinder shown in Figure 6.7, if the four tie rods (A) are 20 mm in diameter, the distance between the heads (B) is 400 mm, the heads (C) are 50 mm thick, the cylinder wall has an OD (D) of 100 mm and an ID (E) of 90 mm, and the studs are tightened to a torque of 100 Nm, determine the following, assuming the tie rods and cylinder are made of steel (assume threads are lubricated during tightening and heads are considered stiff):
    a. Initial force in each tie rod.
    b. Initial stress in each tie rod.
    c. Initial stress in the cylinder wall.
    d. If the cylinder is pressurized to 5,000 kPa, what is the new load in the studs?
    e. What is the new compressive force in the cylinder?

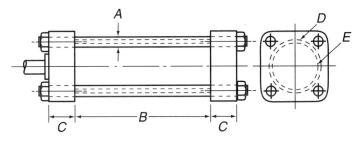

FIGURE 6.7 Cylinder for Problems 8 and 9.

9. In the air/hydraulic cylinder shown in Figure 6.7, if the four tie rods (A) are ½ inch in diameter, the distance between the heads (B) is 10 inches, the heads (C) are 1 inch thick, the cylinder wall has an OD (D) of 2 inches and an ID (E) of 1¾ inches, and the studs are tightened to a torque of 250 in.-lb, determine the following, assuming the tie rods are steel but the cylinder wall is aluminum (assume threads are new and not lubricated, i.e., no traces of cutting fluid):
    a. Initial force in each tie rod.
    b. Initial stress in each tie rod.
    c. Initial stress in the cylinder wall.
    d. If the cylinder is pressurized to 2,500 psi, what is the new load in the studs?
    e. What is the new compressive force in the cylinder?
    f. What is the change in force in the tie rods and cylinder?

10. A 1-inch steel bolt goes through an aluminum ($E = 10 \times 10^6$ lb/in.²) tube as shown in Figure 6.8. The tube and unthreaded bolt are 24 inches long, and UNC threads are used. The tube is 1¼ inches ID and 2¼ inches OD. After tightening and resnugging the nuts, the nut is turned one-half turn. Find the following:
    a. Deflection in each piece.
    b. Stress in each piece.

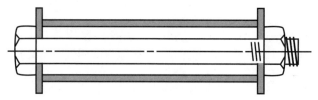

FIGURE 6.8 Bolt and tube for Problem 10.

11. A standard hex cap screw ¾-UNC-10 has a ⁹⁄₁₆-inch-thick head and is subject to a tensile load of 10,000 pounds. Find the stress and factor of safety based on proof strength on the shank if this is an SAE Grade 2 bolt.
12. In Problem 11, find the shear stress on the head and find the factor of safety based on 50 percent of the tensile strength.
13. Explain, based on the results of Problems 11 and 12, why we are not normally concerned about the head strength for standard cap screws.

6.11 CUMULATIVE PROBLEMS

14. For the locomotive in Chapter 2, Problem 11, the steam piston heads are held in place by eight ⅝-inch UNF cap screws. Assume the outside diameter of the cylinder wall is 14 inches and the cap screws are tightened using the turn-of-the-nut method. If a margin factor Q of 3 is desired, find the recommended preload.
15. In Problem 14, using the turn-of-the-nut method, what angle of rotation is needed after snug if the clamped head section is 2 inches thick?
16. What is the stress in the bolts in Problem 15 after tensioning?
17. In Problem 14, if the cylinder and cap screws are tensioned cold and if the cylinder and heads heat up together, do we need to account for the thermal expansion in order to maintain the required preload? Explain why or why not.

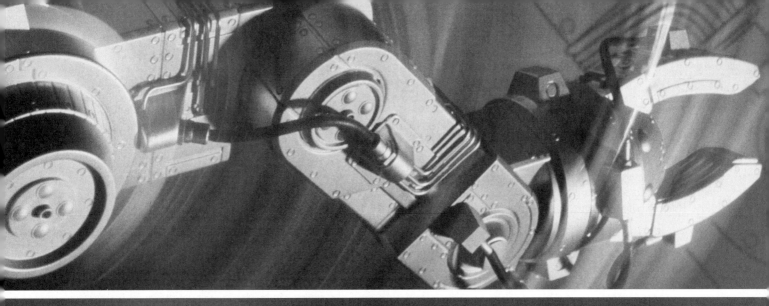

CHAPTER 7

Impact and Energy Analysis

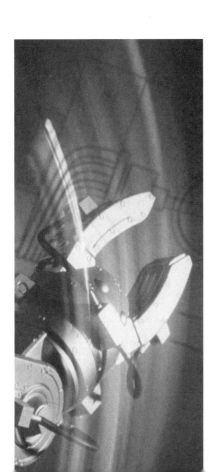

impulse load
A force applied rapidly or of short duration, typically at less than one natural frequency of the receiving part.

potential energy
The energy associated with elevation such as a falling object.

kinetic energy
The energy associated with velocity, as in a moving automobile.

Often in the design of machine parts we have conditions where a load is applied suddenly, either deliberately or incidentally to the design. An example of a suddenly applied or **impulse load** would be a hammer striking a nail or an automobile colliding with a guardrail system. An example of an unintentional case would be the wearing of bearings in a rotating assembly, with the result that clearances become excessive and impact loads are created. An interesting example of impact loads changing the performance of a machine can be seen in an automobile engine, where the main bearings can provide excellent service for up to 200,000 or 300,000 miles. As wear progresses, impact loads become significant due to increased clearances in the bearings. The rate of wear can then accelerate quickly as even relatively small clearances in the presence of high loads can result in significant increases in stress levels.

In this chapter, we are going to analyze impact loading by the use of energy methods where energy given up by the moving object equals energy absorbed by the receiving object. By determining the energy to be given up, which could be in the form of **potential energy** or **kinetic energy**, and setting that energy equal to the absorbed energy in the part through deflection analysis, we can then determine the associated stress in the part and the resultant effect.

7.1 OBJECTIVES

After completing this chapter, you should be able to

- Describe the types of energy that are transmitted by moving or falling objects.
- Understand the principles of energy absorption and how deflections in machine parts absorb energy.
- Understand how energy is transferred into a part and, in some cases, returned (as in a spring).
- Determine the impact-induced stress in a part and how it relates to the part's ability to resist these stresses.

7.2 IMPACT ENERGY

The basic analysis of impact forces involves relating the energy being transmitted to a part to the energy gained by the part. For example, a diver jumping up from a diving board falls a set distance to the board, whereupon the board is deflected downward. The energy, in the form of poten-

tial energy from the diver, deflects the board a set distance which results in that amount of energy being stored in the board, which then returns this energy to the diver, projecting the diver upward. Another example is an automobile hitting a guardrail system where the kinetic energy associated with the velocity of the auto would then bend and/or break the guardrail system, imparting a similar amount of energy to the guardrail. An impact therefore involves energy that is imparted from a moving or falling object and then absorbed by the deflection of a receiving object. We can envision this as being like a spring system. To analyze an impact, we would use the basic relationship of

$$U_{given\ up} = U_{gained}$$

This energy is absorbed in the part by the relationship of energy gained.

$$U = \frac{F\delta}{2} \quad (7.1)$$

where δ is the total **elastic deformation**, or

$$U = \frac{k\delta^2}{2} \quad (7.2)$$

where k is considered to be a **spring constant** of the part. This is similar to the spring constant for springs because any part effectively acts as a spring when subject to a load that causes a corresponding deflection. The energy could be either kinetic or potential energy from either a falling weight or the velocity of a moving mass:

$$U = KE = \tfrac{1}{2}mv^2 \quad (7.3)$$

or potential energy

$$U = PE = mgh \quad (7.4)$$

By rearranging equations 7.1 and 7.2 and by relating the energy given up to the energy gained, the stiffness or force can be determined:

$$U = \frac{k\delta^2}{2} \quad \text{as} \quad k = \frac{F}{\delta} \quad \text{or} \quad F = kx$$

$x = \delta$ for a spring

elastic deformation
The deformation of a part below the elastic limit such that the part returns to its original shape when the load is removed.

spring constant
The equivalent stiffness of a part, similar to a spring rate for a spring, that can be determined based on the stiffness of any potential load-receiving part.

Kinetic energy (7.3)

$U = KE = \tfrac{1}{2}mv^2$

Potential energy (7.4)

$U = PE = mgh$

The energy that can be absorbed in torsion would correspondingly be

$$U = \frac{T\theta}{2} \quad \text{in.-lb or ft-lb}$$

where θ is the **angular deformation** of the bar.

angular deformation
Twist in a round bar; usually expressed in degrees or radians.

We can then relate the energy given up to the energy absorbed through the use of the following string equation:

$$U = w(h + \delta) = mg(h + \delta) = PE = \frac{F\delta}{2} = \frac{SA\delta}{2} = \frac{k\delta^2}{2} = \frac{S^2AL}{2E}$$

$$S = \frac{F}{A} \quad k\delta = SA \quad \delta = \frac{FL}{AE}$$

where $h = $ height from which the weight falls

We can pick any two parts of this string equation to relate the potential energy to the properties we are looking for, either deflection, stress, or a combination thereof, in order to solve for either the stiffness, deflection, or stress in the receiving part. Kinetic energy can be substituted for potential energy and the same process used.

If we have a compressive impact from a dropping weight, we could determine deflection as follows:

Impact-deflection (7.5)

$$\delta = \frac{W}{k} + \frac{W}{k}\left(1 + \frac{2hk}{W}\right)^{1/2} \tag{7.5}$$

where we can determine the stiffness of the part as follows:

Stiffness (7.6)

$$k = \frac{F}{\delta} = \frac{AE}{L} \tag{7.6}$$

or

$$\delta = \frac{FL}{AE} = \frac{SL}{E} \tag{7.7}$$

or, by rearranging this, derive the stress in the part as follows:

$$S = \frac{2W}{A}\left(\frac{h}{\delta}+1\right) \qquad (7.8)$$

Since the relationship of deflection to stress is

$$\delta = \frac{SL}{E} \qquad (7.9)$$

we can also determine the stress from a falling weight as follows:

$$S = \frac{W}{A} + \frac{W}{A}\left(1 + \frac{2hEA}{LW}\right)^{1/2} \qquad (7.10)$$

where the relationship of $\frac{W}{A}$ is the static residual stress. If we have a part such as a stepped shaft, which has multiple stiffnesses, we can simply determine its equivalent stiffness in the same manner that we will later learn to deal with springs in series or in a basic electricity course for resistors in parallel as follows:

$$k = \frac{k_1 k_2}{k_1 + k_2} \qquad (7.11)$$

The following example problem describes a falling object landing on the end of a short shaft and the associated impact stress and deflection of the shaft.

Example Problem 7.1

What is the stress and deflection in the 2-inch-diameter steel bar in Figure 7.1 if the 500-pound weight is dropped 1 inch? Assume the weight and supports are rigid.

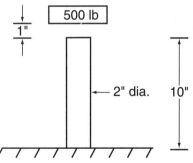

FIGURE 7.1 Steel bar for Example Problem 7.1.

Impact-stress (7.10)

$$S = \frac{W}{A} + \frac{W}{A}\left(1 + \frac{2hEA}{LW}\right)^{1/2}$$

$$S = \frac{W}{A} + \frac{W}{A}\left(1 + \frac{2hEA}{LW}\right)^{1/2} \tag{7.10}$$

$$S = \frac{500 \text{ lb}}{\frac{\pi (2 \text{ in.})^2}{4}} + \frac{500 \text{ lb}}{\frac{\pi (2 \text{ in.})^2}{4}}\left(1 + \frac{2\left[1 \text{ in. } 30 \times 10^6 \text{ lb/in.}^2 \frac{\pi (2 \text{ in.})^2}{4}\right]}{10 \text{ in. } 500 \text{ lb}}\right)^{1/2}$$

$$S = 31,062 \text{ lb/in.}^2$$

$$\delta = \frac{SL}{E} = \frac{31,061 \text{ lb/in.}^2 \; 10 \text{ in.}}{30 \times 10^6 \text{ lb/in.}^2} \tag{7.9}$$

$$\delta = .010 \text{ in.}$$

7.3 VELOCITY AND IMPACT

In the last section, impact loads were analyzed for falling objects such as a hammer hitting a nail or other similar event. We often have, instead of potential energy causing an impact load, kinetic energy of the form

$$KE = \tfrac{1}{2}mv^2 = \tfrac{1}{2}\frac{W}{g}v^2$$

In this case, we have a velocity of an object striking another, such as the previously discussed example of an automobile hitting a guardrail. In this case, the energy part of our string equation simply becomes equal to

$$W\left(\frac{v^2}{2g} + \delta\right) = \frac{F\delta}{2} = \frac{k\delta^2}{2} \tag{7.12}$$

where the velocity is the velocity of the object before impact and the deflection is the deflection in the object after impact occurs. Care should be taken in the use of this formula to make sure that the units are consistent. In some cases, if the object being struck has significant mass, we may wish to include that as, obviously, the mass of the object undergoes an acceleration for a short period of time, and this affects the total energy absorption. In this case, the formula for either the potential energy from a dropping weight or the velocity associated with kinetic energy could be factored in as follows:

Impact (7.12)

$$W\left(\frac{v^2}{2g} + \delta\right) = \frac{F\delta}{2} = \frac{k\delta^2}{2}$$

$$\delta = \delta_{st} + \delta_{st}\left(1 + \frac{2h}{\delta_{st}}\right)^{1/2} = \delta_{st} + \delta_{st}\left(1 + \frac{v^2}{g\delta_{st}}\right)^{1/2} \quad \text{(7.13)}$$

where δ_{st} is the deflection under the static load.

Example Problem 7.2 involves an automobile hitting the end of a bar at low speed in a parking lot and the resulting deflection and stress in the bar. Note that for purposes of this example, we are assuming that the automobile is completely rigid in the analysis so that all of the deflection occurs in the bar. In reality, both objects typically absorb some of the energy. In fact, modern automobile design places a high priority on the ability of the automobile to absorb energy in the case of a collision, leaving less energy to be absorbed by the occupants.

Example Problem 7.2

A rigid 2,000-pound automobile hits a fixed stop constructed of a 36-inch-long, 2-inch-diameter steel bar at a speed of 5 mph. (See Figure 7.2.) If the impact is absorbed only by the bar, what is the deflection and stress in the bar?

$$W\left(\frac{v^2}{2g} + \delta\right) = \frac{k\delta^2}{2} \quad \text{(7.12)}$$

(ignoring δ distance after impact as it is insignificant)

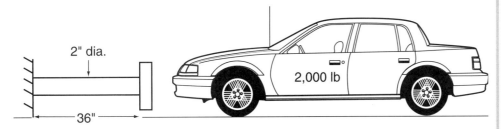

FIGURE 7.2 Example Problem 7.2: Automobile striking a steel bar.

Solve for k.

$$k = \frac{AE}{L} \qquad (7.6)$$

$$k = \frac{\pi\, 2\text{ in.}^2}{4} \frac{30 \times 10^6 \text{ lb/in.}^2}{36 \text{ in.}} = 2.62 \times 10^6 \text{ lb/in.}$$

Converting velocity:

$$5\frac{\text{mi}}{\text{hr}} \quad 5,280\frac{\text{ft}}{\text{mi}} \quad 12\frac{\text{in.}}{\text{ft}} \quad \frac{\text{hr}}{3,600 \text{ sec}} = 88\frac{\text{in.}}{\text{sec}}$$

Substituting:

$$W\left(\frac{v^2}{2g} + \delta\right) = \frac{k\delta^2}{2} \qquad (7.12)$$

$$2,000 \text{ lb} \left(\frac{\left(\frac{88 \text{ in.}}{s}\right)^2}{(2)\, 386\, \frac{\text{in.}}{s^2}} + 0\right) = \frac{2.62 \times 10^6 \text{ lb/in.}\, \delta^2}{2}$$

$$20,060 \text{ in.-lb} = 1.31 \times 10^6 \text{ lb/in.}\, \delta^2$$

Rearranging and solving:

$$\delta^2 = \frac{20,060 \text{ in.-lb}}{1.31 \times 10^6 \text{ lb/in.}}$$

$$\delta = .124 \text{ in.}$$

$$\delta = \frac{SL}{E} \qquad (7.9)$$

or

$$S = \frac{\delta E}{L}$$

$$S = \frac{.124 \text{ in.}\, 30 \times 10^6 \text{ lb/in.}^2}{36 \text{ in.}}$$

$$S = 103,300 \text{ lb/in.}^2$$

Note that for most steels this may cause failure even if we ignore column failure.

7.4 IMPACT ON BEAMS

When an object hits a beam, the energy is absorbed in a similar manner to the axial load shown in the previous two examples, except that we need to calculate the stiffness of the beam. The most straightforward way to do this is to take the deflection from Appendix 2, rearrange this by putting it in $F = kx$ form, and simply solve for k where we would actually determine the equivalent stiffness of our beam system. Example Problem 7.3 shows first the determination of the stiffness of a beam system and then an automobile as in Example Problem 7.2 colliding with this beam.

Example Problem 7.3

A 3,000-lb automobile (assumed to be rigid) hits a guardrail constructed from an 8 in., 23 lb/ft steel I-beam that is 40 ft between supports at its center. The beam's moment of inertia is $I_x = 64.2$ in.^{4}, and it is made of AISI 1137 hot-rolled steel ($S_y = 48,000$ psi). Ignoring the weight of the beam, at what velocity would the beam yield?

FIGURE 7.3 Example Problem 7.3: Automobile hitting a guardrail.

$$S = S_y = \frac{Mc}{I} \qquad M = \frac{FL}{4} \qquad \textbf{(Appendix 2)}$$

$$S = \frac{FL}{4}\frac{c}{I}$$

$$S = \frac{F(480 \text{ in.})(4 \text{ in.})}{4 \cdot 64.2 \text{ in.}^4} = 48,000 \text{ lb/in.}^2$$

$$F = 6,420 \text{ lb}$$

$$\delta = \frac{FL^3}{48EI} \qquad \text{(Appendix 2)}$$

$$\delta = \frac{6,420 \text{ lb } (480 \text{ in.})^3}{48(30 \times 10^6) \text{ lb/in.}^2 \; 64.2 \text{ in.}^4}$$

$$\delta = 7.68 \text{ in.}$$

$$KE = \frac{1}{2}\frac{W}{g}v^2 = \frac{F\delta}{2} \qquad (7.3, 7.1)$$

$$KE = \frac{1}{2}\frac{3,000 \text{ lb}}{32.2 \text{ ft/sec}^2}v^2 = \frac{6,420 \text{ lb } (7.68 \text{ in.})}{2}$$

$$v = 23 \text{ ft/sec}$$

or

$$23\frac{\text{ft}}{\text{sec}} \; \frac{\text{mile}}{5,280 \text{ ft}} \; \frac{3,600 \text{ sec}}{\text{hr}}$$

$$v = 15.7 \text{ miles/hr}$$

7.5 DESIGNING FOR IMPACT

In complicated assemblies, the energy absorbed during an impact is divided between different sections of an assembly. A good example of this is an automobile that hits a pothole on a highway. Some of the impact from hitting the bottom of the pothole is absorbed in the tire and some of the energy is absorbed in the rim, axle, springs, and the frame of the automobile. Finally, some of this energy is absorbed in the seat and then the occupants. If we wished to make this impact as minimal as possible for the occupants, we would try to make each part of the system as "soft" as possible in order to maximize the energy absorption before the impact load reached the occupants. In the design of an automobile, this becomes a trade-off between the car's ride and cornering ability and its ability to absorb shock loading. Another example that will help to explain the trade-offs involved with designing for impact would be a heavy-duty, over-the-road truck. Because of this vehicle's need to carry excessive loads, its springs and frame need to be far stiffer than those of a passenger automobile. In this design we would try to cushion the cab of the vehicle and the seats instead of the tires, springs, and frame.

You can see from these two examples that stiffness design is often a trade-off. Another example of this might be bolting that is used in an

impact environment. By making the unthreaded part of a bolt equal in size to the smaller root diameter of the threaded area, we can actually create a higher-strength bolt as by decreasing the stiffness we increase the load-carrying capability. Example Problem 7.4 shows a bolt subject to a load where, in the first case, we have kept the shank of the bolt at the nominal or major diameter, and in the second case, we make the shank the same size as the minor diameter, improving the load-bearing capability under impact conditions.

Example Problem 7.4

A ¾-UNC bolt with 6 inches of unthreaded shank and 2 inches of thread in engagement is subjected to a falling load of 500 pounds from a height of .030 inches. What is the maximum stress in this bolt?

$$\text{Area of threads: } .334 \text{ in.}^2 \quad \text{(Table 6.1)}$$

$$\text{Area of shank: } \frac{\pi D^2}{4} = \frac{\pi (.75 \text{ in.})^2}{4} = .442 \text{ in.}^2$$

Overall stiffness is

$$k = \frac{k_1 k_2}{k_1 + k_2} \quad (7.11)$$

$$k = \frac{\left(\frac{A_1 E_1}{L_1}\right)\left(\frac{A_2 E_2}{L_2}\right)}{\frac{A_1 E_1}{L_1} + \frac{A_2 E_2}{L_2}}$$

$$k = \frac{\left(\frac{.334 \text{ in.}^2 \left(30 \times 10^6 \text{ lb/in.}^2\right)}{2 \text{ in.}}\right)\left(\frac{.442 \text{ in.}^2 \left(30 \times 10^6 \text{ lb/in.}^2\right)}{6 \text{ in.}}\right)}{\frac{.334 \text{ in.}^2 \left(30 \times 10^6 \text{ lb/in.}^2\right)}{2} + \frac{.442 \text{ in.}^2 \left(30 \times 10^6 \text{ lb/in.}^2\right)}{6 \text{ in.}}}$$

$$k = 1,533,500 \text{ lb/in.}$$

Solve for deflection.

$$\delta = \frac{W}{k} + \frac{W}{k}\left(1 + \frac{2hk}{W}\right)^{1/2} \quad (7.5)$$

$$\delta = \frac{500 \text{ lb}}{1,533,500 \text{ lb/in.}}$$
$$+ \frac{500 \text{ lb}}{1,533,500 \text{ lb/in.}}\left(1 + \frac{2(.030 \text{ in.})\, 1,533,500 \text{ lb/in.}}{500 \text{ lb}}\right)^{1/2}$$

$$\delta = .0048 \text{ in.}$$

as

$$\delta = \frac{S_1 L_1}{E} + \frac{S_2 L_2}{E} \quad \text{where} \quad S_1 = \frac{A_1}{A_2} S_2 = \frac{.334}{.442} S_2$$

$$.0048 \text{ in.} = \frac{\frac{.334}{.444} S_2\, 6 \text{ in.}}{30 \times 10^6 \text{ lb/in.}^2} + \frac{S_2\, 2 \text{ in.}}{30 \times 10^6 \text{ lb/in.}^2}$$

$$.0048 \text{ in.} = 1.504 \times 10^{-7}\, S_2 + 6.666 \times 10^{-8}\, S_2$$

$$S = 22,100 \text{ lb/in.}^2$$

Stiffness, if shank has the same area as threads, is

$$k = \frac{AE}{L} = \frac{.334 \text{ in.}^2\, (30 \times 10^6 \text{ lb/in.}^2)}{8 \text{ in.}} \quad (7.6)$$

$$k = 1,252,500 \text{ lb/in.}$$

$$\delta = \frac{W}{k} + \frac{W}{k}\left(1 + \frac{2hk}{W}\right)^{1/2} \quad (7.5)$$

$$\delta = \frac{500 \text{ lb}}{1,252,500 \text{ lb/in.}} + \frac{500 \text{ lb}}{1,252,500 \text{ lb/in.}}\left(1 + \frac{2(.030 \text{ in.})\, 1,252,500 \text{ lb/in.}}{500 \text{ lb}}\right)^{1/2}$$

$$\delta = .0053 \text{ in.}$$

if $\delta = \frac{SL}{E}$ then $S = \frac{\delta E}{L}$ (7.9)

$$S = \frac{.0053 \text{ in.}\, 30 \times 10^6 \text{ lb/in.}^2}{8 \text{ in.}}$$

$$S = 19,910 \text{ lb/in.}^2$$

Note that the deflection is greater but the stress is lower.

7.6 SUMMARY

After completing this chapter, you should be able to recognize situations where impact loading may affect your design and be able to incorporate this into your analysis. You should also have a sense of the complexity of designing under impact conditions. This chapter only provides an introduction to this field of analysis. You should, however, be able to analyze straightforward impact problems involving falling objects or the impact from horizontal velocity.

7.7 PROBLEMS

1. In Example Problem 7.1, if the 2-inch bar is made from 6061 T-6 aluminum, determine:
   a. Deflection of the bar.
   b. Maximum stress in the bar.
   c. Would this bar fail?
   d. Explain why this stress is less than that calculated in Example Problem 7.1.
2. In Example Problem 7.1, assume that the bar is made from hot-rolled AISI 1040 steel with a diameter of 50 mm and a length of 250 mm. If a rigid 200 N weight falls 10 mm, determine:
   a. Deflection of the bar.
   b. Maximum stress in the bar.
   c. Is this acceptable for a safety factor of 2 based on ultimate strength?
3. In the pulley system shown in Figure 7.4, the 8-pound box is traveling at a rate of 10 ft/sec, the cable is steel and has an effective diameter of $3/16$ inch, and dimension A is 10 ft. If the pulley stops instantly, determine the deflection and stress in the cable. Assume $E = 12 \times 10^6$ lb/in.^{2}
4. In Problem 3, if A is changed to 20 feet, determine:
   a. Stress in the cable.
   b. Deflection in the cable.
   c. Explain why both changed.
   d. If this cable was made from aluminum, without redoing the calculation, estimate the stress in the cable and the deflection.

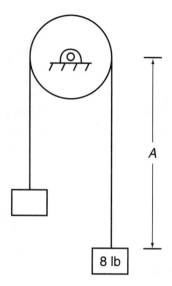

FIGURE 7.4 Pulley system for Problems 3 and 4.

5. A steel bar is 1 inch in diameter and 72 inches long. If a 2,500-pound weight was to drop ⅛ inch, calculate (ignore any column action):
   a. Stress in the bar.
   b. Elongation of the bar.
   c. Static stress in the bar after the weight comes to rest.
   d. Static deflection in the bar.
   e. Recalculate a, b, c, and d for a bar of the same dimensions made from aluminum.

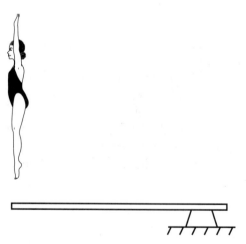

FIGURE 7.5 Diving board for Problem 6.

6. A diver weighing 200 pounds jumps from 1 foot above a diving board that is made from a 1-inch by 16-inch, 5-foot-long aluminum beam (see Figure 7.5). Assume the diver is rigid. Calculate:
   a. Maximum force in the board.
   b. Maximum stress.
   c. Maximum deflection.
7. A 100-pound rigid weight falls 20 inches into the center of a 20-foot-long W14 × 26 wide-flange steel beam that is simply supported. Compute:
   a. Maximum stress in the beam (ignore the weight of the beam).
   b. Maximum deflection.

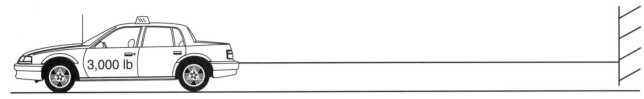

FIGURE 7.6 Cable and car for Problem 8.

8. In a classic movie scene, a cable is tied to a police car, which then speeds away. The police car and the object to which the cable is tied are considered rigid. The police car weighs 3,000 pounds. The cable is 100 feet long and has an area of .2 in.^{2} and a modulus of elasticity of 12×10^6 lb/in.^{2} If the car is going 5 mph when it reaches the end of the cable, calculate:
   a. Deflection in the cable.
   b. Stress in the cable.
   c. Maximum force in the cable.
9. A crankshaft bearing in a large diesel engine receives an impulse load of approximately 300 pounds each time the engine fires. When the engine is new and properly lubricated, no significant clearance exists in the connecting rod bearings. Each connecting rod bearing is 2 inches in diameter and 3 inches long. However, after many years of use, this engine develops an effective bearing clearance of .005 inch.
   a. Calculate the bearing stress when new.
   b. Discuss how this increased clearance affects the bearing stress and the ability of the engine oil to "absorb" the increased bearing stress.
   c. What additional two or three facts or properties about this problem would we need to know to solve for the impact stresses in this situation?

10. A ¾-inch UNC steel bolt with 2 inches of thread and 6 inches unthreaded is subject to an impact load from a 500-pound weight. If the stress is not to exceed 20 ksi, from what distance could this weight be dropped?
11. If the unthreaded shank is turned down to the root diameter in Problem 10, how much would this distance increase?
12. In Problem 10, solve for the distance if the bolt material is aluminum. Explain why this changes.

7.8 CUMULATIVE PROBLEMS

13. In the ongoing truck design problem described in previous chapters, a rear "bumper" for the trailer is needed to absorb the impact from a rigid 4,000-pound car going 10 mph. Assume this bumper is supported by two 24-inch-long steel C5 × 6.7 channel sections from Appendix 5.1. Assume these share the load and act like a cantilever beam. Determine the deflection and stress.
14. Does the design described in Problem 13 appear to be acceptable? If not, design another solution to this problem.

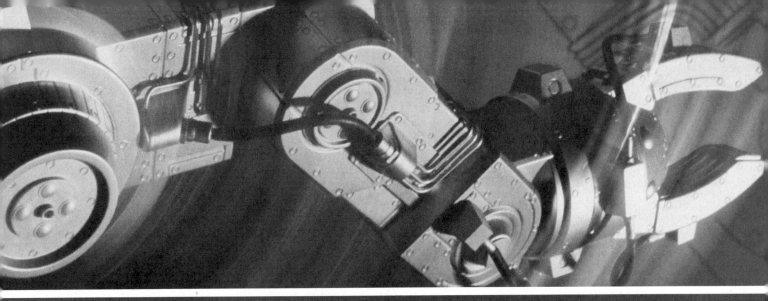

CHAPTER 8

Spring Design

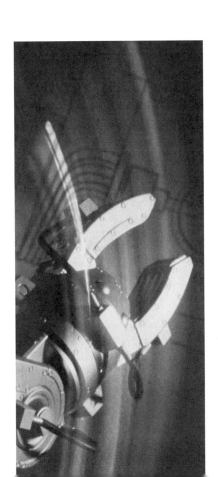

Chapter 8

helical compression spring
A wound spring formed by bending the spring wire into a helical shape, such as the front spring on an automobile.

torque tube
A spring system that relies on a torsional deflection of a solid or hollow tube to act in concert with a lever arm in a spring system.

torsion bar suspension
A torque-tube assembly that also uses the deflection of a bar to create a suspension system, frequently used in the rear wheels of an automobile or truck.

leaf spring
A spring system that uses the principles of a simply supported beam as a spring assembly. Some automobiles and many heavy trucks use spring systems of this type.

constant force spring motor
A spring system that uses a wound-up flat spring to pull on an object in concert with a reel to create a linear force.

extension spring
Very similar to a helical spring with the exception that it is made for retracting as opposed to being compressed.

Springs are in use all around us, absorbing energy, sometimes storing this energy, and then returning this energy to us. Envision the spring on an automobile chassis absorbing the shock of hitting a pothole or an automotive valve spring holding the valve closed until activated by a cam or pushrod. The same concept applies to a cylinder in an enclosed cylinder spring system. For example, this type of cylinder can help us open the door on a hatchback or trunk because it was energized by the previous closing cycle. Springs come in many forms, such as a common **helical compression spring**, a **torque tube** as part of a **torsion bar suspension**, a **leaf spring** in an auto or truck suspension, or a swimming pool diving board. A spring can be as simple as a rubber band or as complicated as the wound coil of a **constant force spring motor.**

In this chapter we will learn how to analyze common types of springs and how to use standard springs from manufacturers' catalogs. We will also continue to expand our understanding of energy absorption.

8.1 OBJECTIVES

After completing this chapter, you should be able to:

- Identify, describe, and understand the principles of several types of springs, including the most common helical compression springs, helical **extension springs,** torsion tubes, and leaf spring systems.
- Design and analyze helical compression springs, including compatibility with allowable stresses; determine deflections; describe the properties of helical springs; and develop the necessary analytical tools for spring design.
- Understand the principles of operation of different types of springs and how to analyze the energy being absorbed by these springs.
- Review the principles of design for other types of springs, such as extension springs and leaf springs.
- Select predesigned springs from manufacturers' catalogs and incorporate them into appropriate designs.

8.2 TYPES OF SPRINGS

Springs can readily be classified by the type of force they can absorb or exert, such as a compression spring, an extension spring, or a torsion spring. Figure 8.1 shows some common types of springs and their classifications.

Spring Design 151

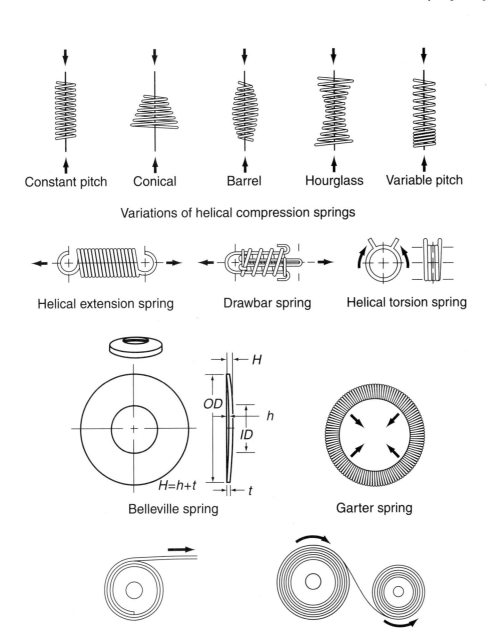

FIGURE 8.1 Common types of springs.

Helical Compression Springs

Helical compression springs are usually made from round wire and are wrapped in a cylindrical form with a fixed pitch similar to a lead screw. Square or rectangular wire can be substituted. These types of springs can often be seen in automobile suspensions, as valve springs in an automotive engine, on a classic pogo stick, and in many other applications. Helical compression springs usually have a uniform pitch and are further classified by how the ends of the springs are manufactured. Figure 8.2 represents different end finishing arrangements for compression springs. Typically, a helical compression spring is considered a linear spring. There is a constant relationship between the force exerted on the spring and the amount of deflection.

We will learn in this chapter how to design typical compression springs and how to select them from a standard manufacturer's catalog.

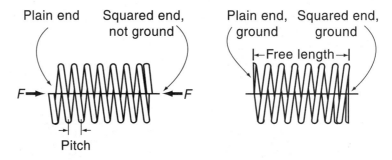

FIGURE 8.2 Helical compression spring types.

Helical Extension Springs

Helical extension springs are very similar to helical compression springs and have a similar cylindrical appearance. The difference between them is that the windings in helical extension springs are typically manufactured with each winding touching the adjacent winding, often with a preset residual load. Helical extension springs are used in the opposite manner from compression springs. They can be manufactured with a wide variety of end types depending on the application. Figure 8.3 shows some types of ends for extension springs. A common example of an extension spring is a hood-closing mechanism on an automobile or the automatic closing mechanism on a storm door.

Twist loop or hook	▭▭ ▭	▭▭ ▭	▭▭ ▭
Cross-center loop or hook		▭▭ ▭	
Side loop or hook		▭▭ ▭ ▭	
Extended hook		▭▭ ▭	

FIGURE 8.3 Extension springs: Types of end hooks.

Torsion Springs

Torsion springs are also similar to helical compression springs with the exception that they normally have an extension off the main body of the spring and are used to resist or create a turning motion. Torsion springs can be manufactured in a left- or a right-hand orientation, depending on the direction of twist desired. Many end types are available for custom usage. An example of a common torsion spring would appear on a guillotine-type paper cutter, where the spring is used to minimize the weight of the shearing arm. Figure 8.4 shows a typical torsion spring.

torsion spring
A spring that looks very similar to a helical spring but whose purpose is to resist rotary motion or to create rotary motion.

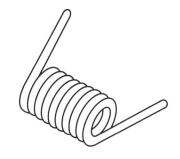

FIGURE 8.4 Torsion spring.

Leaf Springs

Leaf springs are nothing more than simple beams, typically either pinned-end beams or cantilevered beams, that are also used for absorbing energy. They can be made in a wide arrangement of geometries to create either linear spring systems or changing-force systems. Examples are the multileaf springs in a heavy truck or the simple diving board at a swimming pool.

The design of leaf springs uses the principles of stress and deflection, which are covered in other chapters of this book.

Miscellaneous

There are many other special types of spring systems that modify the basic principles of the three primary types of springs. One example is a drawbar spring, which uses a through wire clip to transform a typical compression spring into an extension spring system. Springs such as a constant force spring (which uses a wound-up leaf spring) or a torsion tube system (which is simply a twisting shaft problem) are often used in applications such as automobile suspensions.

8.3 HELICAL COMPRESSION SPRING DESIGN

In order to use helical compression springs, we need to understand the basic dimensional properties of these springs. Figure 8.5 shows the basic diametrical properties of a helical compression spring. Note that the mean diameter (D_m) is the average of the inside and outside diameters of the spring. Figure 8.6 shows the basic length parameters of a typical helical compression spring: L_f is the free length of the spring before installation; L_s is the solid height when the coils of the spring are pressed tightly against each other; L_o is the minimum operating length, which is typically some amount greater than the solid height; and L_i is the installed length, which in some applications may be the free length.

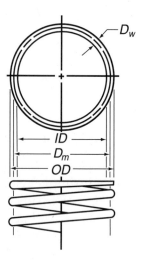

FIGURE 8.5 Compression spring dimensions.

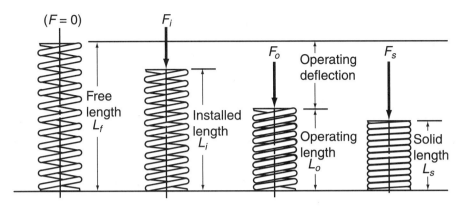

FIGURE 8.6 Compression spring deflections.

To better understand the stresses acting on a compression spring, envision yourself unrolling the coils of a spring into a straight bar. In doing this we create a cantilevered beam system, and the stresses are primarily bending. However, when we take that same beam and wind it in a circular manner, the stresses are now also torsional because of the twisting moment along the entire length. Hence the stiffness of a spring is very much the product of both the overall length of the wire in the spring and the winding diameter, as these properties cause the bending and torsional stresses in the spring. Note, however, that in most spring systems with a reasonable number of windings, the primary stress mechanism is torsion.

8.4 SPRING RATE

The spring rate for standard manufactured springs can often be found in a typical manufacturer's catalog. Appendix 12 includes a selection of standard springs. Often, however, if a standard spring is not available, we will need to calculate the spring rate needed for our design. The spring rate for a spring is the relationship of the change in force to the change in length

$$k = \frac{\Delta F}{\Delta L} \tag{8.1}$$

where the force then exerted by the spring is

$$F = k(L_f - L_o) \tag{8.2}$$

8.5 SPRING INDEX

The ratio of the mean diameter of a spring to the wire from which the spring is constructed is referred to as the spring index:

$$C = \frac{D_m}{D_w} \quad \text{(8.3)}$$

It is generally recommended that for most applications, C should range between 5 and 12. The diameter of the wire from which springs are typically constructed is specified in gage sizes. Appendix 10 summarizes standard wire gages. Note that music wire is the most common gage used for small sizes, whereas for larger wires the U.S. Steel Wire Gage is typically used.

inactive coils
In a helical spring, an often flattened coil on either end that, in some cases, is ground to result in a square end. The inactive coil does not add to the springiness of the system.

8.6 NUMBER OF COILS

The total number of coils in a spring and the number that are active in the spring system are usually different. This is because most springs have square and ground ends or other end features that add a number of **inactive coils** to the spring system. We would adjust the number of active coils in the spring system from Table 8.1.

TABLE 8.1
DIMENSIONAL PROPERTIES OF HELICAL COIL SPRINGS

Type of Ends	Free Length	Total Coils	Solid Height
Plain	$PN_a + D_w$	N_a	$D_w N_a + D_w$
Plain ground	PN_a	N_a	$D_w N_a$
Squared	$PN_a + 3D_w$	$N_a + 2$	$D_w N_a + 3D_w$
Squared and ground	$PN_a + 2D_w$	$N_a + 2$	$D_w N_a + 2D_w$

P = pitch of coils, N_a = number of active coils, D_w = diameter of wire

8.7 PITCH

The pitch of a spring is the center-to-center distance between adjacent coils and is calculated in the same manner as that of a lead screw. It would typically be expressed as the number of coils per inch of length. The associated pitch angle, as seen in Figure 8.7, is simply the angle of winding and is found from

$$\lambda = tan^{-1}\left(\frac{P}{\pi D_m}\right) \qquad (8.4)$$

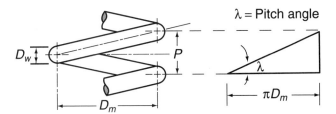

FIGURE 8.7 Pitch of spring.

8.8 TYPES OF MATERIALS

Springs are made from many different materials, but most springs are made from music wire, hard-drawn high-carbon steel, oil-tempered high-carbon steel, or stainless steel. Some light-duty springs are made from copper- or nickel-based alloys. Many different coatings are used to resist corrosion. Most high-strength steel springs are heat-treated. Because of this heat treatment, most of the structural properties of the materials from which springs are made vary significantly with diameter; the heat treatment results in significantly different properties near the surface than near the center core, especially as the wire diameter increases.

There are two common methods for determining the stress allowable in coil spring applications. One technique involves calculating the stress allowable directly off individual curves for each spring material. The second technique uses a formula-based system that factors in the wire size and is generally of the form

$$S_u = \frac{Q}{D_w^x} \qquad (8.5)$$

where Q is the expected ultimate strength of a 1-inch bar and x is a factor for how this strength changes with wire size. Typical values of Q and x can be found in Appendix 11.

It is also necessary to incorporate a loading factor that takes into account the severity of usage, the expected useful life, and the reduced capacity in shear as follows:

Loading Factor

.405   **Light Service:** Static loads or up to 10,000 cycles of loading with a low rate of loading (nonimpact).

.324   **Average Service:** Typical machine design situations; moderate-rate loading and up to one million cycles.

.263   **Severe Service:** Rapid cycling above one million cycles; impact loading; possibility of shock. (Engine valve springs are a good example of severe service.)

Adding the loading factor, Formula 8.5 becomes

$$S_s = (LF)\frac{Q}{D_w^x} \tag{8.6}$$

Many spring designers prefer the formula-based development of the stress allowable because with this method a direct wire sizing can be obtained. The alternative is the more indirect method that uses a graph to determine the allowable stress values. The formula-based application is especially useful when using a computer or calculator program for sizing spring systems, and this method is used in this chapter.

8.9 SPRING STRESSES

As discussed earlier in this chapter, the stress in a spring is primarily a torsional stress. As developed in Chapter 3, this torsional shear would be equal to

$$S_s = \frac{Tc}{J} \tag{3.5}$$

For a round shape (from Appendix 3):

$$J = \frac{\pi D^4}{32}$$

Substituting

$$S_s = \frac{16T}{\pi D_w^3} \tag{8.7}$$

As we assume the force is centrally loaded, then the torque related to the force acting on the spring would be equal to

$$T = \frac{FD_m}{2}$$

with this substitution, the shear stress becomes

$$S_s = \frac{8FD_m}{\pi D_w^3} \qquad (8.8)$$

However, the peak stress in a coil spring is somewhat greater than that derived from Formula 8.8. This is due to the spring curvature, which causes a stress concentration effect on the coil, and thus a combination of transverse shear stress and compressive stress is present in the spring. Research by Wahl created a factor that compensates for the wire size and mean diameter.

$$C = \frac{D_m}{D_w}$$

C is called the spring index. The Wahl factor is then found as follows:

$$K = \frac{4C-1}{4C-4} + \frac{.615}{C} \qquad (8.9)$$

K can also be found from Figure 8.8, which is merely the graphical equivalent of Formula 8.9.

Substituting this factor into our spring stress formula yields

$$S_s = K\frac{8FD_m}{\pi D_w^3} \qquad (8.10)$$

Spring stress (8.10)

$$S_s = K\frac{8FD_m}{\pi D_w^3}$$

We can now complete the spring sizing formula by setting our allowable shear stress greater than or equal to the actual shear stress and solving

$$(LF)\frac{Q}{D_w^x} \geq K\frac{8FD_m}{\pi D_w^3} \qquad (8.11)$$

$$(LF)\frac{Q}{D_w^x} \geq K\frac{8FD_m}{\pi D_w^3} \qquad (8.11)$$

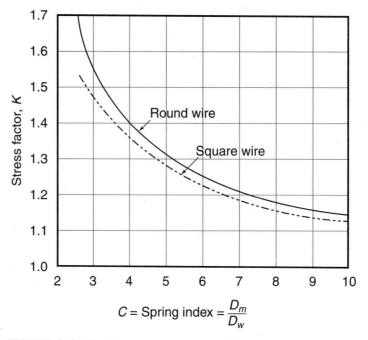

FIGURE 8.8 Wahl stress correction factors.

At this point, we can solve this equation for whatever spring property is desired, typically D_w, or in some cases both D_w and D_m. Example Problem 8.1 illustrates this technique.

Example Problem 8.1

Design a helical compression spring for the following conditions: ¾-inch maximum outside diameter, two inches free length, and a 35-pound load at a ½-inch deflection. Assume severe service and S&G ends made from music wire. Try an outside diameter of .725 inch and solve for D_w. As many properties of the spring depend on the wire size, often a wire size is assumed and then verified. Assume $D_w = .1$ for this first trial.

Finding the Wahl factor:

$$\frac{D_m}{D_w} = \frac{.725 - .1}{.1} = 6.25 \qquad \text{Wahl} \approx 1.25 \qquad \textbf{(Figure 8.8)}$$

$$S_u = \frac{Q}{D_w^x} = \frac{190 \text{ ksi}}{D_w^{.154}} \quad \text{(Appendix 11)}$$

$$(LF) = .263 \quad \text{(From text: Severe Service)}$$

$$(LF)\frac{Q}{D_w^{.154}} \geq \frac{K\,8F\,D_m}{\pi D_w^3} \quad (8.11)$$

$$\frac{D_w^3}{D_w^{.154}} = \frac{K\,8F\,D_m}{(LF)\,\pi\,Q}$$

$$\frac{D_w^3}{D_w^{.154}} = D_w^{2.85} = \frac{1.25\,(8)\,(35)\,(.625)}{.263\,\pi\,(190,000)}$$

$$D_w = .100 \text{ inch} \quad \text{(Appendix 10)}$$

Use U.S. Steel 12-gage wire: $D_w = .105$ inch.

8.10 SOLID HEIGHTS AND SOLID STRESSES

Another concern we have when using springs is that, either during installation or during other events when the spring could be compressed, each coil could touch the adjacent coil where the spring has become "solid." In designs where this is expected only occasionally, we would check the stresses at that point and compare them to the allowable stress for light-duty service. Also note that this light-duty service factor is just slightly above the point at which we would expect permanent yielding; if desired, this could be used as an approximate method for verifying yield as well.

8.11 DEFLECTION OF SPRINGS AND SPRING SCALE

The deflection of a spring can be determined from the following formula where N_a (from Table 8.1) is the number of active coils in the spring:

$$\delta = \frac{8FD_m^3\,N_a}{GD_w^4} = \frac{8FC^3\,N_a}{GD_w} \quad (8.12)$$

G is the shear modulus of elasticity, which is a similar property to the modulus of elasticity (E) and can be determined for most common materials from Table 8.2.

Spring deflection (8.12)

$$\delta = \frac{8F\,D_m^3\,N_a}{G\,D_w^4}$$
$$= \frac{8F\,C^3\,N_a}{G\,D_w}$$

TABLE 8.2
SPRING WIRE MODULUS OF ELASTICITY IN SHEAR (G) AND TENSION (E)

Material and ASTM No.	Shear Modulus (G)		Tension Modulus (E)	
	psi	GPa	psi	GPa
Hard-drawn steel: A227	11.5×10^6	79.3	28.6×10^6	197
Music wire: A228	11.85×10^6	81.7	29.0×10^6	200
Oil-tempered: A229	11.2×10^6	77.2	28.5×10^6	196
Chromium-vanadium: A231	11.2×10^6	77.2	28.5×10^6	196
Chromium-silicon: A401	11.2×10^6	77.2	29.5×10^6	203
Stainless steels: A313				
Types 302, 304, 316	10.0×10^6	69.0	28.0×10^6	193
Type 17-7 PH	10.5×10^6	72.4	29.5×10^6	203
Spring brass: B134	5.0×10^6	34.5	15.0×10^6	103
Phosphor bronze: B159	6.0×10^6	41.4	15.0×10^6	103
Beryllium copper: B197	7.0×10^6	48.3	17.0×10^6	117
Monel and K-Monel	9.5×10^6	65.5	26.0×10^6	179
Inconel and Inconel-X	10.5×10^6	72.4	31.0×10^6	214

Note: Data are average values. Slight variations with wire size and treatment may occur.

By substituting into the formula $F = kx$, which determines the spring rate for the spring (k), we can then modify this formula to determine the spring rate for any spring as follows:

Spring rate (8.13)

$$k = \frac{GD_w^4}{8D_m^3 N_a}$$

(8.13)

as x and δ are the same quantity. This spring rate is the one commonly found in manufacturers' catalogs for standard springs.

Example Problem 8.2 shows the technique for determining the number of active coils and the spring rate.

Example Problem 8.2

Determine the number of coils necessary to meet the design criteria using the spring from Example Problem 8.1.

$$\delta = \frac{8F D_m^3 N_a}{G D_w^4} \quad (8.12)$$

$$N_a = \frac{\delta G D_w^4}{8 F D_m^3}$$

$$N_a = \frac{.5 (11.85 \times 10^6)(.105)^4}{8 (35)(.620)^3} \qquad \frac{\text{in. lb/in.}^2 \text{ in.}^4}{\text{lb-in.}^3}$$

$$N_a = 10.8 \qquad \text{(active coils)}$$

Find total coils.

$$\text{total coils} = N_a + 2$$

$$10.8 + 2 = 12.8 \quad \text{(total coils with squared and ground ends)}$$

Find pitch.

$$L_f = P N_a + 2 D_w \qquad \textbf{(Table 8.1)}$$

$$2 = P 10.8 + 2(.105)$$

$$P = .166 \text{ in.}$$

Find the spring rate for this spring.

$$k = \frac{G D_w^4}{8 D_m^3 N_a} \quad (8.13)$$

$$k = \frac{(11.85 \times 10^6)(.105)^4}{8(.620)^3 \; 10.8} \qquad \frac{\text{lb/in.}^2 \text{ in.}^4}{\text{in.}^3}$$

$$k = 70 \text{ lb/in.}$$

8.12 SPRING BUCKLING

Another concern in the design of helical compression springs is that, when the diameter of the spring is small or the length is long, spring buckling can occur. Spring buckling is the sideways deflection of a spring. When deflection occurs, the spring is no longer stable. This is not a concern with springs that are supported along their length. A good example of this is a ballpoint pen in which the spring is supported in the center by the ink reservoir; a stable spring is unneeded in this system. An unsupported spring, however, needs to be stable. In order to determine a spring's stability, we need to calculate the ratio of the free length (L_f) of the spring to the mean diameter (D_m) and compare it to Figure 8.9. If the answer appears to the left side of the graph, our spring will be stable and not be subject to buckling.

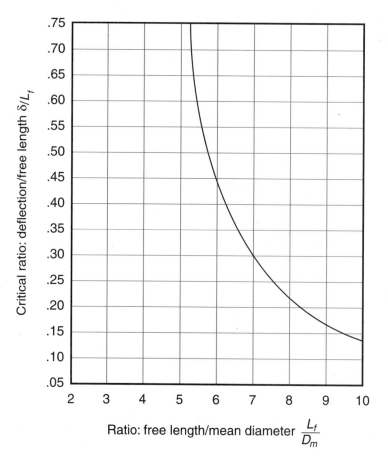

FIGURE 8.9 Spring buckling graph for springs with square and ground ends on a flat surface.

Example Problem 8.3 checks the stability of the spring that was designed in Example Problem 8.1.

Example Problem 8.3

Checking the stability of the spring from Example Problem 8.1:

$$\frac{L_f}{D_m} = \frac{2}{.620} = 3.2$$

From Figure 8.9 for squared and ground ends, this is a stable spring.

8.13 DESIGN PROCESS

In order to effectively design a spring, we typically need to know some characteristics of the space available and the loading conditions. Once these have been determined, the process can be iterative with regard to changes in pitch or number of coils within the set parameters of the space allowed, the material, the mean diameter, and the wire size.

Example Problem 8.4 shows the process whereby we select a standard spring using the same information from Example Problem 8.1.

Example Problem 8.4

From Appendix 12, spring #C0720-096-2000 matches the spring in Example Problem 8.1 fairly closely. It has a free length of 2 inches and an outside diameter of .720 inch. However, its spring rate is slightly higher. The designer would need to determine if this is appropriate for the design criteria. With this spring under the design load of 35 pounds, the deflection would be

$$F = kx$$
$$x = \frac{F}{k} = \frac{35 \text{ lb}}{73.3 \text{ lb/in.}} = .48 \text{ in.}$$

8.14 FLAT SPRINGS

Flat springs can be in the form of a cantilevered beam, a simply supported beam, or any other type of beam such as the typical beams shown in

Appendix 2. A typical example of a flat spring is the leaf spring used in the rear of many automobiles that both provides the necessary cushioning and holds the rear axle in position. A straight leaf spring would be analyzed as a beam. However, when leaf springs have curvature, prestressed multiple leaves, and tapers, the analysis gets far more complicated. To aid in our understanding of springs as well as beams, Example Problem 8.5 demonstrates the design of a diving board as a flat spring system. The stiffness of this system is calculated as well.

Example Problem 8.5

A diver deflects a diving board 4 inches. This diving board is made from an aluminum plate 1 inch thick by 12 inches wide and 72 inches long. What is the force at this point?

$$I = \frac{bh^3}{12} = \frac{12 \text{ in.} (1 \text{ in.})^3}{12} = 1 \text{ in.}^4 \qquad \textbf{(Appendix 3)}$$

$$\delta = \frac{-FL^3}{3EI} \qquad \textbf{(Appendix 2)}$$

$$F = \frac{3\delta EI}{L^3}$$

$$F = \frac{3 \cdot 4 \text{ in.} \left(10 \times 10^6 \text{ lb/in.}^2\right) 1 \text{ in.}^4}{(72 \text{ in.})^3} = 322 \text{ lb}$$

Determining stiffness:

$$F = kx \quad \text{or} \quad F = k\delta$$

$$k = \frac{F}{\delta} = \frac{3EI}{L^3}$$

$$k = 3\frac{10 \times 10^6 \text{ lb/in.}^2 \; 1 \text{ in.}^4}{(72 \text{ in.})^3}$$

$$k = 80 \text{ lb/in.}$$

8.15 ENERGY ABSORPTION

Springs are frequently used to absorb energy. That energy can either be returned immediately or, if the spring is captured in its compressed state, returned at a later time. In the case of the diving board, which was analyzed as a flat spring, we can take this same example and calculate the storage of energy under this deflection. Note that the energy saved will be equal to half of the force times the distance, as the force in the beam increases from zero to the maximum value as follows:

$$U = \frac{F\delta}{2} \tag{8.14}$$

Example Problem 8.6

From Example Problem 8.5, what was the energy from the 4-inch deflection?

$$U = \frac{F\delta}{2} \tag{8.14}$$

$$U = \frac{322 \text{ lb } 4 \text{ in.}}{2} = 644 \text{ in.-lb}$$

8.16 SUMMARY

After completing this chapter, you should be aware of the different types of springs available. You should now understand how to analyze the more common types of springs. In addition to being able to design springs, you should know how to use standard springs from manufacturers and how to select them as this is likely to be the method most commonly used. In addition, other properties of springs, such as the energy that can be absorbed in springs, should now also be better understood.

Your ability to actually design springs will enable you to better understand how minor modifications to springs, such as wire sizes, spring lengths, and pitch, can affect their performance, both in the general design and in the selection of standard items.

Lastly, you should have gained an understanding of how other machine parts, such as cantilevered beams and torsion tubes, can be used in spring systems and how the analysis methods learned elsewhere in this book apply to the use of spring systems.

8.17 PROBLEMS

1. A spring has an overall length of 3 inches when it is not loaded and a length of 1.85 inches when carrying a load of 12 pounds. Compute its spring rate.
2. The following data are known for a spring:
   - Total number of coils = 20
   - Squared and ground ends
   - Outside diameter = .75 inch
   - Wire diameter = .059 inch (25-gage music wire)
   - Free length = 3 inches

   For this spring, compute the spring index, the pitch, the pitch angle, and the solid length.
3. A compression spring with squared and ground ends is made from gage 22 music wire and has an OD of .875 inch. The pitch is .225 inch and the free length is 5 inches. Determine:
   a. Number of active coils
   b. Mean diameter
   c. Spring constant (k)
   d. Solid height
   e. Pitch angle (λ)
   f. The deflection to make this spring solid
   g. The force to make this spring solid
   h. The Wahl factor
   i. The stress at the solid height
4. Assume you have a spring design with a given wire type and a given wire and coil diameter. The spring has proven too stiff and you want to make it softer, that is, reduce k. What would you change if you wanted to retain the same wire diameter?
5. A prototype machine needs a spring to fit inside a tube with an inside diameter of ¾ inch and needs a force of ≈ 16 pounds at a 3-inch length and ≈ 32 pounds at a 2-inch working length.
   a. Find the spring rate of this spring.
   b. What would be the free length?
   c. Select a spring from Appendix 12.
6. Design a helical compression spring for the following conditions: 1⅜-inch maximum outside diameter, 4-inch free length, and 75-pound load at a 1½-inch deflection. Assume average service, S&G ends, and oil-tempered ASTM A229 wire. For your first iteration, assume the Wahl factor (K) = 1.3.

7. Find the spring rate for the spring in Problem 6. Does this match the expected spring rate from the problem statement?
8. In Problems 6 and 7, determine the solid height and determine if the solid stress is acceptable.
9. A valve assembly for an automobile requires 48 N to hold the valve closed. The space available for the compression spring is as follows: ID = 1.5 cm, OD = 2.3 cm, and length at load = 4 cm. Use music wire.
    a. Select a spring for this application from Appendix 12.
    b. For this spring, if the valve moves 1 cm to open, determine the force on the valve from the spring at this position.
    c. For this spring, what was the solid height and solid force?
10. For spring #C0850-0811-1750, determine:
    a. The number of active coils.
    b. Is this a stable spring at the operating length (L_o) if used with the ends fixed?
11. For the steel torque-tube lever arm shown in Figure 8.10 that is part of a vehicle suspension system, if a load of 100 pounds is applied as shown in Figure 8.10, determine the angular deflection of the torque tube.

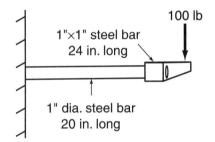

FIGURE 8.10 Torsion arm suspension system for Problems 11, 12, and 13.

12. For the load in Problem 11, determine the deflection of the lever arm.
13. For Problems 11 and 12, determine the total deflection and the equivalent stiffness of the system (ignore bending of the torque tube).
14. A brass spring is constructed of U.S. steel 12-gage wire, has a free length of 3 inches, an outside diameter of 1 inch, squared and ground ends, and 18 total coils. Compute the spring rate for this

spring and the deflection under a load of 5 pounds. Now take this same spring and stretch it to a free length of 4 inches without changing any other properties of the spring. Determine the new spring rate and the new deflection under this same load. Explain why this value did not change, and calculate the force that would be required to make both of these two springs solid.

8.18 CUMULATIVE PROBLEMS

15. For the truck mentioned in the cumulative problems in earlier chapters, use the following constraints:
    a. Each of the axles on the truck body will share one-quarter of the trailer part of the truck, which has a total weight of 40,000 pounds.
    b. The spring should deflect approximately 4 inches under the static load.
    c. Under some road conditions, such as hitting a pothole, the maximum load on the spring increases to double the static load.

    Design a spring for these conditions and include diameter, wire size, pitch, and material. Verify that the stress in the spring is suitable under solid conditions and that buckling is not a concern. Multiple coil springs could be used for each wheel assembly.
16. For the design conditions in Problem 15, design a torsion bar system for this application.

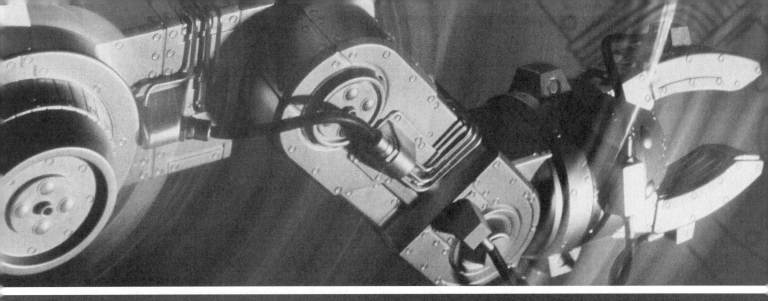

CHAPTER 9

Electric Motors

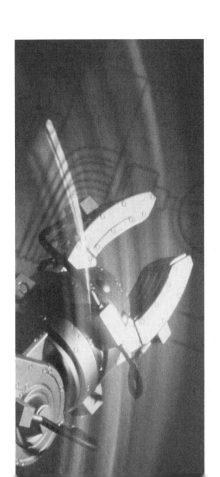

As we look around us, electric motors exist in many products that we use daily. If we take a mental visit through a typical kitchen, we are likely to have a refrigerator with an electric motor that drives the compressor system for cooling; a dishwasher with an electric motor that powers a pump to force water through the system and possibly includes a motor-driven aerator to mix up the water; a blender in which an electric motor drives rotating blades; an exhaust fan; an electric can opener that may have a battery-powered motor; and an electric clock on the wall that could be powered by batteries or plugged into the wall. Further, if we think about our home workshop, we may have power saws and drills that are electrically driven. The designer of these products must understand some of the basics about motors in order to select the appropriate type.

Further, as we think of things like an automobile, we may have electric windows, seats, and a heater fan. The new hybrid electric cars, in which we combine the gasoline engine with an electric motor system, can both drive the automobile and provide regenerative braking in which this same electric drive can recharge batteries for later reuse. You may be sitting in a home office as you study this text where there is, for example, a computer printer that contains a drive system that moves the paper through the system or has some form of electric linear motor for moving the print head. Your office may have many other applications in which electric drive systems are used.

How are all of the drive systems different? A designer involved in the selection of different electric drives needs to understand the attributes and the basic differences between these drive systems. We also need a basic understanding of how power is generated in order to understand the basic principles of electrical drive systems. For instance, the electric power generated by the power company is typically generated as **three-phase power**. A factory may use the power in this three-phase form at a higher voltage than would a household, which may use power from the same source as a single phase and at a lower voltage. If we go back to our factory that is receiving the three-phase power, parts of the factory, such as the lighting systems, probably use only one of these phases. We may also have many products that use DC **(direct current)** power, such as many of our battery-operated devices. In some cases, we may convert the alternating current generated by the power company to direct current for different purposes. The power that we use in the United States is generated at 60 cps (cycles per second); however, Europe uses a frequency of 50 cycles per second. There is no real reason for this difference other than the fact that different people developed these systems at different times in history.

We are now very much accustomed to having a regular power supply, and seldom do our homes, schools, and factories lose power. This is

> **three-phase power**
> Electricity that is generated with three sets of windings in a generator system, such that each revolution adds part of a sine wave cycle to the available power. Three wires are needed to transmit the power.
>
> **direct current**
> A form of electricity in which there is no sinusoidal wave transmitting the power, simply a potential between two separate conductors.

because the generating systems in this country have become much more reliable and also because, for all practical purposes, most of our country is interlinked so that when there is a shortage in one area, this shortage can be made up by taking power from other parts of the country. It wasn't many years ago when each municipality had its own power company. The city or the region would be without power when a power supply problem happened with the local generating facility. The author remembers stories about how, in the small community where he grew up, all the power generated in town came from the local paper company, which used the river as a power source, both for operating the paper mill and for supplying power to the town. When the water level in the river was low and the paper company needed all of the supply, the electric generators could not create enough power to supply both the mill and the town. During drought periods when the mill could not supply its excess power to the town, the lights in town were shut off in order for the mill to have enough water to generate power for the mill. During rainy seasons or high-water periods, plenty of electricity was available for the homes as well as for the paper mill.

The scope of electric supply and all of its nuances are certainly not covered in this text, but it is interesting to recognize how the use of electricity, and corresponding electric motors and battery-operated systems, have changed over the years. As designers, if we have a better understanding of how these systems work, we will be better able to either select the correct systems for our designs or to understand the systems that others, such as electrical engineers, are helping us to select.

9.1 OBJECTIVES

After completing this chapter, you should be able to:

- Understand the different types of electrical input to motors.
- Understand and describe the basic types of motors.
- Understand what types of motors may be appropriate for different types of applications.
- Read a torque-versus-speed curve for a typical motor.
- Recognize frame and mounting types.
- Understand the principles of thrust and overhanging loads as they apply to motor selection.
- Understand the basic principles of operation for **alternating current** (AC) and direct current (DC) motors.
- Describe some other types of specialized motors.

alternating current
The sinusoidal wave form in which most current is generated and transmitted.

9.2 AC AND DC POWER

The power that is generated by American electric utilities has a frequency of 60 hertz (Hz) or 60 cycles per second. Other countries in the world, particularly in Europe, usually use a frequency of 50 Hz. The power that is generated by the utility is created as a three-phase power, which means it is transmitted by three separate wires, along with a ground conductor, and has a waveform as shown in Figure 9.1. As will be seen later in this chapter, this power generation is done using a **synchronous generator.** The different windings in the generator create these three distinctly different phases of the power. This power is then transmitted at high voltage over long distances in high-tension lines. It is then reduced by the use of **transformers** to an intermediate voltage level for transfer to residential and commercial streets and then typically transformed at least one more time down to the voltages that would typically be used in commercial, industrial, and residential buildings. Typically, industrial and large commercial installations use three-phase power for driving motors because generally this is a more efficient form than single-phase power. Figure 9.2 shows single-phase power, which is used in nonindustrial applications. Generally, the power that is available for industrial, commercial, and residential use is a ratio of 120 volts, as shown in Table 9.1. Note that in this table the voltage levels for three-phase power in some cases are slightly different from single-phase or from the system voltage. This is due to factors that are beyond the scope of this chapter. In general, we would try to use as high a voltage as possible because higher voltages can transmit a greater amount of power. The relationship of power is as follows:

$$P = I V \qquad (9.1)$$

where P = the power (typically in watts)
I = amperage
V = voltage

synchronous generator
A common type of generator used in a power plant that typically rotates at the synchronous speed or, on some occasions, a multiple thereof to create (in the United States) 60 cycle or (in Europe) 50 cycle power.

transformer
An electrical device that changes voltage and amperage to different levels.

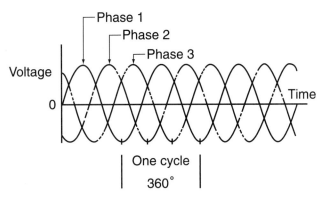

FIGURE 9.1 Three-phase waveform.

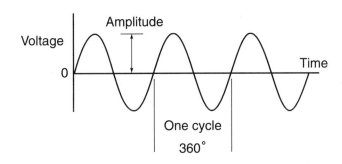

FIGURE 9.2 Single-phase waveform.

TABLE 9.1
AC MOTOR VOLTAGES

System Voltage	Motor Voltage Ratings	
	Single-Phase	Three-Phase
120	115	115
120/208	115	200
240	230	230
480		460
600		575

Speeds of Alternating Current (AC) Motors

synchronous speed
A ratio of the speed that is generated by the power company.

inductive motor
The most common type of electric motor in which inductive forces rotate the motor by means of a magnetic field.

Alternating current electric motors have what is called a **synchronous speed,** which is a product of the frequency of generation of the electric power. As discussed previously, in the United States this is 60 cycles per second or, correspondingly, 3,600 cycles per minute. Therefore, an electric motor with a single pair of magnetic poles under no-load conditions would operate at this speed, as does the generator that created this power. This speed of rotation is considered the synchronous speed of the motor. However, most electric motors in use are **inductive motors,** and there is an inductive slip during rotation under load that causes a full-load speed of slightly less than the synchronous speed. This varies based on many factors. In general, it is typically a few percent lower than the synchronous speed, usually about 5 percent less depending on the type and configuration of the motor design. Table 9.2 presents different synchronous speeds of motors and typical inductive speeds for those motors. Note that the synchronous speed is always a ratio or a product of the 3,600 cycles per minute generation rate because the synchronous speeds are based on the number of pairs of poles in the motor. The number of these poles is often a product of the desired speed of the motor and the necessary usage. In fact, in some motors are created with multiple sets of poles so that different combinations can be activated at different times to create a multiple-speed motor. A three-phase motor uses three sets of poles, placed in different orientations, in order to use the three different phases in sequence and to create a more uniform torque. Also, the three-phase motor, because of this difference and because the power is supplied at more points during the cycle, is typically smaller and slightly more efficient than a single-phase motor. The initial cost of a three-phase motor is often greater than the single-phase, but usually the lifetime cost, including the cost of power along with the increase in efficiency, makes the overall lifetime cost less. The speed of an induction motor varies with load, depending on the configuration.

TABLE 9.2

AC MOTOR SPEEDS (TYPICAL INDUCTION MOTOR)	
Synchronous Speed (rpm)	Full-Load Speed (rpm)
3,600	3,450
1,800	1,725
1,200	1,140
900	850
720	690
600	575

Figure 9.3 is an example of a motor torque curve. This curve shows the synchronous speed, which would be the 100% value, as well as the rated or induction speed, which would be called the full-load torque speed. Also shown is the **breakdown torque,** which is the point at which the curve changes direction. In essence, this is the point at which the motor slows down and the torque decreases instead of increasing. Note that the starting torque is often called the **locked-rotor torque** and is the startup power of the motor system. The shape of this curve will vary depending on the design of the windings, poles, clearances, and many other factors beyond the scope of this chapter.

When selecting a motor for a specific application, you can compare the characteristics of these types of curves for different styles of motors in order to pick the most appropriate type for the application at hand. For example, if we look at other curves, such as those shown in Figure 9.4, we can see that different motors have different slopes, especially at the full-load point. Also, the induction load can be a different percentage of the synchronous speed, and the locked-rotor torque or starting torque can also vary significantly. We may select a type of motor system based on the need for high initial starting torque or another for low or smoother starting torque, or we may choose still another because of the relationship between synchronous and induction speeds.

> **breakdown torque**
> The point at which, when additional load is applied to a motor, the speed may decrease as opposed to increasing.
>
> **locked-rotor torque**
> A term that applies to the torque of a motor to start rotation.

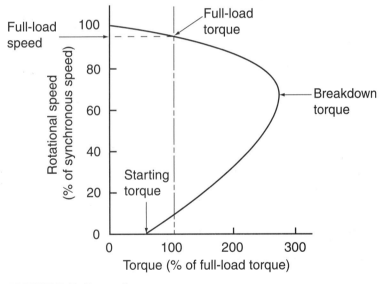

FIGURE 9.3 Typical motor torque curve.

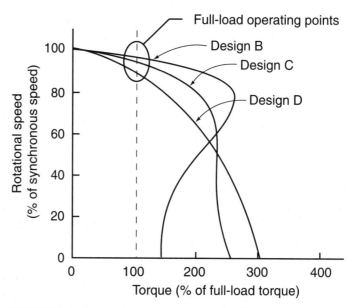

FIGURE 9.4 Typical torque curves for National Electric Manufacturers Association (NEMA) Type B, C, and D motors.

Synchronous Motors

A synchronous motor is similar in design to an induction motor. A synchronous motor runs at the synchronous speed and has no induction slip. These can be useful characteristics for many applications. An obvious case would be an electric clock because, as this type of motor runs at the exact speed generated by the power company, we can make a very accurate clock. Note also that a synchronous motor is, for all practical purposes, the same machine as a generator and, with very minor changes, a typical synchronous motor can be turned into a generator or could be used as both. A good example of using a combined motor generator is the active braking systems used in some electric drive devices, such as the hybrid automobiles discussed in the introduction to this chapter. A synchronous motor operates at the synchronous speed until a certain torque is exceeded. This is called the **pullout torque,** which is a torque that causes the motor to fall out of synchronization. If that torque remains, it generally causes the motor to stop or fail. Hence, it is important when using a synchronous motor not to have torque requirements greater than this synchronous pullout torque value.

pullout torque

When a synchronous motor needs a greater torque than can be created this is the value at which the motor falls out of synchronization or stops.

Universal Motors

A universal motor is very similar to a series-wound DC motor, which is described later in this chapter, because it can operate on either AC or DC power. This type of motor has electric coils that are connected through commutators. A commutator is a type of slip ring that allows power to be transmitted to the center windings. Because of the inherent wear on these slip rings, which make mechanical contact in order to transfer electric energy to the rotating parts of the motors, universal motors are not considered as reliable as other types. However, universal motors have other characteristics that make them ideal for certain applications. A universal motor can run at very high speeds, ranging up to 20,000 rpm in some circumstances, and this results in a very high power-to-weight ratio, as well as a very high power-to-size ratio for this type of motor. This makes it ideal for applications such as handheld electric drills, saws, and food mixers. Such things as sewing machines, vacuum cleaners, and portable leaf blowers typically would use a universal motor. Also, because of their similarities to DC motors, frequency differences between the European 50 cycles per second and the American 60 cycles per second do not significantly affect their performance—generally, voltage and amperage determine motor characteristics. These types of motors, however, are not useful in applications where speed must be closely regulated as, generally, the speed varies significantly with the load. These motors are ideal in applications where variable speeds are needed, as in a variable-speed electric drill. The control system to create variable speeds is quite simple and typically consists of a resistive-type speed control.

DC Motors

DC motors are generally for specialized usage as most industrial, commercial, and residences have only AC power available. AC power can be converted to DC but, in general, DC motors are used where they are either powered by batteries or used in mobile applications such as automobiles, boats, farm equipment, and construction equipment. DC power can be created from AC power by using systems such as **rectifiers.** A silicone control rectifier (SCR) is one common type. DC motors may be used in applications that require rapid starting, stopping, and directional change, such as in a robotic drive system. DC motors can also operate over a wide range of speeds, and the torque can be controlled by simply varying the current supplied to the motor. This can be useful in such applications as tensioning systems or winding materials on a coil or spool. Additionally, **dynamic braking** can be easily incorporated into a DC motor system by reversing or minimizing voltage to control either moderate or direct braking.

> **rectifier**
> An electrical device that converts alternating current into direct current.

> **dynamic braking**
> A system that uses the magnetic field in an electric motor to slow down rotation as opposed to creating rotation.

The most common types of DC motors include shunt-wound, series-wound, compound-wound, and permanent magnet motors, which all have some or all of the attributes discussed previously. Shunt-wound motors have slightly better speed control than the other types and are often used in small fans, such as the heater fan in an automobile. Series-wound motors have a steep speed torque curve with a very high starting torque and provide good service for cranes, hoists, or traction drives in mobile equipment. Compound-wound DC motors have basically the same attributes as the shunt-wound, in combination with the series-wound, and are often used in applications where built-in braking may be desired or good low-speed control is needed. Permanent magnet DC motors use permanent magnets for the armature field and therefore result in a fairly linear speed torque curve. They are useful in such applications as power windows and seats in an automobile. Sometimes they are used with small gear reducers to provide lower-speed, higher-torque output.

Single-Phase Motors

Single-phase motors are similar to three-phase motors with the exception that only one phase of the input power is used. There are many types of single-phase motors; the four most common types are split-phase, capacitor-start, permanent-split capacitor, and shaded-pole. Each of these motor types has slightly different characteristics, but their performance is somewhat similar. The biggest difference generally is in their ability to start under high load. A shaded-pole or permanent-split capacitor type generally has low starting torque, whereas a capacitor-start generally is much higher. In applications where a motor would need to start under load, as in certain types of air conditioning systems, we would pick an appropriate style accordingly. Single-phase motors are most often used in residential and commercial applications that are not wired for three-phase operation. They usually cost less than equivalent-size three-phase motors but tend to be slightly larger and less efficient. They are generally available in most smaller sizes and up to about 10 horsepower (7.5 kw). Figure 9.5 shows the speed versus torque for a typical single-phase motor.

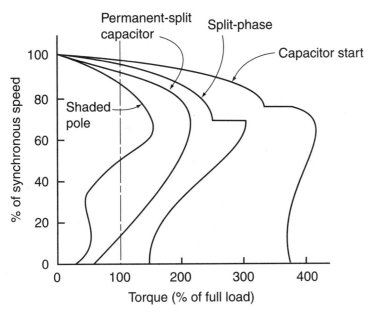

FIGURE 9.5 Common types of single-phase motors: Torque versus speed curves.

9.3 MOTOR MOUNTING AND FRAME TYPES

Due to the various uses of electric motors, there are many standard types of mounting systems. Some of these types are described in the following paragraphs.

Foot Mounted

A foot-mounted motor typically has four mounting holes on a rigid base so that the motor can be bolted to a base. An example is shown in Figure 9.6.

C-Face Mounting

C-face mounting has a standard pattern on the face of the motor, such that a pump, gearbox, or other standard attachment can be bolted directly to the motor. Generally, this attachment would not require a coupling as the attachment typically has an internal keyway to match with a standard motor key. Figure 9.7 shows a C-face motor assembly. The C-face motor

FIGURE 9.6 Foot-mounted motor style. (Courtesy of Boston Gear)

FIGURE 9.7 C-face type motor for direct mounting. (Courtesy of Boston Gear)

FIGURE 9.8 Combination C-face and foot-mounted style allows devices such as a pump or clutch to be direct mounted. (Courtesy of Boston Gear)

may or may not also have a base mounting, depending on whether the attachment to the motor is mounted or the motor is mounted and the attachment is supported by the motor. Figure 9.8 shows a C-face motor assembly with a foot base as well. Note that, as this is a common standard, mating parts could then be selected with the same C-face standard pattern for direct mounting applications.

Special-Purpose Mountings

Many special types of mounting, such as a cushion base, a vertical C-face, or other types used for specialized applications, can either be standard or custom manufactured. Many specialized products such as oil burners for furnaces and fans with integral motors may have a specialized motor mounting created for that specific application.

9.4 MOTOR ENCLOSURES

Different types of enclosures are usually based on the need for weather protection, depending on the use of a motor. The most common types of motor housing assemblies, which are available for each of the motor types discussed in Section 9.3, are discussed in the following paragraphs.

Open

An open motor often has a sheet metal housing, as opposed to a cast-metal housing, and is suitable in applications where water or water spray is not present. An open-frame motor is shown in Figure 9.9.

Drip Proof

A drip-proof motor typically has access for airflow for cooling on the lower parts of the motor assembly, so that in light rain or similar situations, damage will not occur. An example of a drip-proof motor is shown in Figure 9.10.

Totally Enclosed Nonventilated (TENV)

This type of motor has no openings, and cooling is through the housing. In some cases, fins may be cast into the motor housing to increase the surface area in order to provide additional cooling. An example of this is shown in Figure 9.11.

FIGURE 9.9 Open-frame motor. (Courtesy of Baldor Electric Co.)

Electric Motors | **185**

FIGURE 9.10 Drip-proof motor. (Courtesy of Boston Gear)

FIGURE 9.11 Totally enclosed non-ventilated motor (TENV). (Courtesy of Baldor Electric Co.)

Totally Enclosed Fan-Cooled (TEFC)

This type of assembly is very similar to the totally enclosed nonventilated, with the exception that a fan is mounted on one end of the motor and blows air over the assembly in order to aid in cooling. An example of a totally enclosed fan-cooled motor is shown in Figure 9.12.

Explosion Proof (TEFC-XP)

This is also a totally enclosed assembly where the shafts have seals and the electrical connections are more secure to prevent sparks from being generated. An example of an explosion-proof motor is shown in Figure 9.13.

Frame Sizes

To permit replacement of motors in standard applications with motors from different suppliers, the National Electrical Manufacturers Association (NEMA) has established standards for frame sizes. These standards govern diameters, heights, lengths, shaft lengths, key sizes, and mounting arrangements for standard NEMA motor frames so that, when a certain standard

FIGURE 9.12 Totally enclosed fan-cooled motor (TEFC). (Courtesy of Emerson Power Transmission Corp. All rights reserved.)

FIGURE 9.13 Explosion-proof motor (TEFC-XP). (Courtesy of Emerson Power Transmission Corp. All rights reserved.)

frame is specified, a motor from any manufacturer can be used. Table 9.3 and Figure 9.14 show the NEMA standards. Individual manufacturers should be consulted for allowable shaft overhanging and thrust loads, if present.

TABLE 9.3
MOTOR FRAME SIZES

		Dimensions (in.)								
hp	Frame Size	A	C	D	E	F	O	U	V	Keyway
¼	48	5.63	9.44	3.00	2.13	1.38	5.88	.500	1.50	.05 flat
½	56	6.50	10.07	3.50	2.44	1.50	6.75	.625	1.88	3/16 × 3/32
1	143T	7.00	10.69	3.50	2.75	2.00	7.00	.875	2.00	3/16 × 3/32
2	145T	7.00	11.69	3.50	2.75	2.50	7.00	.875	2.00	3/16 × 3/32
5	184T	9.00	13.69	4.50	3.75	2.50	9.00	1.125	2.50	¼ × ⅛
10	215T	10.50	17.25	5.25	4.25	3.50	10.56	1.375	3.13	5/16 × 5/32
15	254T	12.50	22.25	6.25	5.00	4.13	12.50	1.625	3.75	⅜ × 3/16
20	256T	12.50	22.25	6.25	5.00	5.00	12.50	1.625	3.75	⅜ × 3/16
25	284T	14.00	23.38	7.00	5.50	4.75	14.00	1.875	4.38	½ × ¼
30	286T	14.00	24.88	7.00	5.50	5.50	14.00	1.875	4.38	½ × ¼
40	324T	16.00	26.00	8.00	6.25	5.25	16.00	2.125	5.00	½ × ¼
50	326T	16.00	27.50	8.00	6.25	6.00	16.00	2.125	5.00	½ × ¼

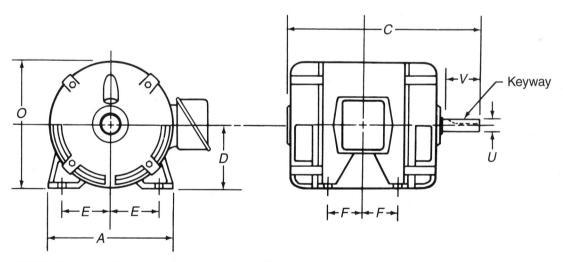

FIGURE 9.14 NEMA standard motor frame dimensions.

9.5 MOTOR CONTROL

There are many aspects to motor control, such as starting the motors, shutdown, braking, and overload protection that are not included in this chapter. Control systems can be purchased or designed. One example is reversing starters that can run motors in both directions by switching windings. This is fairly easy in a three-phase motor assembly, but it can also be incorporated into single-phase or universal motors. Another example is dual-speed motor controls in which multiple windings are created in the motor assembly. By switching between them or combining them, these motors can operate at different speeds. Dynamic braking systems or separate brake motors can be used if needed. Controls that create reduced voltage in order to gradually start a machine or assembly can also be purchased or designed.

It is also usually necessary to provide a motor system with overload protection by using fuses, circuit breakers, or other specialized devices that protect a motor from overheating or, in some cases, from supplying too much power to the machine or assembly.

9.6 OTHER TYPES OF MOTORS

There are many types of motors available to the designer, but only a few common types will be discussed here. Specialized motors and specialized applications of motors can run the gamut for systems that are created for very specific purposes and very specific applications. Some of the more common types follow.

Servomotors

> **servomotor**
> A term that applies to a motor that has feedback loops built into the system for telling a control system the actual motor position and/or speed.

As the field of robotics and accurate positioning devices has evolved, there is a need in many applications to accurately place devices, such as an industrial robot placing a part in a machining operation. Either AC or DC servomotors may be used to control position. A **servomotor** includes feedback loops that determine both position and often, speed, so that the control system can know the position of the motor at any time and the velocity at which it is approaching its target, and can make corrections. A servomotor system may also include braking and be controlled by a computer system to coordinate its activities with other drive units.

Stepper Motors

A stepper motor has a whole series of internal windings so that partial rotation can occur. Some stepper motors may have steps as small as 1.8°. The motor can rotate to or stop at any multiples of these small steps to achieve fairly accurate control with or without the feedback and velocity controls of servomotors. Often these motors are used in applications where feedback may not be required but precise positioning is needed or in applications where small increments are needed. The disadvantages of stepping motors are that they are fairly heavy and have poor torque-to-weight ratios.

Brushless Motors

The typical universal or DC motor discussed earlier in this chapter requires contact through a commutator or slip rings to the shaft of the motor. These become wear items and affect the life of this type of motor. A brushless motor uses solid-state electronics to eliminate the contacts, but in other performance characteristics it is very similar to either the DC or the universal motor.

Linear Motors

Linear motors can take different forms. They can be designed in a similar manner to stepper motors with either integral steps or a brush-type DC motor. The difference between them is that linear motors slide a shaft as opposed to rotating one. Hence, they have a limited stroke but can provide some interesting machine control attributes.

9.7 SUMMARY

After completing this chapter, you should have gained an understanding of the basic principles of different types of motor systems. You should know when different motor systems may be applicable.

You should also understand the advantages of different types of motors as they relate to the relationship between size, power, lifetime, and cost. This is important to gaining a better understanding of which types of motors may be most applicable to the design being undertaken.

9.8 PROBLEMS

1. Answer the following:
   a. What is the standard frequency of AC power in the United States?
   b. In Europe?
2. Answer the following:
   a. How many conductors are required to carry single-phase power?
   b. Three-phase power?
3. Define the following terms:
   a. Synchronous speed for an AC motor
   b. Full-load speed for an AC motor in the United States
   c. Full-load speed for an AC motor in Europe
   d. Locked-rotor torque. What is another term for this parameter?
   e. Breakdown torque
   f. Pullout torque as applied to a synchronous motor
4. A motor nameplate lists the induction speed of a motor as 3,450 rpm. What would be the speed of this motor at zero load?
5. From the following options of electric motors that can be used in an industrial plant:
   120-volt single-phase
   240-volt single-phase
   240-volt three-phase
   440-volt three-phase
   a. Select the motor that would have the smallest overall size for a given power output.
   b. Which of these motors would require the smallest conductors (wire size) for the same power output?
6. Select the type of motor enclosure that should be used in a plant that manufactures flour.
7. Select a motor for use in a meat-processing plant if the motor is to be exposed.
8. Select a style of motor for use in an exposed environment such as in a car wash.
9. For a NEMA-standard, 5 horsepower, 184 T frame, how far is the centerline of the rotating shaft above the base of the motor, and what size shaft and keyway does this motor have?
10. If an inductive motor has a nameplate rating of 5 horsepower and an induction speed of 3,450 rpm, what would be the rated torque at that speed?
11. What torque would a 7 kw motor at a nameplate-rated speed of 3,450 rpm have in both inch-pounds and newton-meters?

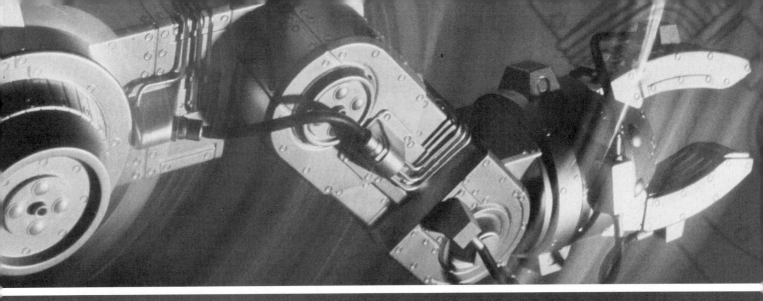

CHAPTER 10

Pneumatic and Hydraulic Drives

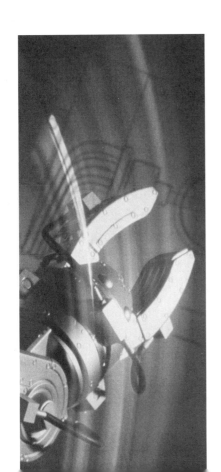

pneumatic
The use of a gas; for example, air or compressed air.

hydraulic
The use of a fluid; for example, water or oil.

Pneumatic and **hydraulic** drives have many unique applications in the machine design field. They are useful because of their ability to create high forces in a relatively small space and because of the ease with which they create direct linear motion. For example, a modern hydraulic excavator uses many types of hydraulic drives, including many hydraulic cylinders for the excavator arm, hydraulic motor drive systems for the track and for rotating the deck assembly. (See Figure 10.1.) The use of hydraulic and pneumatic drives is not limited to mobile construction

FIGURE 10.1 Hydraulic excavator. This Kubota model features a load sensing hydraulic system. (Courtesy of Kubota Tractor Corp.)

equipment. Their unique attributes can be very useful in all sorts of different applications. For example, most elevator systems in low-rise applications use **telescoping** hydraulic cylinders, and many robotic systems use hydraulic drive systems.

In this chapter we will discuss many of the basic principles of hydraulic and pneumatic drive systems, both for linear applications such as cylinders and for rotary drives such as air and hydraulic motors.

10.1 OBJECTIVES

After completing this chapter, you should be able to:

- Describe the principles of operation of both linear and rotary hydraulic and pneumatic drive systems.
- Understand the types of hydraulics and pneumatic cylinders.
- Explain how the principles of force, work, and power apply to these types of systems.
- Describe the types of control accessories available, such as manual and automatic valves, **regulators,** and flow control devices such as **needle valves.**

10.2 PRINCIPLES OF OPERATION

In conventional hydraulic or pneumatic air cylinders, the force that can be created is the product of the pressure and the cross-sectional area of the piston as follows:

$$F = PA \qquad (10.1)$$

Note that the retraction force on a cylinder is slightly different than the extension force because the area of the piston rod must be subtracted from the overall area. The stroke of a typical cylinder is the length of the cylinder assembly minus the width of the piston, as shown in Figure 10.2.

Most types of cylinders can be purchased with any length stroke needed. If the overall extended length is important to the design, many of the tie-rod–type cylinder systems can be modified to any length by the factory before shipping.

---

telescoping
Describes a series of tubes of different sizes that slide within each other to make a longer assembly.

regulator
Typically, a spring-operated bellows or diaphragm assembly that allows pneumatic or hydraulic fluids to pass through at a set pressure.

needle valve
Used in pneumatic or hydraulic systems to control the flow rate.

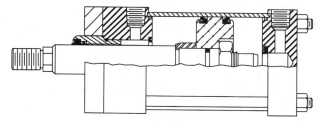
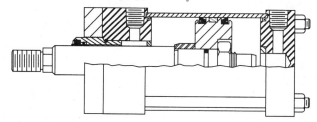

FIGURE 10.2 Cross-sectional view of pneumatic and hydraulic cylinders (note the similarities). (Courtesy of Boston Gear)

The work that is created by the cylinder assembly is the product of force times distance such that

$$W = Fd = PAd \tag{10.2}$$

Conversely, the power that is created is simply the rate of doing work as follows:

$$P = \frac{W}{t} = \frac{Fd}{t} = \frac{PAd}{t} \tag{10.3}$$

The pump or compressor power needed to create this amount of work would have to be determined from the efficiency of the supply system in combination with any frictional loss or inefficiency of the cylinder assembly systems. Typically, the supply system power would be significantly more than the power output of the system to allow for these types of losses. The fluids consumption of a hydraulic system is the product of the volume used times the cycle rate and is calculated for a single cycle as follows:

$$V = Ad \tag{10.4}$$

and the volume rate necessary for this operation is then a product of the time involved as follows:

$$\frac{V}{t} = \frac{Ad}{t} \qquad (10.5)$$

Example Problem 10.1

Select a hydraulic cylinder diameter and largest standard rod size available for a punch press that has 100 psi hydraulic pressure available, needs an extension force of 450 lb, and a retraction force of 400 lb. Verify your results with Figure 10.3.

$$F = PA \qquad (10.1)$$

$$A = \frac{F}{P}$$

$$A = \frac{450 \text{ lb}}{100 \text{ lb/in.}^2} = 4.5 \text{ in.}^2$$

$$A = \frac{\pi D^2}{4}$$

$$D = \sqrt{\frac{4A}{\pi}}$$

$$D = \sqrt{\frac{4\left(4.5 \text{ in.}^2\right)}{\pi}}$$

$$D = 2.39 \text{ in.}$$

Use 2½-inch bore cylinder

For rod size, try a 1-inch rod.

$$A = \frac{\pi D_b^2}{4} - \frac{\pi D_r^2}{4}$$

$$A = \frac{\pi (2.5 \text{ in.})^2}{4} - \frac{\pi (1 \text{ in.})^2}{4}$$

$$A = 4.123 \text{ in.}^2$$

$$F = PA$$

$$F = 100 \text{ lb/in.}^2 \left(4.123 \text{ in.}^2\right)$$

$$F = 412 \text{ lb}$$

This would meet requirements.

Theoretical Push and Pull
Stroke Forces

Hydraulic Cylinders

Cylinder Force and Volume Chart–Hydraulic

Cyl. Bore	Rod Size	Action	Working Area	Stroke Force in Pounds / Operating Pressure (psig)									Fluid Displacement Per Inch of Stroke (U.S. Gallons)
				50	80	100	250	500	750	1000	1500	2000	
1-1/2	—	PUSH	1.767	88	141	177	442	884	1325	1767	2651	3534	.0076
	5/8	PULL	1.460	73	117	146	365	730	1095	1460	2191	2921	.0063
	1	PULL	.982	49	79	98	245	491	736	982	1473	1963	.0042
2	—	PUSH	3.142	157	251	314	785	1571	2356	3142	4712	6283	.0136
	5/8	PULL	2.835	142	227	283	709	1417	2126	2835	4252	5670	.0123
	1	PULL	2.356	118	188	236	589	1178	1767	2356	3534	4712	.0102
2-1/2	—	PUSH	4.909	245	393	491	1227	2454	3682	4909	7363		.0212
	5/8	PULL	4.602	230	368	460	1150	2301	3451	4602	6903		.0199
	1	PULL	4.123	206	330	412	1031	2062	3093	4123	6185		.0178
3-1/4	—	PUSH	8.296	415	664	830	2074	4148	6222	8296	12444		.0359
	1	PULL	7.510	376	601	751	1878	3755	5633	7510	11266		.0325
	1-3/8	PULL	6.811	341	545	681	1703	3405	5108	6811	10216		.0295
4	—	PUSH	12.566	628	1005	1257	3142	6283	9425	12566			.0544
	1	PULL	11.781	589	942	1178	2945	5890	8836	11781			.0510
	1-3/8	PULL	11.081	554	887	1108	2770	5541	8311	11081			.0480
5	—	PUSH	19.635	982	1571	1963	4909	9817	14726	19635			.0850
	1	PULL	18.850	942	1508	1885	4712	9425	14137	18850			.0816
	1-3/8	PULL	18.150	908	1452	1815	4538	9075	13613	18150			.0786
6	—	PUSH	28.274	1414	2262	2827	7069	14137	21206	28274			.1224
	1-3/8	PULL	26.789	1339	2143	2679	6697	13395	20092	26789			.1160
	1-3/4	PULL	25.869	1293	2070	2587	6467	12935	19402	25869			.1120
8	—	PUSH	50.265	2513	4021	5027	12566	25133	37699				.2176
	1-3/8	PULL	48.781	2439	3902	4878	12195	24390	36585				.2112
	1-3/4	PULL	47.860	2393	3829	4786	11965	23930	35895				.2072

FIGURE 10.3 Cylinder forces both for extension and retraction. (Courtesy of Boston Gear)

From Figure 10.3, a 2½-inch bore cylinder with a 1-inch rod has an extension force of 491 pounds and a retraction force of 412 pounds.

If the cylinder has a stroke of 20 inches and a continuous cycle time of 2 seconds, determine the necessary flow rate.

Volume per cycle:

$$V = Ad \qquad (10.4)$$

$$V = A_b d + A_{b-r}\, d$$

$$V = 4.909 \text{ in.}^2 \cdot 20 \text{ in.} + 4.123 \text{ in.}^2 \cdot 20 \text{ in.}$$

$$V = 180.7 \text{ in.}^3$$

Flow rate required:

$$\frac{V}{t} = \frac{180.7 \text{ in.}^3}{2 \text{ sec}}$$

$$\frac{V}{t} = 90.4 \text{ in.}^3/\text{sec}$$

or $\quad 90.4 \dfrac{\text{in.}^3}{\text{s}} \quad \dfrac{\text{ft}^3}{1{,}728 \text{ in.}^3} \quad \dfrac{7.48 \text{ gal}}{\text{ft}^3} \quad \dfrac{60 \text{ s}}{\text{min}}$

$= 23.5 \text{ gal/min}$

10.3 TYPES OF HYDRAULIC AND PNEUMATIC MOTION SYSTEMS

The most common type of hydraulic and pneumatic motion system is a linear cylinder where, due to the pressure of the gas or fluid, motion is developed. The biggest difference in the performance of the two types of systems is due to the compressibility of fluids versus gases. The oil, water, or other fluid in a hydraulic system is considered **noncompressible.** This assumption is valid if the pressure ranges of the fluid are not excessively high. Because of this noncompressibility, a hydraulic system offers better incremental control than a pneumatic system. The gas in a pneumatic system (usually air, but any type of gas can be used) is compressible, so **incremental control** is usually not easily obtained. Hence, a pneumatic system is typically used only for movement between the two end points of the cylinder, whereas a hydraulic system can stop at any point, and small, incremental movements can be realized. Hydraulic systems can be powered directly by a pump-and-reservoir system. They can also be powered by a mixed system such as an air-over-oil or air-over-water system, in which the pressure is created by air acting on the hydraulic reservoir to create pressure in the fluid. This type of mixed system might be used where an air supply is available in order to avoid the complexity of having a separate hydraulic pump system. An example of a mixed fluid system is shown in Figure 10.4 and analyzed in Example Problem 10.2.

noncompressible
A term that is used to describe a fluid that has a high bulk modulus such that little or relatively little change in volume occurs with pressure.

incremental control
When small changes in position are achievable.

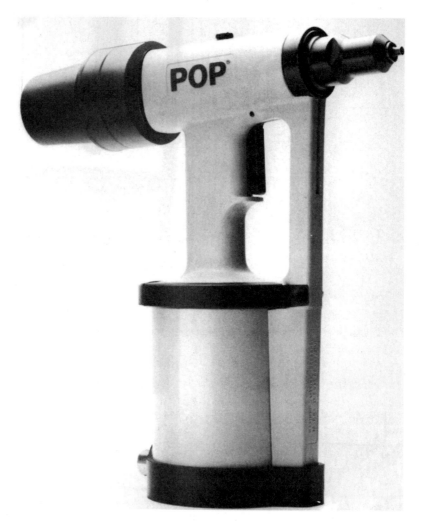

FIGURE 10.4 Pneumatic pop rivet gun.

Example Problem 10.2

For the pop rivet gun shown in Figure 10.5 that utilizes an air/hydraulic force multiplier, determine the oil pressure and force on both piston rods. If the air piston moves four inches, determine the stroke of the top cylinder.

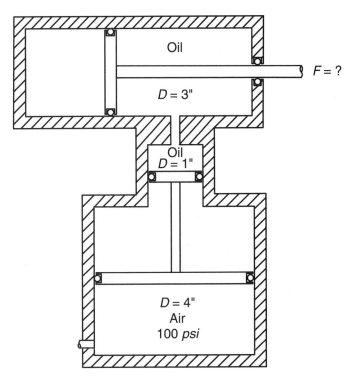

FIGURE 10.5 Sectional view of pneumatic/hydraulic pop rivet gun.

$$F_1 = P_a A_a \tag{10.1}$$

$$F_1 = 100 \text{ lb/in.}^2 \; \frac{\pi (4 \text{ in.})^2}{4}$$

$$F_1 = 1,257 \text{ lb}$$

$$F_1 = P_0 A_0$$

$$P_0 = \frac{F_1}{A_0} = \frac{1,257 \text{ lb}}{\frac{\pi (1 \text{ in.})^2}{4}} = 1,600 \text{ lb/in.}^2$$

$$F_2 = P_0 A_{20}$$

$$F_2 = 1,600 \text{ lb/in.}^2 \; \frac{\pi (3 \text{ in.})^2}{4}$$

$$F_2 = 11,310 \text{ lb}$$

Find the volume in a 1-inch cylinder for the 4-inch travel:

$$V = Ad \qquad (10.4)$$

$$V = \frac{\pi (1 \text{ in.})^2}{4} 4 \text{ in.}$$

$$V = 3.14 \text{ in.}^3$$

Travel for 3-inch cylinder:

$$d = \frac{V}{A}$$

$$d = \frac{3.14 \text{ in.}^3}{\frac{\pi (3 \text{ in.})^2}{4}}$$

$$d = .44 \text{ inch}$$

There are also structural considerations in selecting a hydraulic or pneumatic cylinder assembly, typically involving whether or not the piston rod has appropriate bending or column strength for the application. Most manufacturers offer standard cylinders with what are called oversized rods so that these larger piston rods can be used if the bending load or column action is significant due to the placement of the load on the cylinder assembly. A good example of this is the excavator mentioned earlier, where some of the hydraulic rod systems have their maximum forces during extension and the column compressive bending stresses on the cylinder rod assembly can be significant. In this case, the rod may need to be an oversized model.

Figure 10.6 shows a seal assembly for a typical hydraulic system. Many different types of seals are available, depending on the pressure, the need

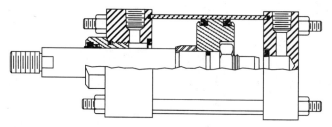

FIGURE 10.6 Seals for a hydraulic cylinder. (Courtesy of Boston Gear)

to prevent leaks, and the expected life. The manufacturer's catalog should be consulted with regard to the appropriate seals for each application. Generally, air cylinders tend to use O-rings, whereas hydraulic cylinders tend to use either cup seals or O-rings. Many types of materials are used, depending on the application.

Many mounting arrangements are possible for air or hydraulic cylinder assemblies, as shown in Figure 10.7. Depending on the application, cylinders can be mounted between clevis mounts, trunion mounts, or one of many other types. Mounts are available in standard configurations from manufacturers. Special designs can be created if necessary.

In order to control these cylinder assemblies, valving is typically used to regulate the pressure and volume supplied to the assembly. Hydraulic and pneumatic systems may use manual or **solenoid valves,** which can be activated by many types of automatic control systems. Needle valves may be used to control the flow rate and thus control the velocity of movement. In pneumatic systems it is also often necessary to control pressure. Thus, in many cases a regulator may be used along with needle valves for velocity control. Some of these types of products are shown in Figure 10.8.

> **solenoid valve**
> An electrically operated valve in which the valve is opened and closed by an electrically operated magnet system.

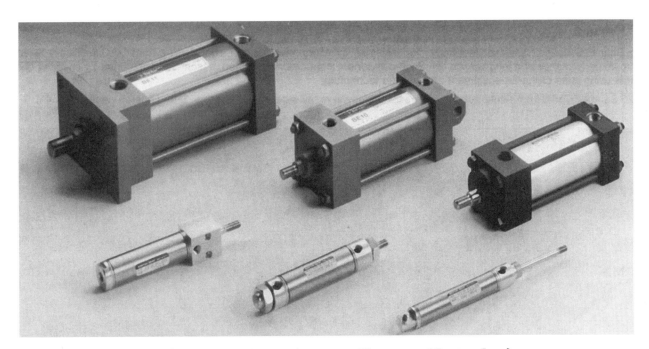

FIGURE 10.7 Different types of cylinders and mounts. (Courtesy of Boston Gear)

FIGURE 10.8 Manual and solenoid valves. (Courtesy of Boston Gear)

10.4 MISCELLANEOUS PNEUMATIC AND HYDRAULIC MOTION SYSTEMS

In addition to hydraulic and pneumatic cylinders, other products are available that use gas- or liquid-drive systems such as air or hydraulic motors to develop rotary motion. The manufacturer's literature should be consulted for power and flow characteristics for these types of systems. Examples of these are shown in Figures 10.9 and 10.10.

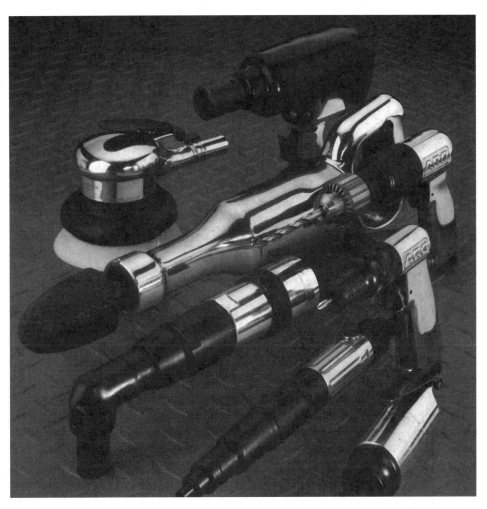

FIGURE 10.9 Air motor hand tools. (Courtesy of Ingersoll-Rand)

FIGURE 10.10 Hydraulic motors. (Courtesy of Webster Electric Co., Inc.)

10.5 SUMMARY

After completing this chapter, you should understand the basic principles of fluid motion systems, including hydraulic and pneumatic cylinder assemblies and air and hydraulic motors. You should understand the basic principles of pressure, work, and power necessary in these systems and be able to select standard items from manufacturers' reference material.

10.6 PROBLEMS

1. An air cylinder lifts a transfer table between two roller track systems. The weight of the table plus the boxes being transferred is 250 lb.

a. If 80-psi shop air were available, what size cylinder would be needed?
b. If you wish to reduce the velocity of this table, what device could be added to the system?
2. An elevator services a five-story building. The distance from the basement to the fifth floor is 75 feet. The weight of the elevator and passengers is 2,500 pounds. If 200 psi hydraulic pressure is to be used (ignore force to accelerate):
   a. Determine the cylinder diameter required.
   b. Determine the volume of reservoir required.
   c. If it is necessary to go from the basement to the fifth floor in 30 seconds, what is the flow rate needed?
   d. Determine the power supplied by the fluid.
   e. Assuming the overall system has an efficiency of 50 percent, what is the horsepower of the motor that would be required?
3. A dump truck uses a telescoping cylinder to lift the body. If the largest section is an eight-inch bore, the second largest is seven inches, and the smallest is six inches, what is the force developed as each section is engaged if the fluid pressure is 600 psi?
4. In Problem 3, solve if the sections are 200 mm, 175 mm, and 150 mm, respectively, and the pressure is 4,000 kPa.
5. An engine lift is needed to lift 500 pounds when the arm is horizontal as shown in Figure 10.11. If a manual pump can deliver 250-psi oil:
   a. Select a cylinder diameter (remember to compensate for the force angle).
   b. How large does the oil reservoir need to be if the stroke is 20 inches?
   c. Is there a way the reservoir volume can be reduced?
6. In Example Problem 10.2, if the stroke of the pop rivet mandrel needs to be increased to ½ inch:
   a. By how much does the air cylinder travel need to increase?
   b. If it were necessary to increase the mandrel force to 15,000 pounds, what air pressure would be required?
7. An automobile lift in a service station uses an air-over-hydraulic lift system. What air pressure would be required to lift the auto under the following conditions:
   a. 3,500-pound automobile; 8-inch-diameter cylinder.
   b. 1,600 kg automobile; 200-mm-diameter cylinder.
8. In Problem 7, should the valve control the input oil or the air supply? Describe how this would affect the lift performance.

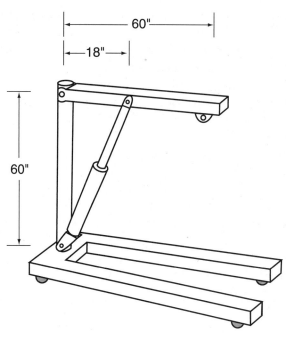

FIGURE 10.11 Automobile engine lift.

9. Select a hydraulic cylinder, including rod size, from Figure 10.3 for an application where 500 psi oil is available and a minimum extension force of 4,000 pounds and a retraction force of 3,300 pounds are needed.

10.7 CUMULATIVE PROBLEMS

10. For the trailer spring system that was designed in Chapter 8, Problem 15, design an air spring system that has a maximum air supply pressure of 250 psi and works in concert with a helical spring where the air spring carries approximately one-half of the load.
11. If, in the air spring used in Problem 10, it is necessary to vary the spring force depending on the load, what device would be used to do this?

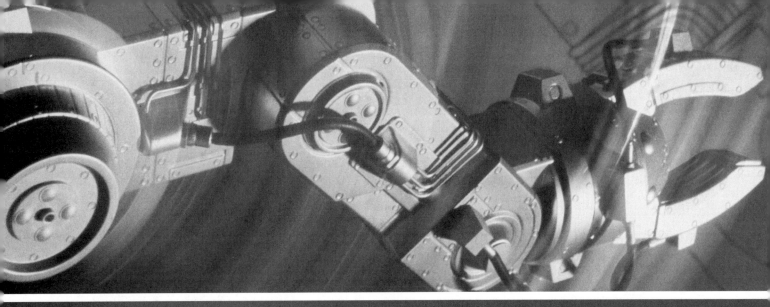

CHAPTER 11

Gear Design

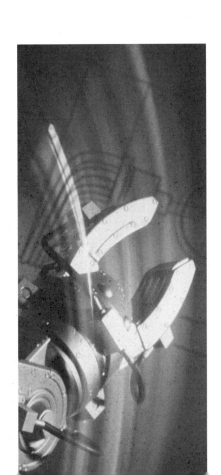

Gears are used in many mechanical devices. If we think of all the gears that are used in an automobile, the applications are numerous. The most obvious is the transmission, where the torque and speed from the engine are modified by reducing the speed in the lower gears, which correspondingly increases the torque. The differential further reduces this speed to the wheels, thereby also increasing the torque. Other gear mechanisms include electric seats, which use electric motors and gears to move the seats both fore and aft and up and down, or electric windows, where the rotational speed of motors is significantly reduced with a corresponding increase in torque, which then typically drives a **rack-and-pinion** gear system to push the windows up or down. In the engine, gears may be used for opening and closing valves and for many other purposes. Think of other mechanisms that involve the use of gears. Gears can be very small, such as those used in paper drives in computer printers. They can be very large, such as those that lower or rotate bridges to allow ships to pass by. From all this we can begin to understand the multitude of applications in which gearing systems are used. Because gear systems either increase or decrease rotational velocity, they have the inverse effect on the torque being transmitted. The principle of gear systems is that the power transmitted through a set of gears is constant, with the exception of minor frictional losses. The ratio of the gear system results in a change in the output rotational speed, and the torque changes by the inverse of that same ratio.

As you learn the principles in the next few chapters of how gear systems work and how to design with them, you will begin to realize how useful they can be. You will learn how gear systems can be applied in many cases, and you will learn how to design for the appropriate lifetime for gear systems. You will learn to select necessary attributes of the gearing such as pitch, size, quality of manufacture, and materials. You will also learn about the effects of such things as impact on gearing systems, how to determine what types of gears to select for specific applications, and how to select appropriate ratios.

11.1 OBJECTIVES

This chapter is intended to give you an understanding of the different types of gears and how to calculate velocity ratios, and it outlines some of the physical characteristics of different gearing systems. After completing this chapter, you should be able to:

- Understand the basic principles of gearing.
- Understand gear trains and how to calculate ratios.

---

rack-and-pinion gear
Uses a conventional pinion gear in concert with a straight gear to convert rotary to linear motion.

- Recognize different types of gearing systems and the relative advantages and disadvantages between them.
- Understand the geometry of different types of gears and their dimensional properties.
- Begin to recognize many of the different principles of gearing and also recognize the many unorthodox ways gears can be used in different types of motion systems.

11.2 TYPES OF GEARS

If we think of a gear system as two cylinders rolling together without slippage (as shown in Figure 11.1), the **surface velocity** of the two gears is always the same. This, in a nutshell, is really the principle of most forms of gearing—that is, we create a zero-slippage wheel system. The rotational velocity and torque are a product of either the number of teeth or the radii of the gears. The force between mating teeth is always equal and opposite, with the surface velocity of the mating gear systems correspondingly identical. Figure 11.2 shows a variety of different gear types, ranging from the most common spur gear systems through more specialized gears. Examples shown include **bevel** or **miter** gears, **helical** gears, rack-and-pinion gears such as those in the steering system of an automobile, **internal gears,** and **worm gears.** Some of the advantages and disadvantages of different types of gear systems and their basic principles are explained in the following sections.

surface velocity
The surface speed of the gear or other rotating object at the pitch line. For example, for the tire of an automobile it would be the speed that the car is moving.

bevel gear
An angular gear used to join perpendicular shafts that is similar but different in profile from a spur gear.

miter gear
A bevel gear in which both gears have the same number of teeth and, hence, a one-to-one ratio.

helical gear
A gear that uses the same general form as a spur gear but is cut on an angle.

internal gear
A gear profile made on the inside of a circle.

worm gear
A gear system that uses a lead-screw–type shape to turn a gear, which results in high ratios.

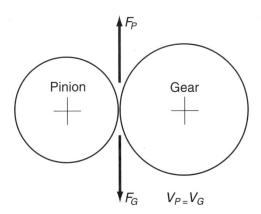

FIGURE 11.1 As if gears were friction wheels: Note that surface speed will be the same for both.

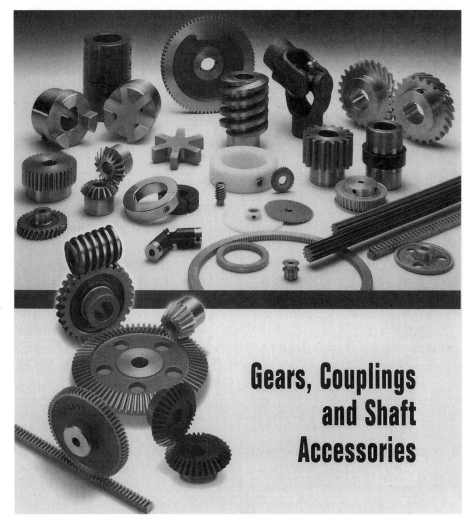

FIGURE 11.2 Common gear types. (Courtesy of Boston Gear)

Spur Gears

spur gear
The most common form of gearing; uses a shaped tooth to transmit and sometimes change rotational speed.

Spur gears are the most common form of gearing. Their involute shape results in high efficiency, and relatively high ratios can be achieved. They are ideally suited for low- to moderate-speed applications because, in general, they cost less than other forms of gearing. Spur gears are available in materials ranging from hardened steel to brass, bronze, cast iron, and plastic. In low-torque applications such as clocks, they may even be made out

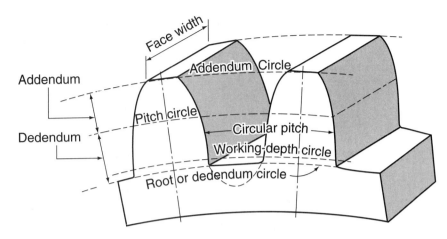

FIGURE 11.3 Spur gear tooth features.

of sheet-metal stampings. Many manufacturers stock spur gears in typical sizes, and they are therefore readily available. A spur gear uses parallel shafts. The basic geometries of a spur gear are shown in Figure 11.3.

Helical Gears

Helical gears use the same basic profile as spur gears except that they are cut on an angle. They also typically use parallel shafts. Because the teeth are cut on an angle, they engage gradually during the rotation process. This results in greater strength for high-velocity applications because the impact loading that may be present in a spur gear system is minimized in helical gear systems. The disadvantage of helical gears is that their relative cost is greater than that of spur gears; however, in applications where space is limited and high loads and velocities are present, this increased cost may be justified. In one modified form of helical gears, called a **cross-helix** the gear systems are cut at a 45-degree angle and the shafts are perpendicular. The gears in this type of application are typically subjected to considerable side load and friction. Helical gears are shown in Figure 11.4.

cross-helix
A specialized form of a helix gear in which the gears are perpendicular to each other.

Bevel and Miter Gears

Bevel gear systems (Figure 11.5) are used for applications where the shafts are perpendicular to each other. This type of gear system is ideal for

FIGURE 11.4 Helical gears. (Courtesy of Boston Gear)

FIGURE 11.5 Miter and bevel gears. (Courtesy of Boston Gear)

FIGURE 11.6 Bevel and miter gears: straight and spiral types. (Courtesy of Boston Gear)

assemblies that must corners. A bevel gear set can change the ratio. A miter gear, which is a one-to-one ratio bevel gear, is used to simply turn the corner. The principles of bevel gears are very similar to spur gears with one minor difference. The pitch of a bevel gear is not constant; the tooth profile increases in size as it goes outward on the bevel gear face. Other than this principle and the angles involved, the analysis of bevel gears and spur gears is quite similar, as are their advantages and disadvantages.

We can add the advantages of helical gears to a bevel or miter gear system by the use of spiral miter gears or spiral bevel gears, where an angle or a contoured angle is cut into the face of the gear teeth to minimize impact loading (Figure 11.6). Along with this advantage is a significant increase in cost due to the complex profile of the teeth and the complexities in manufacturing.

Rack-and-Pinion Gearing

A rack-and-pinion gear system can be envisioned as simply taking one of the gears from a spur gear system and unrolling its teeth into a flat profile (Figure 11.7). Thus, rotary motion is converted to linear motion. A very common application of this type of gear system is in the steering mechanism for many automobiles, where the rotation of the steering wheel converts to a linear motion to turn the steering system. Many other applications exist where it is necessary to convert rotary motion to a reciprocating or other type of linear motion. The basic principles of spur gears apply to rack gears with the exception that, due to the profile of the gear teeth, the spacing of the rack gear teeth needs to be modified slightly unless clearances are increased.

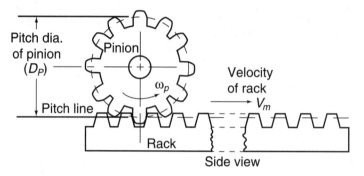

FIGURE 11.7 Rack-and-pinion gear system.

Worm Gear Assemblies

The profile of the worm gear assembly is only somewhat similar to a spur gear (see Figure 11.8). The principles of operation for worm gears really are closer to a lead screw assembly in the sense that, unlike a spur gear, one rotation of the worm assembly moves the mating gear one incremental tooth-width. Furthermore, a worm gear assembly relies on rotating sliding friction much like a lead screw assembly, and therefore many of the principles of gearing do not apply. However, worm gears are studied along with gearing systems because they have similar functional properties such as changing rotational speed and torque.

FIGURE 11.8 Worm gear system with C-face motor mounting. (Courtesy of Peerless-Winsmith, Inc.)

11.3 CATEGORIES OF GEARS

It is often useful to categorize gear assemblies according to whether their mounting shafts are parallel or perpendicular. For the different types of gears that we have learned about already in this chapter, we have discussed whether the shafts are parallel or perpendicular. This is because, in most cases, gear systems are used both to change rotational velocity and torques and also to turn corners. We can break the gear types that we have reviewed so far into these categories:

Parallel Shaft
Spur gears
Helical gears

Perpendicular Shaft
Bevel and miter gears
Cross-helix
Worm gears

Other Types
Rack-and-pinion

The differentiation of gears into these categories can be useful when we have design constraints where either parallel shafts or perpendicular shafts are needed. We can further differentiate gears into whether or not speed and torque need to be modified.

11.4   VELOCITY RATIOS AND GEAR TRAINS

For a single set of gears, the velocity ratio is defined as the ratio of the rotational speed of the input gear to that of the output gear. We often generalize this as also referring to a **gear train** value, which is the ratio between the input and output speed of one or more pairs of gears. With multiple sets of gears, the ratios become multiplicative for each set of gears in the gear train. This enables us to calculate the individual parts of a gear train value as mathematical factors of the overall ratio. For instance, if we needed an overall ratio of 27:1, an individual gear train with a 3:1 ratio could be mated with a second set of the same ratio, which could then be mated to a third set, giving us an overall ratio of $3 \times 3 \times 3$ or the 27:1 overall ratio.

As the motion in a gear train can be thought of as tooth-to-tooth with no slippage, the individual ratio between two gears is the ratio of the number of teeth on the gears, or the radii, diameters, or speeds of these gears such that the velocity ratio for a pair of gears would be

$$V_r = \frac{N_g}{N_p} = \frac{D_g}{D_p} \tag{11.1}$$

where V_r = velocity ratio

N_p = number of teeth on pinion

N_g = number of teeth on gear

D_p = pitch diameter of pinion

D_g = pitch diameter of gear

In most cases, gear systems are used to reduce speed and increase torque. However, there are certainly exceptions to this where we are

> **gear train**
> Two or more sets of gears combined to result in higher ratios.

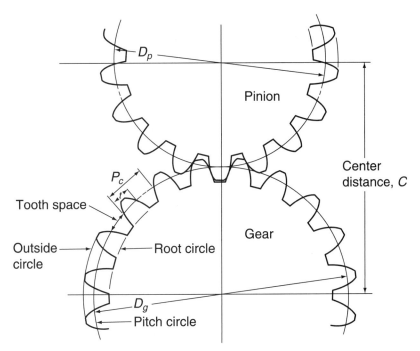

FIGURE 11.9 Spur gears in mesh.

increasing the speed with a corresponding decrease in torque. Two gears in mesh are shown in Figure 11.9 where it can be seen that the velocity of the mating teeth is equal and the rotation speed of the gear is correspondingly reduced, as reflected in Formula 11.1, in order to obtain the rotational output speed of the gear system. Note that the torque for this gear would correspondingly increase by the same ratio. However, the power transmitted does not change, assuming the system is 100 percent efficient.

When we have multiple sets of gears in series, we would refer to this as a gear train. In a gear train, the ratio is multiplied for each set of gears. Example Problem 11.1 shows a gear train analysis for a drive system, assuming 100 percent efficiency. Note that in any gear train system, the rotation is reversed through each set of gears. Often an idler is added to yield the desired direction of rotation.

Example Problem 11.1

For the set of four gears shown in Figure 11.10, calculate the output speed, output torque, and horsepower for both input and output conditions and the overall velocity ratio.

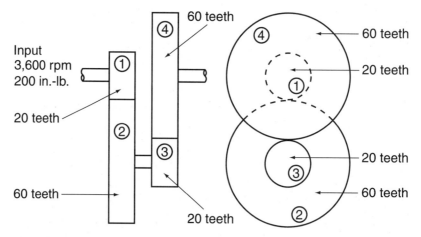

FIGURE 11.10 Double-reduction spur gear set for Example Problem 11.1.

$$V_r = \frac{N_2}{N_1} \cdot \frac{N_4}{N_3} \tag{11.1}$$

$$V_r = \frac{60}{20} \cdot \frac{60}{20} = \frac{9}{1}$$

Output speed:

$$n_4 = \frac{n_1}{V_r}$$

$$3{,}600 \text{ rpm} \cdot \frac{1}{9} = 400 \text{ rpm}$$

Output torque:

$$T_4 = T_1 V_r$$

$$T_4 = 200 \text{ in.-lb} \cdot \frac{9}{1} = 1{,}800 \text{ in.-lb}$$

Input horsepower:

$$hp = \frac{Tn}{63,000} \quad (2.6)$$

$$hp = \frac{200 \text{ in.-lb } 3,600 \text{ rpm}}{63,000}$$

$$hp = 11.4$$

Output horsepower:

$$hp = \frac{1,800 \text{ in.-lb } 400 \text{ rpm}}{63,000}$$

$$hp = 11.4$$

Example Problem 11.2

For the gear train shown in Figure 11.11, determine the train value, output speed, output direction, output torque, and output power.

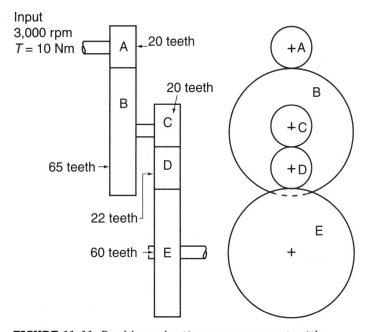

FIGURE 11.11 Double-reduction spur gear set with idler for Example Problem 11.2.

Train value:

$$V_r = \frac{N_B}{N_A} \cdot \frac{N_D}{N_C} \cdot \frac{N_E}{N_D}$$ (11.1)

$$V_r = \frac{65}{20} \cdot \frac{60}{20} = 9.75/1$$

Note: The idler cancels out and has no effect on the overall train value.

Output speed:

$$n_E = \frac{n_A}{V_r} = \frac{3,000 \text{ rpm}}{9.75/1} = 307.7 \text{ rpm}$$

Output torque:

$$T_E = T_A V_r$$
$$T_E = 10 \text{ Nm } (9.75/1) = 97.5 \text{ Nm}$$

Direction:
If Gear A is clockwise,
    Gear B is counterclockwise.
    Gear C is counterclockwise.
    Gear D is clockwise.
    Gear E is counterclockwise.

Output power:

$$P = Tn_E = 97.5 \text{ Nm } 307.5 \frac{\text{rev}}{\text{min}} \frac{\text{min}}{60 \text{ sec}} \pi$$

$$P = 1,571 \text{ Nm/sec or } \frac{J}{s} \text{ or W}$$

$$P = 1.57 \text{ kW}$$

11.5 SPUR GEAR STYLES

As spur gears are by far the most common form of gearing, many options are available for materials, mounting, and hub design, depending on the application. Figure 11.12 shows a series of different types of spur gears

FIGURE 11.12 Types of spur gears. (Courtesy of Boston Gear)

ranging from a large spoke model, where spokes are used to reduce the weight of the gear system, through a more typical smaller, solid-hub model. Many styles are available as standard items from manufacturers. They often include an integral **keyway** for mounting. The rack gear shown, which is in essence a spur gear, looks almost as if you had taken a spoke gear, cut off the outer rim, and straightened it. The tooth profile for a rack gear is similar to that of a spur gear. Many gear forms are available in many grades of plastics, which, in addition to being light, can absorb impact loads well. Plastic gears can be manufactured by injection molding at a relatively low cost. Low-cost stamped metal spur gears can be cut from sheet metal. These might be found in an inexpensive clock in combination with many plastic gears. In some cases, a small gear may even be manufactured as part of the shaft itself (as shown in Figure 11.13), where the gear tooth form is machined directly into the shaft to eliminate the need for a separate assembly. Spur gear forms may be used in applications such as **spline** assemblies, such as those used in the driveshaft of an automobile. A spur gear profile is also used in an internal gear (as shown in Figure 11.14), where the gear teeth are manufactured on the inner radius for use in applications such as spherical gear reduction systems. Some very large gear systems may be built so that sections of individual teeth can be replaced when worn rather than replacing the whole gear assembly.

keyway
A slot cut into a shaft and a mating hub that uses a square or other-shaped key to transmit rotational motion.

spline
A device that allows a change in length of a shaft while transmitting rotary motion.

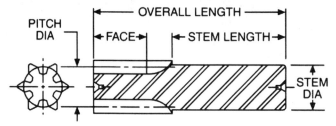

FIGURE 11.13 Pinion gear (stem pinion). (Courtesy of Boston Gear)

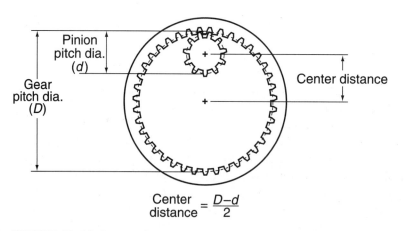

Center distance $= \dfrac{D-d}{2}$

FIGURE 11.14 Internal gear system.

As we study spur gear design in Chapter 12, we will see applications where plastic gears or composite resin gears may actually provide higher-strength service than steel gears due to their ability to absorb shock loading. Gears may be made out of cast iron, bronze, brass, aluminum, steel, and many plastic compositions to meet the requirements of different applications.

To differentiate between the large and the small gear in a spur gear system, we commonly refer to the input, which is often the smaller gear, as a **pinion** and the larger gear, or the output, as a gear. We will use that terminology in this text to differentiate between the two.

11.6 SPUR GEAR GEOMETRIES

So that we can use spur gears from different manufacturers interchangeably, spur gears have common dimensions and symbols. The American Gear Manufacturers Association (AGMA) sets certain standards for the manufacture and dimensions of spur gears. These standards and definitions are summarized in this section.

Pitch Diameter (D_p)

Pitch diameter is the most important consideration in the placement of spur gear systems. If we imagine gears as being solid rollers, as discussed earlier in this chapter, their mating point would be called the pitch diameter. This pitch diameter is often also referred to as the "working circle" or "pitch circle." We use this dimension not only for the placement of our gears but also as the point at which the forces and power are transmitted. Figure 11.15 shows the imaginary solid roller mating point and Figure 11.16 shows the forces acting on a spur gear system. If you look carefully at Figure 11.16, you will notice that the shape of the gear tooth is somewhat curved. This is called the **involute curve** of the gear tooth system. This involute shape is what causes a rolling action when the spur gear teeth mate with each other. Correctly placed spur gears result in a smooth rolling action with no slipping between the gear teeth as they come into and out of engagement. This characteristic of spur gears is important. If there is slippage between the teeth as they engage and disengage, there will be wear on the teeth. Obviously, a minor amount of this can occur in any gear system, as gear teeth are never made or placed exactly. However, we typically lubricate the teeth to minimize the effect of any sliding when they come into or out of engagement, to cushion the teeth as they come into engagement, and

> **pinion**
> The term that is applied to the smaller gear in a gear-mating system, as opposed to the larger gear, which is referred to only as the gear.

> **pitch or diametral pitch**
> An index of the size of the teeth that is found by dividing the total number of teeth by the diameter of the gear.

> **involute curve**
> The special shape of a gear profile that allows for rolling action when mated with a similar involute shaped gear.

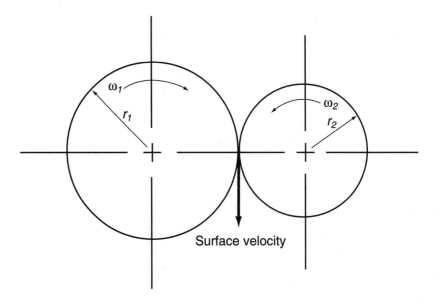

FIGURE 11.15 Spur gears shown as solid wheels: Note surface velocity is the same for both.

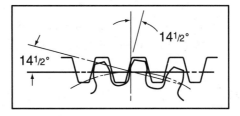

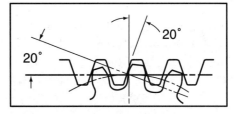

FIGURE 11.16 14½° and 20° pressure angle teeth.

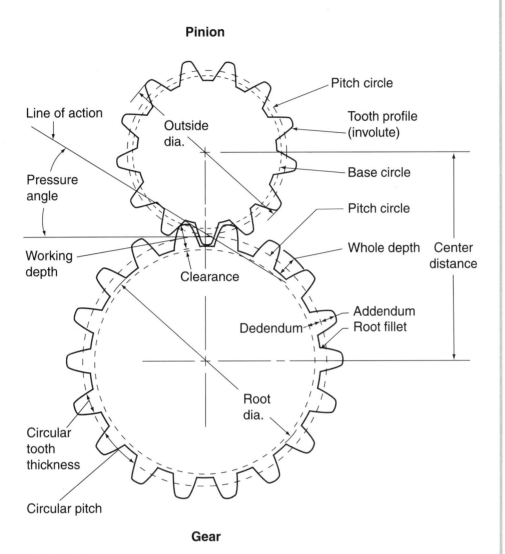

FIGURE 11.17 Mating spur gears.

often to cool the teeth as well. Figure 11.17 shows two gear teeth in engagement, mating at the point of their pitch circle. It can be seen from this relationship that the center-to-center distance would be equal to

$$C \text{ to } C = \frac{D_{pg}}{2} + \frac{D_{pp}}{2} = \frac{N_g + N_p}{2P_d} \quad (11.2)$$

where the two different pitch circles or pitch radii would then meet.

Circular Pitch (P_c)

Circular pitch relates to the physical size of the gear teeth and can be thought of as the tooth-to-tooth distance along the circumference of the circle. Circular pitch is a similar quality then to the lead distance in a lead screw. The larger the value, the larger the physical size of the tooth. Therefore

$$P_c = \frac{\pi D_p}{N} \tag{11.3}$$

where N is the number of total teeth on the gear. The circular pitch of two gears in mesh must be the same.

Diametral Pitch (P_d)

The **diametral pitch** is the most common system in use for signifying the size of gear teeth. It is, however, somewhat confusing. The pitch number, or diametral pitch, is the relationship of the number of teeth divided by the diameter of the gear as measured at the pitch diameter. Figure 11.18 shows different pitches of gears. Diametral pitch is calculated as follows:

$$P_d = \frac{N_g}{D_g} = \frac{N_p}{D_p} \tag{11.4}$$

We would therefore refer to the diametral pitch of a gear system as being a 4 pitch or a 20 pitch, for example. Unlike circular pitch, the smaller number would be for larger teeth.

As gears are generally manufactured by standard tooling, there are certain standard pitches available, and we usually use these standards. When we need to convert from circular pitch to diametral pitch, we do it as follows:

$$D_p = \frac{N P_c}{\pi} \tag{11.5}$$

and since

$$D_p = \frac{N}{P_d}$$

---

circular pitch
Relates to the physical size of the gear tooth but, unlike diametral pitch, is measured along the circumference of the gear circle.

Circular pitch (11.3)

$$P_c = \frac{\pi D_p}{N}$$

diametral pitch
Commonly referred to as simply "pitch," it is the size of the gear tooth and is measured as the number of teeth per inch of diameter.

Diametral pitch (11.4)

$$P_d = \frac{N_g}{D_g} = \frac{N_p}{D_p}$$

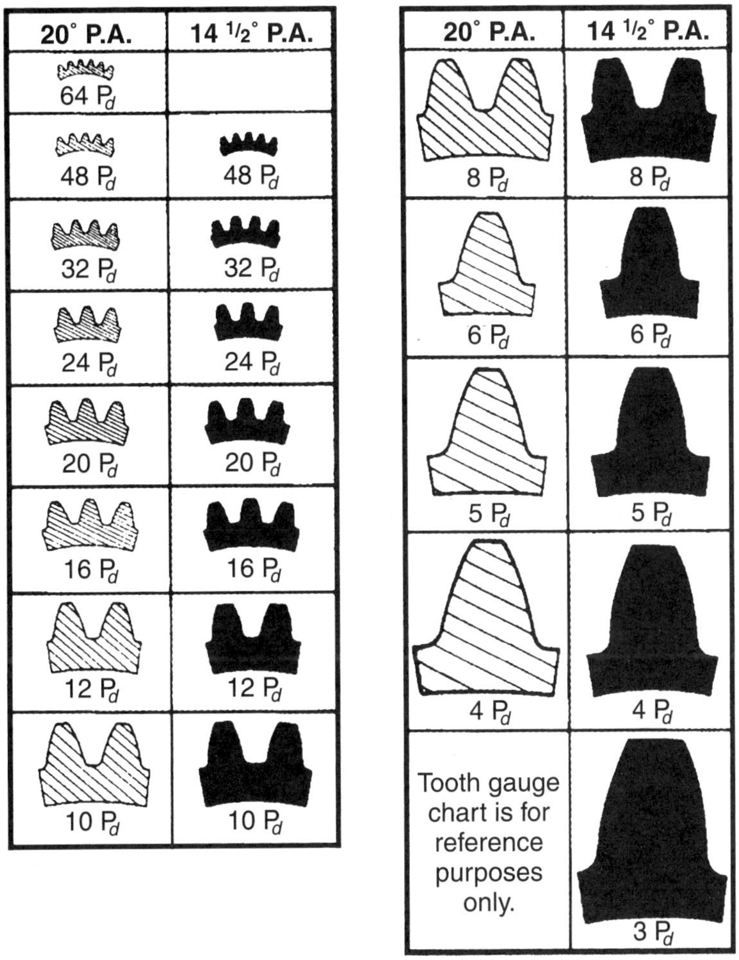

20° P.A.	14 ½° P.A.		20° P.A.	14 ½° P.A.
64 P_d			8 P_d	8 P_d
48 P_d	48 P_d		6 P_d	6 P_d
32 P_d	32 P_d		5 P_d	5 P_d
24 P_d	24 P_d		4 P_d	4 P_d
20 P_d	20 P_d		Tooth gauge chart is for reference purposes only.	3 P_d
16 P_d	16 P_d			
12 P_d	12 P_d			
10 P_d	10 P_d			

FIGURE 11.18 Actual tooth sizes. (Courtesy of Boston Gear)

and combining

$$\frac{N}{P_d} = \frac{N P_c}{\pi}$$

then

$$P_d = \frac{\pi}{P_c} \qquad (11.6)$$

metric module
A very similar characteristic to circular pitch; used in the metric SI system for specifying gear-tooth size.

In the metric system, a module system is used. To obtain the **metric module** (m), we would use the relationship of

$$m = \frac{D_g}{N_g} = \frac{D_p}{N_p} \qquad (11.7)$$

Therefore

$$m = \frac{1}{P_d}$$

Or converting from inches to millimeters using 25.4 mm per inch:

$$m = \frac{25.4}{P_d} \text{ where } P_d \text{ is in inches} \qquad (11.8)$$

Note that we would never intermix a metric module system with a diametral pitch as the standards are different; hence the teeth would not match. However, you should recognize the metric module system's similarity to diametral pitch so that you will know what metric module teeth would be similar in size. Table 11.1 lists standard metric modules with their

TABLE 11.1
STANDARD MODULES

Module (mm)	Equivalent P_d	Closest Standard P_d (teeth/in.)
.3	84.667	80
.4	63.500	64
.5	50.800	48
.8	31.750	32
1	25.400	24
1.25	20.320	20
1.5	16.933	16
2	12.700	12
2.5	10.160	10
3	8.466	8
4	6.350	6
5	5.080	5
6	4.233	4
8	3.175	3
10	2.540	2.5
12	2.117	2

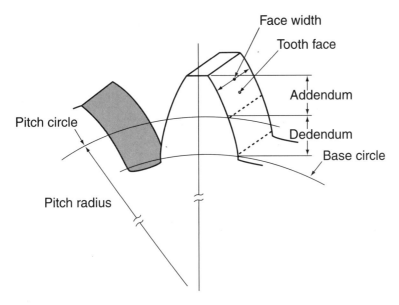

FIGURE 11.19 Tooth profile showing addendum and dedendum.

equivalent diametral pitches and the closest standard pitches. Note that the metric module system is not the same as the diametral pitch system but is very close.

Other gear teeth features we may be concerned with are shown in Figure 11.19. These are somewhat self-explanatory, such as the **addendum** and the **dedendum,** which are the amounts added or subtracted from the pitch circle. The pitch diameter minus the dedendum is also commonly called the root diameter. The addendum is very similar to the outside diameter or the clearance diameter. We also need to know the width of the gear teeth because that becomes important when we analyze the strength of gear teeth in Chapter 12. In addition, we will often be concerned with **backlash,** especially for coarser gear teeth. Backlash is the radial distance at which two mating teeth can move relative to each other. Obviously, the greater the backlash, the greater the potential for impact, especially under high-speed loading. The amount of backlash is a product of the size of the gear teeth and the accuracy or quality of manufacturing. Its effect can be influenced greatly by the type of lubrication as the lubrication cushions the gear teeth.

addendum
Could be thought of as the outside diameter of a gear. It relates to the clearance needed for gears or the maximum diameter of a gear system. It is the amount that is added to the pitch diameter.

dedendum
The distance from the pitch circle to the root of the gear or the base circle.

backlash
The clearance between gear teeth or the amount of clearance if the gears were to reverse direction.

Pressure Angle

pressure angle
A property of the gear-face profile related to the involute angle.

The **pressure angle** is the angle between the tangent of the pitch circles and a line drawn perpendicular to the surface of the teeth (see Figure 11.20). The pressure angle is relevant because it affects the shape of the gear teeth and, as gear teeth are transmitting a force, there is a component of this force trying to push apart the two shafts on which the gears are mounted. This is the separation force, and it is the product of this pressure angle. There are three different standard pressure angles in general use: 14½°, 20°, and 25°. The shape of these teeth varies slightly with these different pressure angles as shown in Figure 11.21. The advantages and disadvantages of different pressure angles are related to the strength of the teeth, the interference between teeth, and the separating force. A 25° pressure angle gear generally has a slightly wider base that results in a slightly stronger tooth. However, this is offset by the fact that the shaft separation force is larger.

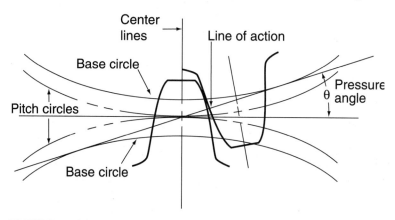

FIGURE 11.20 Gear tooth pressure angle.

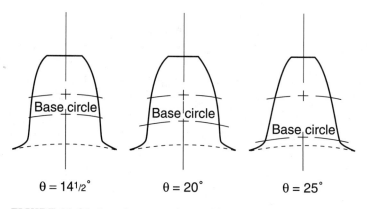

FIGURE 11.21 Involute tooth profiles.

Example Problem 11.3

For the set of gears shown in Figure 11.17, if the diametral pitch is 8, find the pitch diameter, circular pitch, and shaft center-to-center distance. The pinion has 16 teeth and the gear has 32 teeth.

Pitch diameter:

$$D_p = \frac{N_p}{P_d} \quad \text{or} \quad \frac{N_g}{P_d} \tag{11.4}$$

Pinion:

$$D_p = \frac{16}{8} = 2 \text{ inches}$$

Gear:

$$D_p = \frac{32}{8} = 4 \text{ inches}$$

Circular pitch:

$$P_c = \frac{\pi D_p}{N_p} \tag{11.3}$$

$$P_c = \frac{\pi \, 2 \text{ in.}}{16} = .393 \text{ inch}$$

Circular pitch would be the same for both the pinion and the gear.

Centerline distance:

$$C \text{ to } C = \frac{D_p}{2} + \frac{D_g}{2} \quad \text{or} \quad C \text{ to } C = \frac{N_p + N_g}{2P_d} \tag{11.2}$$

$$C \text{ to } C = \frac{16 + 32}{2(8)}$$

$$C \text{ to } C = 3 \text{ inches}$$

11.7 HELICAL GEAR GEOMETRIES

One disadvantage of spur gears, due to their straight-tooth geometry, is that impact loads are created, especially at high surface speeds. Especially with coarse series of spur gears, an impact load is imparted to the teeth that can accelerate the deterioration of the tooth systems and require the designer to de-rate the gear assembly. Helical gears reduce this problem because their angled design causes each tooth to come into engagement gradually. Hence, impact loading is reduced in a correctly designed and installed helical gear system.

Because of the gradual engagement of the teeth, helical gears are quieter and generally stronger than spur gears and are recommended for higher surface velocities. However, because the teeth are not parallel to the shaft, a thrust load is created that needs to be accommodated for in the design of the shafts. Each of the two mating gears creates a thrust load in opposite directions. In some cases, such as in a multiple-gear assembly, a left- and a right-hand helical gear could be mounted on the same shaft to eliminate this thrust load. Sometimes left- and right-hand helical gears are made as a combined assembly. This is called a **herringbone gear**. Figure 11.22 shows a set of helical gears along with a crossed helix, which is quite different in principle but identical in construction.

herringbone gear
A left- and right-hand helical gear combined in a single assembly to eliminate thrust loads on the shaft.

FIGURE 11.22 Helical gears: Note both parallel and perpendicular shafts. (Courtesy of Emerson Power Transmission Corporation. All Rights Reserved.)

Properties such as pitch, diametral pitch, and circular pitch are similar for helical gears as for spur gears; the one exception is the helix angle, which creates an angled force transfer and the associated thrust load. The helix angle typically ranges from 15° to 45°, with the 45° angle being used if the shafts are perpendicular in a cross-helix arrangement.

11.8 BEVEL GEAR GEOMETRIES

Bevel gears are commonly used to transfer motion between perpendicular shafts. If we think again of the two mating cylinders in the discussion on spur gears and replaced the cylinders with two cones, we can envision a similar mating surface that would transmit torque and the associated power around a corner. Furthermore, we can change the ratio of the system in the same manner as we did with spur gears. One significant difference, however, is that the diametral pitch of a bevel gear is not a constant value. Figure 11.23 shows a pictorial representation of the friction cones, and Figure 11.24 reflects the basic dimensions of a bevel gear system. Note in Figure 11.24 that the pitch diameter is dependent on where it is measured. Generally, the maximum pitch diameter value is found at the rear hub of a bevel gear system. The mounting distance is usually a product of the length of the hub and not directly a product of the gears. The bevel angle is

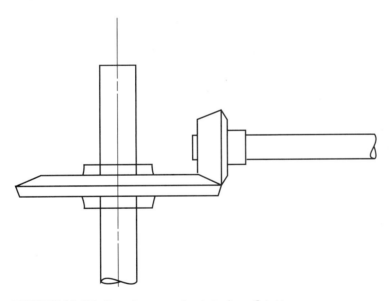

FIGURE 11.23 Bevel gears depicted as friction cones.

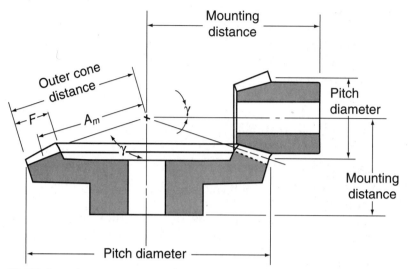

FIGURE 11.24 Bevel gear mounting dimensions.

FIGURE 11.25 Bevel gears. (Courtesy of Boston Gear)

dependent on the ratio between the gears, but the sum of the bevel angle for the two mating gears would always be 90° if they are on perpendicular shaft systems. A miter gear is a special kind of bevel gear that does not change the ratio between the gearing systems. It has a bevel angle of 45°; both gears in this case are identical. Most bevel gear sets are made as pairs because of some special considerations with regard to angle clearances and undercut considerations to minimize interference. Figure 11.25 shows a photo of a straight bevel gear system.

FIGURE 11.26 Spiral bevel gears. (Courtesy of Boston Gear)

Note that there are a few specialized types of bevel gear systems, such as a spiral bevel, a Zerol® bevel, and a hypoid gear system. These are similar conceptually to how a helical gear is a modification of the spur gear system (see Figure 11.26). A spiral bevel has a spiral on the teeth, and the Zerol® system incorporates a curved tooth. The impact loading for these gears is reduced in the same way that the helical gears have reduced impact loads when compared to straight spur gears. Also, special applications sometimes require bevel gear systems for angles other than 90°.

In the standard straight bevel, spiral, or Zerol® bevel system, the apexes of both pitch cones must meet where the shaft centerlines intersect. Alignment is fairly critical for smooth operation. The gear ratio is determined in the same manner as for spur gears: it is the ratio of the number of teeth on each of the mating gears.

Bevel gears are typically made in pairs and are not interchangeable in the same way as spur gears. The alignment of bevel gears is fairly important as the backlash increases significantly with minor differences in alignment. However, this can be adjusted fairly easily by moving either of the gears toward the centerline. This type of system also develops a thrust load. Because of this thrust loading and the inability to place a bearing in the apex area of most bevel gears, shaft size and placement take on greater importance. Generally, bevel gears have a ratio of 4:1 or less. Because of some special considerations with regard to clearances, some manufacturers modify the tooth profile slightly to reduce possible interference.

Example Problem 11.4

If a bevel gear system has 10 teeth on the pinion and 24 teeth on the mating gear, and the pinion is driven, find the output torque and speed if the pinion is driven by a 1½-hp motor at 3,450 rpm.

Torque in input shaft:

$$\text{hp} = \frac{Tn}{63,000} \tag{2.6}$$

$$T = \frac{\text{hp}\, 63,000}{n}$$

$$T = \frac{1\frac{1}{2}(63,000)}{3,450}$$

$$T = 27.4 \text{ in.-lb}$$

Output torque:

$$T_{out} = \frac{N_g}{N_p} T_{in}$$

$$T_{out} = \frac{24}{10}(27.4 \text{ in.-lb})$$

$$T_{out} = 65.8 \text{ in.-lb}$$

Output speed:

$$n_{out} = \frac{N_p}{N_g} n_{in}$$

$$n_{out} = \frac{10}{24}(3,450 \text{ rpm})$$

$$n_{out} = 1,437.5 \text{ rpm}$$

Note that because of the speed, this might be an application where straight bevel gears may be a poor choice.

11.9 WORM GEAR GEOMETRIES

Worm gearing, unlike bevel gearing, is used to transmit motion and power between nonintersecting shafts. Like bevel gears, the input and output shafts for worm gears are typically at right angles. A significant advantage of worm gearing is that very high ratios can be achieved, and worm gearing can be quite smooth and quiet when properly lubricated. An inherent disadvantage of worm gearing is that, because of the rubbing motion between the worm and the worm gear, greater friction is created than with most other forms of gearing. Hence, efficiency is typically lower. The basic dimensions applicable to worm gear systems are shown in Figure 11.27. It can be seen from this figure that, in a single-start worm gear system, the ratio between the worm gear and the worm causes the worm gear to progress one tooth for each revolution of the worm. This is why very high ratios can be achieved. Often a **multiple-start worm** is used if lower ratios are needed. This type of worm can also extend slightly the amount of engagement surface area between the worm and the worm gear. This is a significant concern with worm gears as the force that can be transmitted by a worm gear set is often limited by the surface area of engagement between the two gears. Because there is sliding friction between the gears, the surface area is important. If the lubricated area between the two surfaces is too small, the lubricant tends to be pushed out from between the surfaces and

multiple-start worm
A worm gear that contains two or more concentric thread faces resulting in a progression of a similar number of teeth on the main gear, often resulting in slightly higher operating efficiencies.

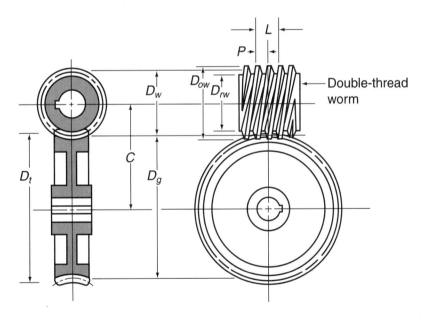

FIGURE 11.27 Worm gear set.

FIGURE 11.28 Worm gear reducer. (Courtesy of Peerless-Winsmith, Inc.)

significant friction increases occur. To minimize this problem (as depicted in Figure 11.27), the face of the worm gear can be made with a concave shape to increase the surface area, or the worm can be created in a convex shape to further increase the surface area.

Typically, the worm and worm gears are sold as a set, especially when concave worm gears and convex worms are used. This is because the geometries in these cases can get quite complicated. Many manufacturers sell worm gear assemblies prepackaged in a case, as shown in Figure 11.28. These prepackaged assemblies can be attached with a coupling to a motor or use a **C-face** or other mounting arrangement for direct motor coupling.

Self-Locking Worm Gear Sets

One of the advantages of most worm gear sets and, specifically, almost all single-start worm gear sets is their ability to be **self-locking.** Self-locking means that, because of the lead angle, the frictional force between the worm threads and the worm gear teeth is high enough that no matter how much torque is applied to the output shaft, the input shaft will not turn. This is often an advantage in cases such as a winch assembly where, once the drive is turned off, locking in position is required. Often this is why worm gear assemblies are selected.

C-face

A term applied to the face of a motor system to which gearboxes or other assemblies can be bolted directly to the end of the motor.

self-locking

A property of certain mechanical elements, often applied to worm gears, when the output shaft will not rotate except in the presence of rotation of the input shaft. Can be used as a built-in brake system.

Mechanical versus Thermal Ratings

Because worm gears that are in continuous operation generate a fair amount of heat, the manufacturer's rating for a worm gear set may be stated in terms of a mechanical versus thermal rating. Sometimes the phrase "continuous versus intermittent duty" is used. The mechanical rating is based on the ability of the gear train to transmit the load without mechanical failure. However, in many cases if this load were to be applied on a continuous basis, the heat generated by the friction would increase the oil temperature to a point that would cause deterioration of the lubricating properties. Therefore, the continuous or thermal rating of a gearbox assembly may be lower than the mechanical rating. It is based on the ability of the outside of the gearbox to dissipate heat. This rating should be used when it is necessary to use a system on a continuous basis at maximum loads. It is interesting to note from Figure 11.29 that different thermal ratings are given for these gearbox assemblies based on the lubrication selected, whereas the gears themselves are identical. This is merely a difference in the maximum temperatures at which the different lubricants can provide adequate service.

11.10 GEAR TRAIN CONFIGURATIONS

The design of a gear system typically involves determining an appropriate ratio between the input and the output speeds. In a single set of gears, this ratio then becomes the ratio between the number of teeth on either gear, the pitch diameter, or the rotational speed as follows:

$$V_r = \frac{N_g}{N_p} = \frac{n_{input}}{n_{output}} = \frac{D_g}{D_p} \qquad (11.9)$$

You can see that there are many different properties you can use to determine this ratio. Often the easiest one to use is simply the number of teeth on each gear because (with the exception of multiple-start worm gear systems) this becomes the direct ratio. In the case of multiple-start worm gear systems, we would divide the number of teeth on the worm gear by the number of starts as well. When we have a series of gears that become a gear train or multiple reduction system, we can determine the overall ratio and each part of the gear train is then a factor of the total ratio. For example, if we wanted a 16:1 ratio in a double-reduction set, each of the individual ratios could be 4:1. If we needed a ratio such as 64:1 and had a triple

1,750 input rpm
Service factor–1.0

Output RPM	Ratio	Series/Size		F726 Lubricant		F732 Lubricant		F732 Fan Cooled Lubricant		F738 Lubricant		F738 Fan Cooled Lubricant	
				Mobil SHC634	Mobil 600W	Mobil SHC634	Mobil 600W	Mobil SHC634	Mobil 600W	Mobil SHC634	Mobil 600W	Mobil SHC634	Mobil 600W
350	5	Input HP		4.95	4.27	4.97	4.97	6.28	5.82	—	—	—	—
		Output	Torque	800	677	1620	1572	2050	1850	—	—	—	—
			HP	4.44	3.76	4.50	4.36	5.69	5.14	—	—	—	—
175	10	Input HP		3.23	2.97	3.59	3.53	4.49	4.41	7.26	7.20	8.65	8.05
		Output	Torque	1050	940	1688	1610	2110	2010	2350	2280	2800	2550
			HP	2.91	2.61	3.12	2.98	3.91	3.72	6.52	6.33	7.77	7.08
116.7	15	Input HP		2.47	2.14	3.37	3.02	4.21	3.86	4.87	4.82	6.08	6.05
		Output	Torque	1150	968	2080	1800	2600	2300	2307	2220	2884	2780
			HP	2.13	1.79	2.89	2.50	3.61	3.19	4.27	4.11	5.34	5.15
87.5	20	Input HP		2.04	1.84	2.81	2.56	3.51	3.25	4.97	4.65	5.92	5.48
		Output	Torque	1250	1086	2110	1850	2640	2350	3100	2800	3686	3300
			HP	1.73	1.51	2.34	2.05	2.93	2.61	4.30	3.89	5.12	4.58
70	25	Input HP		1.75	1.54	2.44	2.26	3.07	2.85	—	—	—	—
		Output	Torque	1290	1100	2140	1900	2700	1400	—	—	—	—
			HP	1.43	1.22	1.98	1.76	2.50	2.22	—	—	—	—
58.3	30	Input HP		1.50	1.36	1.85	1.69	2.32	2.16	3.50	3.40	4.27	3.99
		Output	Torque	1300	1120	2080	1800	2600	2300	3200	2900	3800	3400
			HP	1.20	1.04	1.44	1.25	1.81	1.50	2.96	2.68	3.51	3.14
43.8	40	Input HP		1.13	1.03	1.57	1.46	1.96	1.86	2.76	2.63	3.28	3.10
		Output	Torque	1250	1086	2080	1800	2600	2300	3100	2800	3686	3300
			HP	.87	.75	1.15	1.00	1.44	1.27	2.15	1.94	2.56	2.29
35	50	Input HP		.95	.88	1.38	1.28	1.73	1.65	2.21	2.24	2.76	2.65
		Output	Torque	1250	1075	1995	1750	2500	2250	2960	2800	3700	3320
			HP	.69	.60	.92	.81	1.16	1.04	1.64	1.55	2.05	1.84
29.2	60	Input HP		.83	.77	—	—	—	—	1.97	1.89	2.42	2.36
		Output	Torque	1182	1032	—	—	—	—	2900	2620	3552	3270
			HP	.55	.48	—	—	—	—	1.34	1.21	1.64	1.51
Overhung Load (No Thrust)				1000		1300		1300		2000		2000	
Output Shaft Thrust Load				900		1100		1100		1300		1300	

Maximum Torque (Lb. Ins.)
Overhung load is at centerline of output shaft projection, and with no thrust.

FIGURE 11.29 Thermal capacity of worm gear reducer based on lubricant type. (Courtesy of Boston Gear)

reduction, each individual set could be 4:1. The relationship between teeth on each mating set of gears would also then be 4:1. For instance, if we had a 12-tooth gear, it would mate with a 48-tooth gear. When we are designing gear trains, it is often easiest to develop the overall ratio, estimate how many sets of gears are needed to accomplish this, and then try to determine the factors. Note that it is not necessary for these to be whole number factors, but they must be exact factors. We can then summarize an overall gear train value as follows:

$$V_r = \frac{N_{tg1}}{N_{tp1}} \cdot \frac{N_{tg2}}{N_{tp2}} \cdot \text{etc.}$$

For example, Figure 11.30 reflects a double-reduction gear train system with an overall ratio of 12:1. Note that the input speed has this same relationship to the output and, assuming that the gear train is 100 percent efficient, the torque input would be multiplied by the overall train value to obtain the output torque. Note that power in equals power out. The following example problem shows this relationship.

Example Problem 11.5

The gear train shown in Figure 11.30 is driven by a 1½-hp induction motor at 1,750 rpm. For this gear train, determine gear train value, input and output torque, and speed. If gear A rotates clockwise, determine gear D direction.

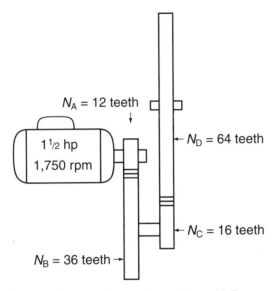

FIGURE 11.30 Example Problem 11.5.

Gear train value:

$$V_r = \frac{N_B}{N_A} \cdot \frac{N_D}{N_C} \quad (11.1)$$

$$V_r = \frac{36}{12} \cdot \frac{64}{16} = 12:1 \text{ ratio}$$

Motor torque:

$$T = \frac{63,000 \text{ hp}}{n} \qquad (2.6)$$

$$T = \frac{63,000 \, (1.5)}{1,750}$$

$$T = 54 \text{ in.-lb}$$

Output torque:

$$T_{out} = T_{in} \, V_r$$

$$T_{out} = 54 \text{ in.-lb} \, \frac{12}{1} = 648 \text{ in.-lb}$$

Output speed:

$$n_{out} = \frac{n_{in}}{V_r}$$

$$n_{out} = \frac{1,750 \text{ rpm}}{\frac{12}{1}}$$

$$n_{out} = 145.8 \text{ rpm}$$

Directions: Gear A is clockwise.

Gear B is counterclockwise.

Gear C is counterclockwise.

Gear D is clockwise.

As a check of the output torque and speed, verify power.

$$\text{hp} = \frac{Tn}{63,000} = \frac{648 \text{ in.-lb} \; 145.8 \text{ rpm}}{63,000}$$

$$\text{hp} = 1½ \text{ OK}$$

Designing gear trains can best be done after a great deal of practice and experience. The following checklist will aid in determining gear trains.

1. Determine the overall ratio between the input and the output.
2. Determine whether the output direction is important.
3. Determine whether there are any angles that need to be turned (that is, do the output and input shafts need to be parallel or perpendicular to each other?).
4. Remember that each set of gears needs to have the same diametral pitch or metric module, but that different sets can use a different pitch and in many cases do, based on changing speeds and torque.
5. If the output direction is not appropriate, an idler may be needed. Remember that an idler has no effect on the magnitude of the train value but merely changes direction.

The following example problem illustrates these principles.

Example Problem 11.6

If an output speed of 20 rpm is needed and the input is 2,400 rpm from the gears shown in Figure 11.31, devise a gear train to accomplish this.

Overall velocity ratio:

$$V_r = \frac{n_{in}}{n_{out}} \quad (11.9)$$

$$V_r = \frac{2,400}{20} = \frac{120}{1}$$

Consider factors of 120 if two gear sets, 10 and 12, are factors. This ratio of gears is not available from Figure 11.31. Try three sets of gears.

If 4, 5, and 6 are factors:

- 20 tooth and 80 tooth result in 4:1
- 20 tooth and 100 tooth result in 5:1
- 20 tooth and 120 tooth result in 6:1

This is a possible solution. Note many others are also possible.

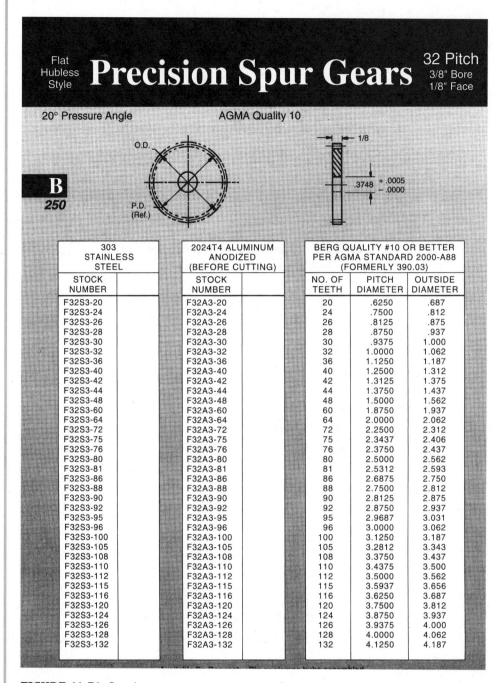

FIGURE 11.31 Stock precision spur gears. (Courtesy of W.M. Berg, Inc.)

11.11 SUMMARY

You should now be familiar with the different types of gearing, the basic principles and dimensional properties of gearing, and when different types of gears may be appropriate. Some of the attributes and their advantages and disadvantages of the many types of gears available should now be recognized.

You should also understand the principles of devising gear trains, how multiple ratios can be calculated, and how to determine the different gears that may be needed in order to obtain the appropriate speed and other design conditions.

11.12 PROBLEMS

1. If a gear has 48 teeth, a 14¼° involute profile, and is a 16 pitch, find the following properties:
   a. Pitch diameter
   b. Circular pitch
   c. Equivalent module
   d. Closest standard module
   e. The center-to-center distance if mated with an 18-tooth gear of the same pitch
2. A 12-pitch pinion with 15 teeth is mated with a 28-tooth gear. Find the following if the pinion rotates at 2,400 rpm and transmits 2 hp:
   a. Pitch diameter of both gears
   b. Circular pitch
   c. Center-to-center distance
   d. Gear ratio
   e. Speed of gear
   f. Torque of both pinion and gear

For Problems 3 through 7, determine gear ratio, output speed, direction of rotation, and output torque for the following input conditions. Check your work by calculating output power and comparing it to input.

3. In Figure 11.32, use the following parameters:
   P_{in} = 3 hp
   n = 1,725 rpm clockwise
   N_A = 20
   N_B = 46

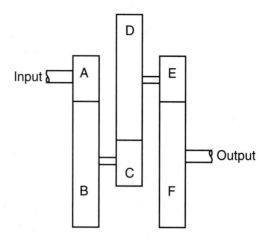

FIGURE 11.32 Gear train for Problems 3 and 4.

$N_C = 16$
$N_D = 48$
$N_E = 20$
$N_F = 50$

4. In Figure 11.32, use the following parameters:
$P_{in} = 7$ kw
$n = 1,450$ rpm counterclockwise
$N_A = 18$
$N_B = 40$
$N_C = 18$
$N_D = 40$
$N_E = 22$
$N_F = 40$

5. In Figure 11.33, use the following parameters:
$P_{in} = 5$ hp
$n = 3,600$ rpm clockwise
$N_A = 15$
$N_B = 20$
$N_C = 48$
$N_D = 20$
$N_E = 60$

In this problem, if the gears have a diametral pitch of 12, find the centerline distance from gear A to gear C.

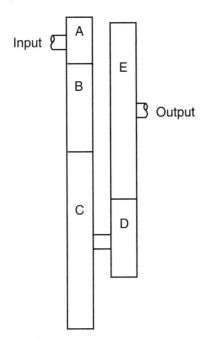

FIGURE 11.33 Gear train for Problem 5.

6. In Figure 11.34, use the following parameters:
   $P_{in} = 2$ hp
   $n = 1,080$ rpm clockwise
   $N_A = 16$
   $N_B = 48$
   $N_C = 20$
   $N_D = 60$
   $N_E = 20$
   $N_F = 80$

7. Figure 11.35 is a double-reduction worm gear system. Both worms are single-start.
   $P_{in} = 5$ hp
   $n = 3,600$ rpm clockwise
   $N_B = 80$
   $N_D = 75$

8. Devise a gear train using spur gears to reduce a motor output speed from 1,750 rpm to 87.5 rpm. Select gears from Figure 11.31.

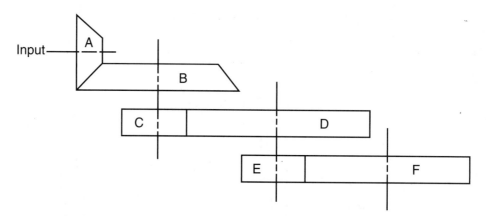

FIGURE 11.34 Gear train for Problem 6.

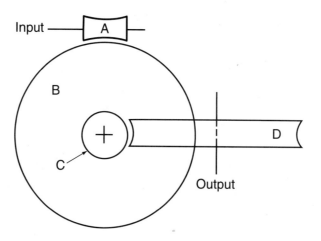

FIGURE 11.35 Gear train for Problem 7.

9. Design a gear train with the input shaft 90° from the output shaft, an input speed of 900 rpm and 20 hp, and a desired output speed of 10 rpm. What is the output torque?
10. Using gears from Figure 11.31, design a gear train to reduce an input speed of 3,450 rpm to 50 rpm.

11.13 CUMULATIVE PROBLEMS

11. For the truck that has been designed in previous chapters, the steering system uses a rack-and-pinion gear system. If, for a movement of 12 inches in the rack, the steering wheel is to turn 360°, decide upon a diametral pitch and pitch diameter.
12. In Problem 11, if the force on the rack is 200 pounds and the steering wheel is 18 inches in diameter, find the torque on the wheel and the force the driver needs to apply to the steering wheel.
13. The transmission for this truck uses 12-pitch gears. If the following ratios are needed, determine the approximate number of teeth in each gear. The input pinion has 36 teeth.
    First gear 3.5:1
    Second gear 2.8:1
    Third gear 1.9:1
    Fourth gear 1:1
14. If the maximum input torque is 450 foot-pounds at 2,800 rpm in Problem 13, find the output torque and speed in each gear.
15. For Problems 13 and 14, sketch a possible configuration for this transmission.

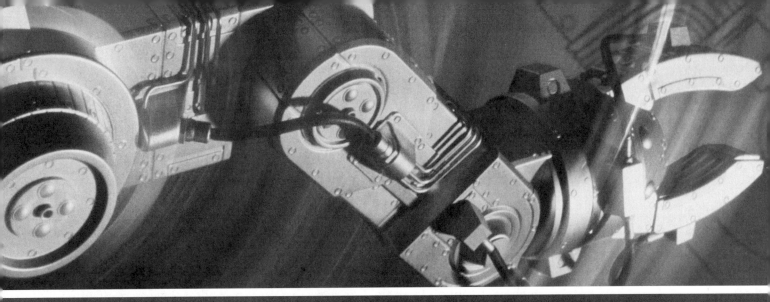

CHAPTER 12

Spur Gear Design and Selection

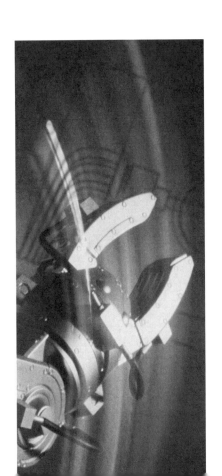

Chapter 12

The physical properties of spur gears were discussed in Chapter 11, including such aspects as pitch, alignment, and centerline spacing. In this chapter, we will learn how to actually design the spur gear systems, including the analysis of how the allowable load on a spur gear system is determined, the load on the teeth, and how to factor in the impact and fatigue effects on the gears. The load on the teeth of spur gears is analyzed like a cantilevered beam that is loaded repeatedly but not reversed. This load acts as a fatigue loading if the gears are used for an extended period. The clearances or backlash inherent in gearing systems also creates an impact load on the teeth. This impact load depends on how accurately the gears are made, their placement, and the pitch or size of the teeth, as well as the involute angle. It is necessary to factor all of these aspects into the design process.

In addition to the loading on the gear teeth and the factors added for clearances, impact, and fatigue, we also need to be concerned with wear on the teeth. Even if the gear teeth are strong enough to carry a repeated load, wear can be a consideration if the surfaces of the teeth are not hard enough and properly lubricated to resist these effects.

This chapter will also briefly review the selection of standard gears from manufacturers' catalogs because that is often part of the design and/or usage of gears.

12.1 OBJECTIVES

After completing this chapter, you should be able to:

- Apply the principles learned in Chapter 11 to the actual design and selection of spur gear systems.
- Calculate the forces on the teeth of spur gears, including the impact forces associated with velocity and clearances.
- Determine the allowable force on the gear teeth, including the factors necessary due to the angle of the involute of the tooth shape and the materials selected for the gears.
- Design actual gear systems, including specifying materials, manufacturing accuracy, and other factors necessary for a complete spur gear design.
- Understand and determine the necessary surface hardness of the gears in order to minimize or prevent surface wear.
- Understand how lubrication can both cushion the impact on gearing systems and cool them.
- Select standard gears available from stocking manufacturers or distributors.

12.2 FORCES ON SPUR GEAR TEETH

As power is transmitted between mating gears, there are equal and opposite forces between the mating teeth. However, these forces are not tangent to the intersection point of the two gears because the gear teeth have an involute shape. From Chapter 11, we learned this was necessary in order to prevent sliding between the surfaces of the teeth when they engage. This involute shape of the teeth results in a force that is normal to the involute shape at the point of contact. The angle of this point of contact is the angle of involute from the tangent point of the gear mating surfaces. Figure 12.1 shows this resultant force (F_r) between the mating gears at the involute angle from the tangent to the pitch circle at the mating point and a third force labeled F_n. Because the resultant force is at a different angle from the **transmitted force** (F_t), there is a separating force trying to push the two gears apart. The relationship between these forces is as follows, and the separating force would be equal to

$$F_r = F_t \tan \theta \qquad (12.1)$$

The resultant force, which is also the maximum force, can be found from

$$F_r = \frac{F_t}{\cos \theta} \qquad (12.2)$$

where F_t = transmitted force, which would be tangent to the pitch circle
F_n = separating force
F_r = resultant force (also the force on the mounting shafts)
θ = pressure angle of the tooth form

> **transmitted force**
> A force that would act tangent to the running circle for gears, sprockets, or pulleys. This is the value that would be used in the power analysis.

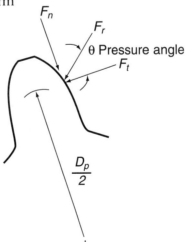

FIGURE 12.1 Forces acting on a gear tooth.

Remember, from your study of physics, that these forces are the same for both the pinion and the mating gear as each force needs to have an equal and opposite resisting force in order to be balanced. We can determine the force transmitted (F_t) from the relationship of torque and power (from Chapter 2) as follows:

$$P = \frac{Tn}{63,000} \quad (2.6)$$

Therefore

$$T = \frac{63,000P}{n}$$

The tangential force, then, would therefore be equal to

$$F = \frac{T}{d} \quad (2.5)$$

$$d = \frac{D_p}{2}$$

$$F_t = \frac{2T}{D_p} \quad (12.3)$$

The following example problem illustrates this.

Example Problem 12.1

A 20-tooth, 8 pitch, 1-inch-wide, 20° pinion transmits 5 hp at 1,725 rpm to a 60-tooth gear. Determine the driving force, separating force, and maximum force that would act on the mounting shafts.

$$P = \frac{Tn}{63,000} \quad (2.6)$$

$$T = \frac{63,000P}{n}$$

$$T = \frac{(63,000)5}{1,725} = 183 \text{ in.-lb}$$

Transmitted force (12.3)

$$F_t = \frac{2T}{D_p}$$

Find pitch circle.

$$D_p = \frac{N_p}{P_d} \qquad (11.4)$$

$$D_p = \frac{20 \text{ teeth}}{8 \text{ teeth/in. diameter}} = 2.5 \text{ in.}$$

Find transmitted force.

$$F_t = \frac{2T}{D_p} \qquad (12.3)$$

$$F_t = \frac{(2)183 \text{ in.-lb}}{2.5 \text{ in.}} = 146 \text{ lb}$$

Find separating force.

$$F_n = F_t \tan \theta \qquad (12.1)$$
$$F_n = 146 \text{ lb} \tan 20°$$
$$F_n = 53 \text{ lb}$$

Find maximum force.

$$F_r = \frac{F_t}{\cos \theta} \qquad (12.2)$$
$$F_r = \frac{146 \text{ lb}}{\cos 20°}$$
$$F_r = 155 \text{ lb}$$

Surface Speed

As the forces between mating gears are always equal and opposite, the surface speed (V_m) of the mating gears is also the same and, therefore, it is useful for gears to use this quantity for defining speed. This surface speed is often referred to as the **pitch-line speed.** It can be found from the relationship

$$V_m = \frac{D_p}{2} \cdot \omega \qquad (12.4)$$

where D_p = the pitch diameter
ω = the angular velocity of the gear

pitch-line speed

The same value as surface speed or surface velocity. This would be the speed at which the gear teeth come into engagement or the rotational speed for a wheel or belt speed for a belt system.

To convert this to a more useable form, we can use quantities that are typically known, such as speed in revolutions per minute, and we can convert the pitch-line speed to feet per minute. We can thus modify this relationship as follows:

$$V_m = \frac{D_p}{2} \cdot \omega = \frac{D_p \text{ in.}}{2} \cdot \frac{n \text{ rev}}{\text{min}} \cdot \frac{2\pi \text{ rad}}{\text{rev}} \cdot \frac{1 \text{ ft}}{12 \text{ in.}}$$

$$V_m = \frac{\pi D_p n}{12} \text{ ft/min} \tag{12.5}$$

From Chapter 11, we can determine the pitch diameter from the relationship

$$P_d = \frac{N_g}{D_g} = \frac{N_p}{D_p} \tag{11.4}$$

or

$$D_p = \frac{N_p}{P_d} = \frac{N_g}{P_d}$$

As using the surface speed in feet per minute is often convenient, it is useful to be able to determine our transmitted force (F_t) from this relationship as follows:

Transmitted force (12.6)

$$F_t = \frac{33,000 \text{ hp}}{V_m} \tag{12.6}$$

Note in this relationship that V_m should be in feet per minute. V_m can be determined from the relationship

$$V_m = \pi D n \tag{12.7}$$

where D is in feet.

  n is in revolutions per minute (rpm).

Example Problem 12.2

In Example Problem 12.1, determine the surface speed:

$$V_m = \pi D n \tag{12.7}$$

or

$$V_m = \frac{\pi D_p n}{12} \qquad (12.5)$$

$$V_m = \pi\ 2.5\text{ in.}\ 1{,}725\text{ rpm}\ \frac{\text{ft}}{12\text{ in.}}$$

$$V_m = 1{,}129\text{ ft/min}$$

12.3 STRENGTH OF GEAR TEETH

There are numerous techniques for determining the appropriate strength or force allowable in gearing systems. All yield somewhat similar results. In this text, we will use a method that uses the **Lewis form factor** for determining an appropriate force that can be applied to the gear teeth. The strength of gear teeth depends on many factors, the first of which is the physical shape and size of the teeth. Figure 12.2 shows different involute shapes of teeth. It can be noted from these that, as the angle of involute increases, the base of the tooth gets wider and typically results in a stronger tooth. The width of the tooth has a significant effect, as does the material from which the teeth are made. Certainly a high-strength steel tooth will have a significantly higher strength than a cast iron or aluminum tooth. The diametral pitch (P_d) is also critical to the strength of the teeth as a lower–pitch number tooth will be larger and thus will have a greater cross-sectional area. The number of teeth in the gear will also be relevant. Therefore, we can determine an appropriate allowable force for gear teeth as follows:

$$F_s = \frac{SbY}{P_d} \qquad (12.8)$$

> **Surface speed** (12.5)
>
> $$V_m = \frac{\pi D_p n}{12}$$
>
> **Lewis form factor**
> A factor used in the design of gears to compensate for the sharpness of the inner root of the teeth, the shape of the teeth, and other factors.

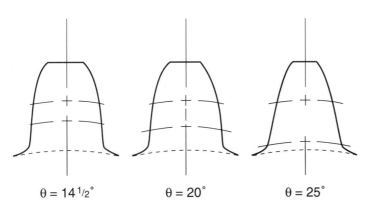

FIGURE 12.2 Standard gear tooth shapes.

However, because gears are almost always subjected to fatigue loading, we would use the endurance limit value for the stress allowable. Thus, the Lewis equation would be modified to be

$$F_s = \frac{S_n bY}{P_d} \tag{12.9}$$

Force capability (12.9)

$$F_s = \frac{S_n bY}{P_d}$$

where S_n = endurance limit for the material
 b = width of tooth
 Y = modified Lewis form factor (Table 12.1)
 P_d = diametral pitch

In some gear design methods, a strength reduction factor is added to this equation, along with a value of the Lewis form factor that assumes the load is applied at the tooth center. However, the technique used in this chapter uses a modified Lewis form factor that assumes the load is applied at the tip of the tooth, which negates the need for the strength reduction factor.

Table 12.1 gives values of the Lewis form factor.

Example Problem 12.3

In Example Problem 12.1, determine the force allowable (F_s) on these teeth if the pinion is made from an AISI 4140 annealed steel, the mating gear is made from AISI 1137 hot-rolled steel, and long life is desired.

Pinion

$$S_n = .5 S_u = .5\,(95 \text{ ksi}) = 47.5 \text{ ksi}$$

$$F_s = \frac{S_n bY}{P_d} \tag{12.9}$$

Find the Lewis form factor (Y) from Table 12.1, assuming full-depth teeth.

$$Y = .320$$

$$F_s = \frac{47,500\,(1)\,.320}{8}$$

$$F_s = 1,900 \text{ lb}$$

TABLE 12.1
TOOTH FORM FACTOR (Y) (LOAD APPLIED AT TOOTH TIP)

Number of Teeth	14½° Full Depth Involute	20° Full Depth Involute
10	.176	.201
11	.192	.226
12	.210	.245
13	.223	.264
14	.236	.276
15	.245	.289
16	.255	.295
17	.264	.302
18	.270	.308
19	.277	.314
20	.283	.320
22	.292	.330
24	.302	.337
26	.308	.344
28	.314	.352
30	.318	.358
32	.322	.364
34	.325	.370
36	.329	.377
38	.332	.383
40	.336	.389
45	.340	.399
50	.346	.408
55	.352	.415
60	.355	.421
65	.358	.425
70	.360	.429
75	.361	.433
80	.363	.436
90	.366	.442
100	.368	.446
150	.375	.458
200	.378	.463
300	.382	.471
Rack	.390	.484

Gear:
$$S_n = .5\,(88\text{ ksi}) = 44\text{ ksi}$$
$$Y = .421$$
$$F_s = \frac{44{,}000\,(1)\,.421}{8}$$
$$F_s = 2{,}316\text{ lb}$$

(Table 12.1)

Use $F_s = 1{,}900$ lb for design purposes.

12.4 CLASSES OF GEARS AND MANUFACTURING METHODS

There are many methods of manufacturing spur gears, ranging from simple sheet-metal stampings for very low-load applications such as inexpensive clocks to plastic injection-molded gears and powder metal gears. For manufactured metallic gears under typical loading conditions, we will consider four basic types: precision, carefully cut, commercial, and hobbed or shaved gears. We place gears in these different categories because the differences in the manufacturing methods, yield different levels of precision. The differences in the accuracy of manufacturing have a significant effect on the amount of clearance or backlash in a gear set and, correspondingly, on the level of impact loading on the gear teeth. Many distributors of gears may carry more than one grade, depending on the application. In many applications, a precision gear may actually be less expensive than a less accurate class because, when the accuracy is greater, the lower clearance and correspondingly lower impact load may enable a smaller gear or gear width to be specified. You will see the difference in the loading factors based on these different accuracies of manufacturing in Section 12.5.

12.5 FORCE TRANSMITTED AND DYNAMIC LOADS

dynamic load factor
A value based on the surface speed of gears and the manufacturing accuracies; used to compensate for the clearances or impact loads inherent in gear systems.

As was discussed in Section 12.4, the effect of the transmitted load will vary depending on the accuracy of the gears. The associated clearance and/or backlash and the speed at which the gears operate determine the impact load. To compensate for these factors, a **dynamic load factor** is determined that incorporates manufacturing accuracy, velocity of impact, and the force transmitted from Section 12.2. The factors for the four basic types of gears are

$$F_d = \frac{600 + V_m}{600} F_t \quad \text{Commercial} \quad (12.10)$$

$$F_d = \frac{1,200 + V_m}{1,200} F_t \quad \text{Carefully cut} \quad (12.11)$$

$$F_d = \frac{78 + V_m^{\frac{1}{2}}}{78} F_t \quad \text{Precision} \quad (12.12)$$

$$F_d = \frac{50 + V_m^{\frac{1}{2}}}{50} F_t \quad \text{Hobbed or shaved} \quad (12.13)$$

Commercial gear dynamic force (12.10)

$$F_d = \frac{600 + V_m}{600} F_t$$

Carefully cut gear dynamic force (12.11)

$$F_d = \frac{1,200 + V_m}{1,200} F_t$$

Precision gear dynamic force (12.12)

$$F_d = \frac{78 + V_m^{\frac{1}{2}}}{78} F_t$$

Hobbed or shaved gear dynamic force (12.13)

$$F_d = \frac{50 + V_m^{\frac{1}{2}}}{50} F_t$$

12.6 DESIGN METHODS

The design of gears typically involves an iterative process because there are many interlinked variables. Material, pitch, pitch diameter, class of gears, and width of the gear face can all be varied depending on the design constraints. Generally the easiest technique is to select a possible solution and verify that the selection is acceptable. You could also leave one variable out, such as pitch or width, and solve for that quantity. The basic technique involves setting the strength of the gear teeth equal to or greater than the dynamic force as follows:

$$F_s \geq F_d$$

At this point, a factor of safety would usually be incorporated, so this formula would become

$$\frac{F_s}{N_{sf}} \geq F_d$$

For gearing, however, instead of a conventional safety factor we often use a **service factor** determined by the type of loading and the potential for shock, especially when selecting standard gears from manufacturers. These guidelines are also very useful in determining factors that may be used in the design of gears. Table 12.2 lists service factors that are often used. Note, however, that unknowns in the design with regard to level of loading or unexpected occurrence may require the use of even larger safety or service factors.

service factor

Similar to a factor of safety but based on specific usage of systems, such as whether the loads would be smoothly applied or other factors inherent in the end usage of gears or other products.

TABLE 12.2
FREQUENTLY USED SERVICE FACTORS

$1 < N_{sf} < 1.25$	Uniform load without shock; centrifugal machines, hoisting machinery, belt-driven machine tools, textile machinery, smoothly running conveyors.
$1.25 < N_{sf} < 1.5$	Medium shock; frequent starts; reciprocating compressors and pumps, pneumatic tools, well-drilling machinery, portable electric tools, heavy-duty conveyors, machine tools.
$1.5 < N_{sf} < 1.75$	Moderately heavy shock; dredging machinery, road machinery, railway motor cars, single-cylinder compressors, ore or stone crushers.
$1.75 < N_{sf} < 2$	Heavy shock; rolling mills, rock crushers.

Example Problem 12.4

If in Example Problem 12.1, the gears are commercial grade, determine the dynamic load. Based on the allowable force from Example Problem 12.3, is this an acceptable design if a factor of safety of 2 is desired? (Use the surface speed and force transmitted from Example Problems 12.2 and 12.3.)

Dynamic load:

$$F_d = \frac{600 + V_m}{600} F_t \qquad (12.10)$$

$$F_d = \frac{600 + 1{,}129}{600} (146 \text{ lb})$$

$$F_d = 421 \text{ lb}$$

Comparing to the allowable force:

$$\frac{F_s}{N_{sf}} \geq F_d$$

$$\frac{1{,}900 \text{ lb}}{2} \geq 421$$

$$905 \text{ lb} > 421 \text{ lb}$$

This design meets the criteria.

Also to be considered in the design of gears is the relationship between the width of the teeth and the pitch of the gears. As a general rule of thumb, wider gear teeth with a low pitch may not yield satisfactory service due to the difficulty of accurate alignment. Generally, the face width of gears should follow the guideline

$$\frac{8}{P_d} < b < \frac{12.5}{P_d} \tag{12.14}$$

It should be noted that this is a general guideline, not an absolute, but you should follow it unless there is a compelling reason to do otherwise. Example Problems 12.5 and 12.6 illustrate the basic technique of gear design.

Example Problem 12.5

The spur gears from the catalog page shown in Figure 12.3 are made from a .2 percent carbon steel with no special heat treatment. What factor of safety do they appear to use in this catalog? (Try a 24-tooth at 1,800 rpm gear for example purposes.) From Appendix 4, an AISI 1020 hot-rolled steel would have .2 percent carbon with an $S_y = 30$ ksi and $S_u = 55$ ksi.

Therefore

$$S_n = .5\, S_u$$
$$S_n = .5\,(55 \text{ ksi}) = 27.5 \text{ ksi}$$

Find D_p.

$$D_p = \frac{N_p}{P_d} \tag{11.4}$$

$$D_p = \frac{24}{12} = 2 \text{ in.}$$

Find V_m.

$$V_m = \frac{\pi D_p n}{12} \tag{12.5}$$

$$V_m = \pi\, 2 \text{ in.}\,(1{,}800 \text{ rpm})\,\frac{\text{ft}}{12 \text{ in.}}$$

$$V_m = 942 \text{ ft/min}$$

SPUR GEARS

APPROXIMATE HORSEPOWER AND TORQUE* RATINGS FOR CLASS I SERVICE (Service Factor = 1.0)

12 DIAMETRAL PITCH STEEL **14½° PRESSURE ANGLE** **3/4" FACE**

No. Teeth	25 RPM H.P.	Torque	50 RPM H.P.	Torque	100 RPM H.P.	Torque	200 RPM H.P.	Torque	300 RPM H.P.	Torque	600 RPM H.P.	Torque	900 RPM H.P.	Torque	1200 RPM H.P.	Torque	1800 RPM H.P.	Torque	3600 RPM H.P.	Torque
11	.05	119	.09	118	.18	116	.35	112	.51	108	.93	97	1.27	88.6	1.55	81.5	2.00	70	2.82	49.4
12	.05	130	.10	129	.20	126	.38	120	.55	116	.99	104	1.35	94.4	1.64	86.3	2.10	73.6	2.92	51.1
13	.06	149	.12	147	.23	144	.44	138	.63	132	1.12	118	1.51	106	1.83	96.3	2.33	81.6	3.19	55.9
14	.07	169	.13	167	.26	163	.49	156	.71	149	1.25	131	1.68	118	2.03	107	2.56	89.6	3.46	60.6
15	.07	189	.15	186	.29	181	.55	172	.78	164	1.37	144	1.83	128	2.20	116	2.76	96.6	3.69	64.6
16	.08	209	.16	206	.32	200	.60	190	.86	180	1.50	157	1.99	139	2.38	125	2.96	103	3.91	68.5
18	.10	249	.19	245	.38	237	.71	224	1.01	242	1.73	182	2.28	159	2.70	142	3.32	116	4.31	75.4
20	.11	289	.23	284	.44	274	.82	257	1.15	215	1.95	205	2.54	178	2.99	157	3.64	127	4.65	81.4
21	.12	310	.24	304	.47	293	.87	274	1.22	257	2.06	216	2.67	187	3.14	164	3.80	133	4.81	84.3
22	.13	328	.26	322	.49	310	.92	288	1.28	270	2.15	226	2.78	195	3.25	171	3.92	137	4.93	86.2
24	.14	365	.28	357	.54	343	1.01	317	1.41	296	2.33	245	2.98	209	3.47	182	4.14	145	5.14	90
30	.19	483	.37	470	.71	447	1.29	407	1.78	373	2.85	299	3.57	250	4.09	215	4.78	167		
32	.21	529	.41	514	.77	488	1.40	442	1.92	403	3.05	320	3.80	266	4.32	227	5.02	176		
36	.24	605	.46	586	.88	552	1.57	495	2.13	448	3.33	349	4.09	286	4.62	243	5.31	186		
40	.27	682	.52	659	.98	617	1.74	547	2.34	492	3.59	377	4.37	306	4.90	257				
42	.28	715	.55	689	1.02	644	1.80	568	2.42	509	3.69	387	4.46	312	4.99	262				

12 DIAMETRAL PITCH CAST IRON **14½° PRESSURE ANGLE** **3/4" FACE**

No. Teeth	25 RPM H.P.	Torque	50 RPM H.P.	Torque	100 RPM H.P.	Torque	200 RPM H.P.	Torque	300 RPM H.P.	Torque	600 RPM H.P.	Torque	900 RPM H.P.	Torque	1200 RPM H.P.	Torque	1800 RPM H.P.	Torque	3600 RPM H.P.	Torque
48	.20	497	.38	477	.70	441	1.22	384	1.62	340	2.41	253	2.88	201	3.19	167				
54	.23	571	.43	545	.79	500	1.36	430	1.79	377	2.62	275	3.09	216	3.40	178				
60	.25	631	.48	600	.87	546	1.47	463	1.91	402	2.74	288	3.21	224						
64	.27	683	.51	647	.93	586	1.56	493	2.02	425	2.87	301	3.33	233						
72	.30	762	.57	718	1.02	644	1.69	533	2.17	455	3.01	316	3.46	242						
84	.36	896	.66	837	1.17	739	1.90	598	2.40	503	3.24	340								
96	.40	1014	.74	938	1.30	817	2.06	649	2.56	538	3.39	356								
108	.46	1148	.84	1054	1.44	906	2.24	706	2.76	579	3.58	376								
112	.47	1187	.86	1087	1.47	929	2.29	720	2.80	588	3.62	379								
120	.50	1264	.91	1150	1.55	975	2.37	748	2.89	607	3.69	387								
144	.59	1487	1.06	1333	1.75	1103	2.61	821	3.11	654	3.86	406								
168	.69	1745	1.22	1541	1.98	1248	2.87	905	3.38	710										

12 DIAMETRAL PITCH NON-METALLIC **14½° PRESSURE ANGLE** **3/4" FACE**

No. Teeth	25 RPM H.P.	Torque	50 RPM H.P.	Torque	100 RPM H.P.	Torque	200 RPM H.P.	Torque	300 RPM H.P.	Torque	600 RPM H.P.	Torque	900 RPM H.P.	Torque	1200 RPM H.P.	Torque	1800 RPM H.P.	Torque	3600 RPM H.P.	Torque
15							.13	41	.20	42	.34	36	.45	31	.56	29	.72	25	1.15	20
18							.19	60	.26	54	.43	45	.56	39	.68	36	.90	31	1.44	25
21							.23	72	.31	65	.50	52	.66	46	.81	42	1.06	37	1.80	31
24							.27	85	.36	76	.58	60	.76	53	.92	48	1.22	42	2.07	36
30							.34	107	.44	92	.73	76	.95	66	1.14	60	1.56	54	2.64	46
36							.40	126	.52	109	.82	86	1.13	79	1.36	71	1.83	64	3.18	56
48							.51	161	.66	138	1.05	110	1.39	97	1.70	89	2.33	81		
60							.60	189	.80	168	1.29	135	1.72	120	2.13	112	2.95	103		

Ratings are based on strength calculation. Basic static strength rating, or for hand operation of above gears is approximately 3 times the 100 RPM rating.

NOTE: 1. Ratings to right of heavy line are not recommended, as pitch line velocity exceeds 1000 feet per minute. They should be used for interpolation purposes only.

2. Non-metallic gears are most commonly used for the driving pinion of a pair of gears, with mating gear made of Cast Iron or Steel, where pitch line velocities exceed 1000 FPM and are not subjected to shock loads.

*Torque Ratings (Lb. Ins.).

FIGURE 12.3 Stock spur gears. (Courtesy of Boston Gear)

Find F_s.

$$F_s = \frac{S_n bY}{P_d} \quad (12.9)$$

$Y = .302$ **(from Table 12.1)**

$$F_s = \frac{27{,}500\,(\tfrac{3}{4})\,.302}{12}$$

$F_s = 519$ lb

Set $F_s = F_d$ and solve for F_t.

$$F_d = F_s = \left(\frac{600 + V_m}{600}\right) F_t$$

$$519 \text{ lb} = \left(\frac{600 + 942}{600}\right) F_t$$

$F_t = 202$ lb

$$T = F\left(\frac{D_p}{2}\right) \quad (12.3)$$

$$T = 202 \text{ lb}\left(\frac{2 \text{ in.}}{2}\right)$$

$T = 202$ in.-lb

$$P = \frac{Tn}{63{,}000} = \frac{202\,(1{,}800)}{63{,}000} = 5.8 \text{ hp} \quad (2.6)$$

or

$$P = \frac{F_t V_m}{33{,}000} = \frac{202\,(942)}{33{,}000} = 5.8 \text{ hp} \quad (12.6)$$

Compared to catalog:

$$N_{sf} = \frac{\text{hp-calculated}}{\text{hp-catalog}}$$

$$N_{sf} = \frac{5.8}{4.14} = 1.4$$

This appears to be a reasonable value. The manufacturer may also have reduced its rating for wear purposes as these are not hardened gears.

Example Problem 12.6

A pair of commercial-grade spur gears is to transmit 2 hp at a speed of 900 rpm for the pinion and 300 rpm for the gear. If class 30 cast iron is to be used, specify a possible design for this problem.

Note that the following variables are unknown: P_d, D_p, b, N_t, N_{sf}, θ. As it is impossible to solve for all simultaneously, try the following: $P_d = 12$, $N_t = 48$, $\theta = 14\tfrac{1}{2}°$, $N_{sf} = 2$, and solve for face width b.

Miscellaneous properties:

$$D_p = \frac{N_p}{P_d} = \frac{48}{12} = 4 \text{ in.} \tag{11.4}$$

$$N_g = N_p \, V_r = 48(3) = 144 \text{ teeth}$$

Surface speed:

$$V_m = \frac{\pi D_p n}{12} \tag{12.5}$$

$$V_m = \pi \frac{4 \text{ in. } 900 \text{ rpm}}{12 \text{ in./ft}}$$

$$V_m = 943 \text{ ft/min}$$

Finding force on teeth:

$$F_t = \frac{33,000 \text{ hp}}{V_m} \tag{12.6}$$

$$F_t = \frac{33,000 \, (2)}{943}$$

$$F_t = 70 \text{ lb}$$

Dynamic force:

$$F_d = \left(\frac{600 + V_m}{600}\right) F_t \tag{12.10}$$

$$F_d = \left(\frac{600 + 943}{600}\right) 70$$

$$F_d = 180 \text{ lb}$$

Since width b is the unknown,

$$\frac{F_s}{N_{sf}} \geq F_d$$

and

$$F_s = \frac{S_n\, b\, Y}{P_d} \tag{12.8}$$

$$\frac{S_n\, b\, Y}{N_{sf}\, P_d} \geq F_d$$

Class 30 CI; $S_u = 30$ ksi; $S_n = .4\, S_u$ (.4 is used because gear is cast iron).

$$S_n = 12 \text{ ksi}$$
$$Y = .344 \qquad \text{(Table 12.1)}$$

Substituting:

$$\frac{S_n\, b\, Y}{N_{sf}\, P_d} = F_d \tag{12.9}$$

$$\frac{12{,}000\, b\, .344}{2\,(12)} = 180$$

$$b = 1.0 \text{ inch}$$

Check ratio of width to pitch.

$$\frac{8}{P_d} < b < \frac{12.5}{P_d} \tag{12.14}$$

$$\frac{8}{12} < 1 < \frac{12.5}{12}$$

$$.66 < 1 < 1.04$$

This is an acceptable design. Note that many other designs are also possible.

12.7 BUCKINGHAM METHOD OF GEAR DESIGN AND EXPECTED ERROR

Buckingham method
An alternative method for the analysis of impact loads or velocity factors in gearing systems. This method uses actual expected clearances of the gears to determine a dynamic load.

Another method often used for the design of gear teeth is the **Buckingham method**. This method is often substituted for the procedure discussed previously because it sometimes offers greater flexibility. In addition, the expected error can be factored in based on different-pitch teeth, or the fact that the error would tend to increase with low-pitch teeth can be included. The Buckingham method will typically result in a more conservative design, and lower factors of safety are often used with this method. To use the Buckingham equation, we can calculate this expected error by first finding the expected error for the class of gear teeth selected from Figure 12.4, take that value and multiply it by 1,000, and multiply that by the values found in Table 12.3.

We could then find a new dynamic force from the following equation:

$$F_d = F_t + \frac{.05 \, V_m \, (bC + F_t)}{.05 \, V_m + (bC + F_t)^{1/2}} \tag{12.15}$$

It is also useful to have a guideline for recommended permissible error based on the velocity of our system. Figure 12.5 shows the recommended

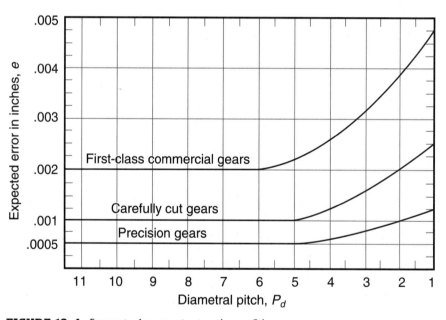

FIGURE 12.4 Expected error in tooth profiles.

TABLE 12.3
VALUES OF C FOR e = 0.001 INCH

Material	14½°	20°
Gray iron and gray iron	800	830
Gray iron and steel	1,100	1,140
Steel and steel	1,600	1,660

For other values of e, multiply the value given by the number of thousandths. For other materials, use $C = k E_g E_p / (E_g + E_p)$. Values of C for bronze are virtually the same as for the gray iron (moduli of elasticity about the same).

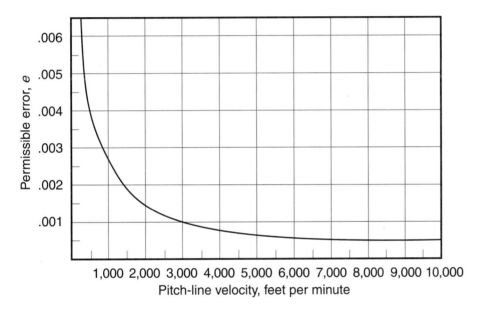

FIGURE 12.5 Recommended maximum error in gear teeth.

permissible error for different pitch-line velocities. Note that this graph is offered as a recommended guideline and should be followed unless a valid reason exists to do otherwise.

An example of the use of Buckingham's equation will clarify this method.

Example Problem 12.7

A pair of steel gears is made from annealed AISI 3140. Each gear has a surface hardness of BHN = 350. The pinion is a precision, 16 pitch, 20° involute, with 24 teeth 1 inch wide. The gear has 42 teeth. To transmit 3 hp at a speed of 3,450 rpm with a safety factor of 1.4, is this a suitable design?

$$S_u = 95 \text{ ksi} \quad \text{(Appendix 4)}$$

$$S_n = .5\, S_u = .5(95 \text{ ksi}) = 47.5 \text{ ksi}$$

$$D_p = \frac{N_p}{P_d} = \frac{24}{16} = 1.5 \text{ in.} \quad (11.4)$$

Find torque.

$$P = \frac{Tn}{63,000} \quad (2.6)$$

$$T = \frac{P(63,000)}{n}$$

$$T = \frac{3\,(63,000)}{3,450}$$

$$T = 55 \text{ in.-lb}$$

Find force transmitted.

$$F_t = \frac{2T}{D_p} \quad (12.3)$$

$$F_t = \frac{2\,(55 \text{ in.-lb})}{1.5 \text{ in.}}$$

$$F_t = 73 \text{ lb}$$

Find surface speed.

$$V_m = \frac{\pi D_p n}{12} \quad (12.5)$$

$$V_m = \frac{\pi\, 1.5 \text{ in.}\, 3,450 \text{ rpm}}{12 \text{ ft/in.}}$$

$$V_m = 1,355 \text{ ft/min}$$

Find force allowable (F_s).

$$F_s = \frac{S_n \, b \, Y}{P_d} \qquad (12.9)$$

$$Y = .337 \qquad \text{(from Table 12.1)}$$

$$F_s = \frac{47,500 \, (1) \, .337}{16}$$

$$F_s = 1,000$$

Expected error from Figure 12.4:

$$e = .0005$$

The value of C from Table 12.3 for steel and steel and 20° involute angle gears is 1660.

$$C = .0005 \, (1,000) \, 1,660$$

$$C = 830$$

Solve for dynamic load using Buckingham's equation.

$$F_d = F_t + \frac{.05 \, V_m \, (bC + F_t)}{.05 \, V_m + (bC + F_t)^{1/2}} \qquad (12.15)$$

$$F_d = 73 + \frac{.05 \, (1,355) \, (1 \bullet 830 + 73)}{.05 \, (1,355) + (1 \bullet 830 + 73)^{1/2}}$$

$$F_d = 699 \text{ lb}$$

$$\frac{F_s}{N_{sf}} \geq F_d$$

$$\frac{1,000}{1.4} \geq 699$$

$$714 > 699$$

This meets the criteria.

12.8 WEAR OF GEARS

In addition to verifying that the allowable force is greater than the dynamic force on a gear tooth, we often also need to insure that wear is not a concern. A typical spur gear tooth comes into and out of engagement many times in its lifetime, so the potential for wear can be significant. As was discussed in Chapter 11, if the gear's placement is exact and the gear is manufactured perfectly, there should theoretically be no sliding action between the gear teeth. In practice, however, that is impossible to achieve. Therefore, because of the slight sliding action as the teeth come into and out of engagement, there is the potential for wear on the gear teeth. To minimize wear, we must factor in the hardness of the gear teeth and several other factors as follows:

$$F_w = D_p\, b\, Q\, K_g \qquad (12.16)$$

Q in this formula can be calculated as follows:

$$Q = \frac{2 D_g}{D_g + D_p} = \frac{2 N_g}{N_g + N_p} = \frac{2 m_g}{m_g + 1} \qquad (12.17)$$

K_g is a product of the hardness of the material and can be found in Appendix 13.

For the gear teeth to resist both the repeated loading and the potential for surface wear, both F_s and F_w must be greater than the dynamic force (F_d), including an appropriate safety or service factor. The equation shown previously would be modified as follows:

$$\text{both } \frac{F_s}{N_{sf}} \text{ and } \frac{F_w}{N_{sf}} \text{ should be } \geq F_d$$

If both of these criteria are met, the design should be suitable. Note that the safety factor for wear is often lower than for stress. The following example problem verifies that the design in Example Problem 12.7 meets these criteria.

Example Problem 12.8

In Example Problem 12.7, verify that the surface is suitable for wear considerations. (For wear use $N_{sf} = 1.2$.)

Wear formula: $$F_w = D_p \, b \, Q \, K_g \qquad (12.16)$$

Find Q.
$$Q = \frac{2 N_g}{N_g + N_p} \qquad (12.17)$$

$$Q = \frac{2(42)}{42 + 24}$$

$$Q = 1.27$$

$$K_g = 270 \qquad \text{(from Appendix 13)}$$

Substituting into Equation 12.16:

$$F_w = 1.5 \, (1) \, 1.27 \, (270)$$

$$F_w = 514$$

This would not be suitable. Try the same calculations if the surfaces each had a BHN = 450.

$$K_g = 470 \qquad \text{(from Appendix 13)}$$

$$F_w = 1.5 \, (1) \, (1.27) \, (470)$$

$$F_w = 895$$

$$\frac{F_w}{N_{sf}} \geq F_d$$

$$\frac{895}{1.2} > 699$$

$$746 > 699$$

This would now be acceptable if the gear teeth were hardened to a BHN of 450.

12.9 SUMMARY

You should now recognize that there are different recommended techniques for the analysis of gears. Two of the most common were included here. Often you will simply select gears from a manufacturer's catalog, but in some cases the intended usage will not be standard, and specialized gears will need to be designed. Especially in mass-production environments, a standard gear may be unavailable, or other nonstandard features may be needed. The problem sets at the end of this chapter include examples where you will select gears from a manufacturer's catalog and analyze them to see if the manufacturer's power ratings are conservative.

After completing this chapter, you should also recognize that there are many variables in the design of gears, including material, size, tooth style and size, width, and manufacturing accuracy, to name only a few.

12.10 PROBLEMS

1. A pair of 20° spur gears transmits 5 hp from an electric motor that turns at 1,750 rpm. These gears are 12 pitch and 1 inch wide, and the pinion has 24 teeth. The mating gear needs to turn at 700 rpm. Compute the following:
   a. The velocity ratio of the set of gears
   b. The pitch diameter of each gear
   c. The surface speed of the gears
   d. The centerline distances between the mounting shafts
   e. The torque on both shafts
   f. The transmitted force
   g. The separating force
   h. The resultant force (the force acting on the mounting shafts)

2. A set of 14½° spur gears transmits 15 hp at 3,450 rpm from an electric motor. The pinion is a 12 pitch with 20 teeth and is ⅞ inch wide. The mating gear has 108 teeth. Compute the following:
   a. The velocity ratio of the gear set
   b. The surface speed
   c. The pitch diameter of both gears
   d. The centerline distances between the mounting shafts
   e. The output torque
   f. The output power if the gears are 100 percent efficient
   g. The transmitted force

3. In Problem 1, assume the gears are both made from AISI 4140 annealed steel and long life is desired. Compute the following:
   a. The Lewis form factor (Y) for both gears
   b. The force allowable for both gears
   c. As both gears are made from the same material, which value should be used in determining the allowable force for this gear train?

4. In Problem 2, assume the pinion is made from cold-drawn AISI 1050 steel, the gear from phosphor bronze (B139C), and long life is desired. Compute the following:
   a. The Lewis form factor for each gear
   b. The force allowable for both gears
   c. The force allowable for this combination
5. If the gear set in Problems 1 and 3 are commercial grade, determine
   a. The dynamic force
   b. If a safety factor of 2 is desired, is this an acceptable design?
   c. Is the width in the recommended range?
6. For the gear set in Problems 2 and 4, if the gears are precision class, determine
   a. The dynamic force
   b. If a safety factor of 1.5 is desired, is this an acceptable design?
   c. Assuming a safety factor of 2 is desired, how wide would the gears need to be?
   d. Is this in the recommended width range?
7. In Problems 1, 3, and 5, if the gears were the same except made by the hobbing technique, what minimum width would now be needed?
8. Determine the power that could be transmitted by a gear with the following properties: $P_d = 24$, $D_p = 3$, $n = 1{,}500$ rpm, 20° involute, commercial grade, $b = 1\frac{1}{4}$ inch, material is 6061 T6 aluminum, and long life is desired. Use a safety factor of 1.5 for this problem.

Use Figure 12.3 for Problems 9, 10, and 11.

9. Input 2 hp at 600 rpm; output 300 rpm. Select gears to meet these criteria.
10. Input ¾ hp at 1,800 rpm; output 300 rpm. Select a nonmetallic and a cast iron gear for this problem.
11. For the 20-tooth steel gear at 1,200 rpm, do you agree with the catalog rating? These are precision gears, and the material is similar to an AISI 1020 hot-rolled steel.
12. A pair of precision gears is to transmit 20 hp from a speed of 1,750 rpm to an output at 875 rpm. The gears are to be made of AISI 4140 annealed steel and long life is needed. Design a set of gears for this problem.
13. A steam turbine transmits 20 hp at a speed of 4,200 rpm to a generator that must turn at 3,600 rpm to generate power. The centerline spacing of the shafts should not exceed 8 inches because of space constraints. Design a set of gears for this problem.
14. Reanalyze Problem 5 using the Buckingham method. Is this design still acceptable?

15. Reanalyze Problem 6 using the Buckingham method. How does this change the answers to parts a, b, c, and d?
16. Determine the necessary surface hardness to prevent premature surface wear for the gears in Problem 12.
17. If the gears in Problems 1, 3, and 5 each have a surface hardness of BHN = 400, would you expect them to fail prematurely because of surface wear?
18. Design a double-reduction gear system to reduce the speed of an electric motor drive from 1,800 rpm to 120 rpm. Select material, diametral pitch, pitch diameter, pressure angle, tooth numbers, and other factors for a complete design. Use a factor of safety of 1.5. Include a sketch of the layout that includes shaft centerline spacing.

12.11 CUMULATIVE PROBLEMS

19. The truck for which we have designed many aspects in previous chapters now needs a transmission. The motor creates 350 hp at 2,400 rpm and requires a selectable gear ratio of 4:1, 3:1, and 2:1 for the first step in the reduction. Design three sets of spur gears.
20. The second step in the reduction in Problem 19 requires a 2:1 reduction. Design a gear set for this (remember the torque and speed have now changed).
21. Prepare a sketch of the gear layout for the gear system that was designed in Problems 19 and 20.

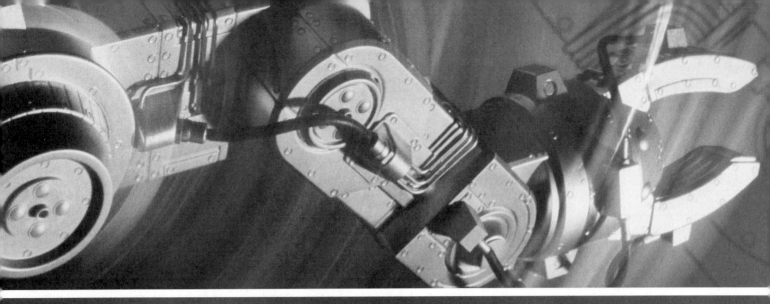

CHAPTER 13

Helical, Bevel, and Worm Gears

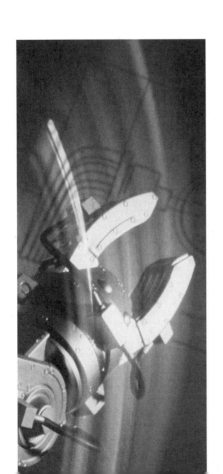

Helical, bevel, and worm gears are similar to spur gears in many ways. However, as discussed in Chapter 11, they have different advantages and disadvantages, so they are typically used in different types of applications. Helical gears, for example, have many of the same space and geometry properties as spur gears, but they are superior in high-velocity applications due to their lower impact loads. Bevel and worm gears have the inherent advantage of being able to turn at a 90° angle in their most typical forms. Bevel gears, in particular, have many similarities to spur gears, especially in how the teeth are formed and in the basic shape and profile of the tooth. Bevel gears can change shaft speed, similar to a spur-gear system, and they are used when it is necessary to turn a corner. Worm gears are also utilized in applications where the shafts are 90° apart. Worm gears, however, have very differing geometries and principles from the other forms of gearing in that they use a lead screw–type sliding motion. Their most significant advantage is that very high ratios can be realized in a relatively small space.

This chapter will focus on the principles involved in using these types of gears. It includes an introduction to the stresses that can be carried in the teeth. Most of this chapter will focus on how these three types of gears are used and the forces associated with them so that you will know how to design the shafts and bearings to support these types of gears.

This chapter will emphasize the selection of helical, bevel, and worm gears from manufacturers' catalogs as, in practice, engineers are more likely to select standard gears than to design custom gears. When this type of gearing does need to be designed, the principles presented in Chapter 12 can be applied.

13.1 OBJECTIVES

In order to use helical, bevel, and worm gears effectively, you must understand their basic attributes and the associated forces, torques, and power they can deliver. After completing this chapter, you should be able to:

- Understand helical-gear dimensions and geometries.
- Calculate the forces on helical gears along with the stresses inherent in gears of this type.
- Determine the allowable forces and actual forces on helical gears.
- Select standard helical-gear systems from manufacturers' catalogs.

- Understand bevel-gear geometries and dimensions.
- Calculate the forces on bevel gears and the associated reaction forces on the supporting shafts and bearings.
- Select standard bevel-gear systems from manufacturers' catalogs.
- Understand worm-gear dimensions and geometries.
- Calculate the forces on worm gears and their associated shafts and bearings.
- Select standard worm-gear systems from manufacturers' catalogs.
- Understand the distinction between thermal and mechanical ratings for standard worm-gear packages.

13.2 HELICAL GEARS

In helical gears, due to the angularity of the tooth face, the distinction between diametral pitch and the effective diametral pitch can be confusing. Figure 13.1 shows the **normal plane diametral pitch,** which is similar to the diametral pitch in a spur gear, but is important only in the manufacture of the gear teeth. For most purposes, the axial plane diametral pitch, or simply diametral pitch, is specified. This pitch is the same dimension used in spur gears to determine the effective spacing and diameter of the gear systems. Figure 13.1 reflects the normal circular pitch and the circular pitch and how they relate to the **helix angle.**

In the design of helical gear systems, we are also concerned with the axial force that is created. The angular characteristics of helical gears cause a thrust force that must be resisted by the bearings on the shaft. This axial force can be calculated as follows:

$$F_a = F_t \tan \psi \quad (13.1)$$

where ψ is the helix angle.

The radial force is calculated in the same manner as for spur gears. This radial force is the force that pushes the gears apart. It is calculated as follows:

$$F_n = F_t \tan \varphi_t \quad (13.2)$$

normal plane diametral pitch
The diametral pitch in an imaginary plane perpendicular to the face of the gear teeth, which would be used in the manufacture of helical gears.

helix angle
The angle at which the tooth form is oriented with respect to the centerline of the gear face. The helix angle would be perpendicular to the normal plane.

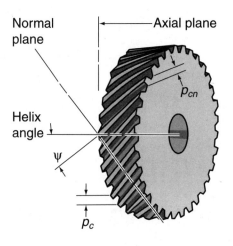

p_{cn} = Normal circular pitch
p_c = Circular pitch

FIGURE 13.1 Helical gear dimensions.

transverse pressure angle
The corrected pressure angle for use in helical gearing that corrects for the helix angle.

where φ_t is the **transverse pressure angle,** which is found as follows:

$$\varphi_t = \tan^{-1}\left(\frac{\tan \theta}{\cos \psi}\right) \qquad (13.3)$$

where θ = normal plane pressure angle

The transverse pressure angle is similar to the involute angle in spur gears, which is a characteristic of the construction angle of the gear.

The corresponding force transmitted is the same as for spur gears.

$$F_t = \frac{33,000 \text{ hp}}{V_m} \qquad (12.6)$$

Example Problem 13.1

A helical gear has a (axial plane) diametral pitch of 12, has 24 teeth, a pressure angle of 20°, a helix angle of 15°, and a width of 1½ inches. It transmits 5 hp at 1,750 rpm. Find the pitch-line velocity, transmitted force, axial force, and separating force.

Pitch-line velocity:

$$D = \frac{N_t}{P_d} = \frac{24}{12} = 2 \text{ in} \quad (11.4)$$

$$V_m = \frac{\pi D_p n}{12} \quad (12.5)$$

$$V_m = \frac{\pi\, 2\,(1,750)}{12}$$

$$V_m = 916 \text{ ft/min}$$

Transmitted force:

$$F_t = \frac{33,000 \text{ hp}}{V_m} \quad (12.6)$$

$$F_t = \frac{33,000(5)}{916}$$

$$F_t = 180 \text{ lb}$$

Axial force:

$$F_a = F_t \tan \psi \quad (13.1)$$

$$F_a = 180 \text{ lb} \tan 15°$$

$$F_a = 48 \text{ lb}$$

Separating force:

$$\varphi_t = \tan^{-1}\left(\frac{\tan \theta}{\cos \psi}\right) \quad (13.3)$$

$$\varphi_t = \tan^{-1}\left(\frac{\tan 20°}{\cos 15°}\right)$$

$$\varphi_t = 20.6°$$

$$F_n = F_t \tan \varphi_t \quad (13.2)$$

$$F_n = 180 \text{ lb} \tan 20.6°$$

$$F_n = 68 \text{ lb}$$

13.3 HELICAL GEAR STRESSES

The stresses and stress allowables and the corresponding force and force allowable on helical gears are analyzed in a manner very similar to spur gears. The force allowable (F_s) is calculated as follows:

$$F_s = \frac{S_n b Y}{P_{dn}} \tag{13.4}$$

where P_{dn} = normal plane pitch
Y = modified Lewis form factor load at tip

It should be noted that this is basically the same formula as that used for spur gears, with the minor exception that the normal plane pitch is found as follows:

$$P_{dn} = \frac{P_d}{\cos \psi} \tag{13.5}$$

and the number of teeth used to find the Lewis form factor is modified as follows:

$$N_e = \frac{P_{dn} D_p}{\cos^2 \psi} = \frac{N}{\cos^3 \psi} \tag{13.6}$$

where N = actual number of teeth

The dynamic load factor is the same as was used for spur gears. For example, for commercial gears, the factor would be

$$F_d = \left(\frac{600 + V_m}{600} \right) F_t \tag{12.10}$$

The material selection and other attributes of helical gears are the same. The differentiating quantity is merely the form factor, which is a product of pressure angle, helix angle, and other factors. The following example shows an analysis of a helical-gear system along with a comparison from a standard manufacturer's catalog.

Example Problem 13.2

For a 16-diametral-pitch, ½-inch-wide, 45° helix angle, and 20° pressure angle gear with 24 teeth at 600 rpm, estimate a power rating and compare it with the catalog rating. Use a stress value of 30,000 psi and assume commercial grade construction.

Find P_{dn}.

$$P_{dn} = \frac{P_d}{\cos \psi} = \frac{16}{\cos 45°} \quad (13.5)$$

$$P_{dn} = 22.6$$

Lewis form factor from Table 12.1 (only approximate for helix gears):
Find:

$$N_e = \frac{N}{\cos^3 \psi} \quad (13.6)$$

$$N_e = \frac{24}{\cos^3 45°}$$

$$N_e = 68$$

$$Y = .427$$

$$F_s = \frac{S b Y}{P_{dn}} \quad (13.4)$$

$$F_s = \frac{30,000(.5).427}{22.6}$$

$$F_s = 280 \text{ lb}$$

$$V_m = \frac{\pi D n}{12} = \frac{\pi \frac{24}{16} 600}{12}$$

$$V_m = 236 \text{ ft/min} \quad (12.5)$$

Set F_s equal to the dynamic force and solve.

$$F_s = F_d = \left(\frac{600 + V_m}{600}\right) F_t \quad (12.10)$$

$$280 \text{ lb} = \left(\frac{600 + 236}{600}\right) F_t$$

$$F_t = 200 \text{ lb}$$

$$P = \frac{F_t V_m}{33,000} \quad (12.6)$$

$$P = \frac{200(236)}{33,000}$$

$$P = 1.4 \text{ hp}$$

The result is only slightly higher than the 1 hp found in Figure 13.2 for a SF = 1.

HELICAL GEARS

APPROXIMATE HORSEPOWER AND TORQUE* RATINGS FOR CLASS I SERVICE (Service Factor = 1.0)

No. Teeth	25 RPM H.P.	25 RPM Torque	50 RPM H.P.	50 RPM Torque	100 RPM H.P.	100 RPM Torque	200 RPM H.P.	200 RPM Torque	300 RPM H.P.	300 RPM Torque	600 RPM H.P.	600 RPM Torque	900 RPM H.P.	900 RPM Torque	1200 RPM H.P.	1200 RPM Torque	1800 RPM H.P.	1800 RPM Torque	3600 RPM H.P.	3600 RPM Torque
24 DIAMETRAL PITCH – 33.94 NORMAL DIAMETRAL PITCH HARDENED STEEL																			.250-.375" FACE	
8	.01	13.5	.01	13.5	.02	13.4	.04	13.2	.06	13.0	.12	12.5	.17	12.0	.22	11.6	.31	10.8	.51	8.9
10	.01	18.0	.01	17.9	.03	17.8	.06	17.4	.08	17.1	.16	16.3	.22	15.5	.28	14.8	.39	13.6	.62	10.9
12	.01	22.5	.02	22.3	.04	22.1	.07	21.6	.10	21.1	.19	20.0	.27	18.9	.34	17.9	.46	16.2	.72	12.6
15	.01	29.1	.02	28.9	.05	28.5	.09	27.8	.13	27.1	.24	25.2	.34	23.5	.42	22.0	.56	19.6	.84	14.8
18	.01	23.6	.02	23.4	.04	23.1	.07	22.4	.10	21.7	.19	19.9	.26	18.4	.33	17.1	.43	15.0	.62	10.9
20	.01	26.8	.02	26.5	.04	26.1	.08	25.2	.12	24.3	.21	22.2	.29	20.4	.36	18.8	.47	16.3	.67	11.7
24	.01	32.6	.03	32.3	.05	31.6	.10	30.3	.14	29.2	.25	26.1	.34	23.7	.41	21.6	.53	18.5	.73	12.8
30	.02	41.3	.03	40.8	.06	39.7	.12	37.8	.17	36.0	.30	31.6	.40	28.1	.48	25.3	.60	21.1	.81	14.1
36	.02	49.9	.04	49.1	.08	47.6	.14	44.9	.20	42.4	.35	36.4	.46	31.9	.54	28.4	.66	23.3	.86	15.1
48	.03	67.0	.05	65.6	.10	63.0	.19	58.3	.26	54.3	.43	45.0	.55	38.4	.64	33.5	.76	26.6	.94	16.5
60	.03	83.8	.06	81.6	.12	77.6	.22	70.7	.31	64.9	.50	52.0	.62	43.4	.71	37.3	.83	29.0	1.00	17.5
72	.04	101	.08	97.7	.15	92.1	.26	82.5	.36	74.8	.56	58.3	.68	47.8	.77	40.5	.89	31.0	1.00	18.2
20 DIAMETRAL PITCH – 28.28 NORMAL DIAMETRAL PITCH HARDENED STEEL																			.375-.563" FACE	
8	.01	29.2	.02	29.1	.05	28.8	.09	28.4	.13	27.9	.25	26.6	.36	25.4	.46	24.3	.64	22.3	1.00	18.0
10	.01	37.7	.03	37.5	.06	37.1	.12	36.3	.17	35.6	.32	33.5	.45	31.7	.57	30.1	.78	27.2	1.20	21.2
12	.02	48.5	.04	48.2	.08	47.5	.15	46.4	.22	45.2	.40	42.2	.56	39.5	.71	37.1	.95	33.2	1.44	25.1
15	.02	62.7	.05	62.2	.10	61.2	.19	59.3	.27	51.6	.50	52.8	.70	48.8	.86	45.4	1.14	39.8	1.66	29.0
20	.02	57.7	.05	57.1	.09	55.9	.17	53.7	.25	51.6	.44	46.2	.60	41.9	.73	38.3	.93	32.7	1.30	22.7
25	.03	73.8	.06	72.8	.11	70.9	.21	67.4	.31	64.3	.54	56.4	.72	50.2	.86	45.2	1.08	37.7	1.44	25.2
30	.04	89.1	.07	87.6	.13	85.0	.25	80.0	.36	75.7	.62	65.0	.81	56.9	.96	50.7	1.19	41.5	1.54	27.0
40	.05	120	.09	118	.18	113	.33	104	.46	97.2	.77	80.5	.98	68.7	1.14	59.9	1.36	47.7	1.69	29.6
50	.06	151	.12	147	.22	139	.40	127	.55	117	.89	93.4	1.11	78.0	1.27	66.9	1.49	52.1	1.79	31.4
60	.07	180	.14	175	.26	165	.47	147	.64	134	.99	104	1.22	85.4	1.38	72.3	1.58	55.4	1.86	32.5
16 DIAMETRAL PITCH – 22.63 NORMAL DIAMETRAL PITCH HARDENED STEEL																			.500" FACE	
12	.03	67.2	.05	66.6	.10	65.6	.20	63.6	.29	61.7	.54	56.6	.75	52.3	.93	48.6	1.20	42.6	1.80	31.1
16	.04	93.4	.07	92.4	.14	90.5	.28	86.9	.40	83.5	.71	74.8	.97	67.8	1.18	62.0	1.51	52.9	2.10	36.7
20	.05	120	.09	118	.18	115	.35	110	.50	104	.87	91.5	1.16	81.5	1.40	73.4	1.75	61.3	2.34	41.0
24	.06	146	.11	144	.22	139	.42	131	.59	124	1.00	107	1.33	93.3	1.58	83.0	1.94	68.1	2.52	44.2
32	.08	197	.15	193	.29	185	.54	172	.76	160	1.26	132	1.61	113	1.87	98.4	2.24	78.4	2.78	48.7
40	.10	249	.19	242	.37	230	.67	210	.92	193	1.47	154	1.84	129	2.11	111	2.46	86.2	3.00	51.8
48	.12	298	.23	289	.43	273	.77	244	1.05	221	1.64	173	2.02	141	2.28	120	2.62	91.8	3.08	53.9
12 DIAMETRAL PITCH – 16.97 NORMAL DIAMETRAL PITCH HARDENED STEEL																			.750" FACE	
12	.07	179	.14	177	.27	173	.53	166	.76	160	1.36	143.2	1.85	130	2.26	119	2.89	101	4.01	70.2
15	.09	231	.18	228	.35	222	.67	211	.96	201	1.68	176	2.24	157	2.69	142	3.37	118	4.51	79.0
18	.11	281	.22	277	.43	268	.80	253	1.14	239	1.95	205	2.57	180	3.05	160	3.75	131	4.86	85.1
24	.15	387	.30	379	.58	364	1.07	337	1.49	313	2.47	260	3.16	222	3.68	193	4.39	154	5.45	95.5
30	.19	489	.38	477	.72	453	1.31	413	1.80	379	2.89	304	3.62	254	4.14	218	4.84	170	5.82	102
36	.23	589	.45	571	.85	538	1.53	482	2.08	437	3.24	341	3.99	279	4.50	237	5.17	181	6.08	106
10 DIAMETRAL PITCH – 14.14 NORMAL DIAMETRAL PITCH HARDENED STEEL																			.875" FACE	
8	.07	181	.14	179	.28	176	.54	171	.79	165	1.44	151	1.78	139	2.45	129	3.20	112	4.62	80.9
10	.10	240	.19	238	.37	233	.71	223	1.02	215	1.83	193	2.49	174	3.03	159	3.88	136	5.39	94.4
12	.12	300	.23	296	.46	288	.87	275	1.25	262	2.20	231	2.95	206	3.55	186	4.46	156	6.01	105
15	.15	387	.30	381	.59	369	1.10	348	1.56	329	2.69	282	3.53	247	4.19	220	5.16	181	6.69	117
20	.21	533	.41	522	.79	501	1.47	464	2.05	432	3.40	357	4.35	305	5.06	266	6.05	212	7.51	131
25	.27	680	.53	662	1.00	630	1.82	573	2.50	526	4.01	422	5.03	352	5.75	302	6.72	235	8.09	142
30	.32	818	.63	793	1.19	747	2.12	669	2.89	606	4.50	473	5.54	388	6.25	328	7.18	252	8.44	148
40	.44	1097	.84	1053	1.55	975	2.69	849	3.58	751	5.32	559	6.36	445	7.04	370	7.89	276	8.97	157
8 DIAMETRAL PITCH – 11.31 NORMAL DIAMETRAL PITCH HARDENED STEEL																			.750" FACE	
8	.10	242	.19	239	.37	234	.71	225	1.03	216	1.85	194	2.51	176	3.06	160	3.91	137	5.43	95
10	.13	321	.25	317	.49	309	.93	293	1.33	280	2.53	245	3.12	218	3.74	197	4.69	164	6.27	110
12	.16	400	.31	394	.61	382	1.14	360	1.62	340	2.78	292	3.65	256	4.34	228	5.33	187	6.92	121
16	.22	555	.43	543	.83	521	1.53	483	2.14	447	3.54	372	4.53	318	5.27	277	6.30	221	7.82	137
20	.28	710	.55	692	1.04	658	1.90	599	2.62	550	4.20	441	5.26	368	6.01	316	7.03	246	8.45	148
24	.34	862	.66	836	1.25	787	2.24	706	3.04	639	4.75	499	5.84	409	6.59	346	7.57	265	8.90	156
32	.46	1160	.88	1113	1.64	1031	2.85	897	3.78	794	5.63	591	6.72	471	7.44	391	8.34	292		
40	.58	1454	1.10	1383	2.00	1259	3.39	1068	4.41	927	6.32	664	7.39	517	8.07	424	8.88	311		
48	.69	1737	1.30	1636	2.33	1466	3.85	1214	4.93	1036	6.85	719	7.87	551	8.50	447				

Ratings are based on strength calculation. Basic static strength rating, or for hand operation of above gears is approximately 3 times the 100 RPM rating.

NOTE: Ratings to right of heavy line exceed 1500 Feet per Minute and should be used for interpolation purposes only.

*Torque Rating (Lb. Ins.)

FIGURE 13.2 Catalog ratings for Example Problem 13.2. (Courtesy of Boston Gear)

13.4 BEVEL GEAR FORCES

As bevel and miter gears are typically designed as a set, the structural analysis of the gear teeth is omitted from this chapter. However, it should be noted that this type of analysis is very similar to the process used for both spur and helical gears. Of more importance to the selection of miter or bevel gears are the associated shaft-loading characteristics. As can be seen from Figure 13.3, due to the angle of the bevel gears, there is both a separating load and a thrust load in addition to the transmitted load. The separating load causes a potential deflection of the shaft, and the force created is calculated as follows:

$$F_n = F_t \tan \theta \cos \gamma \tag{13.7}$$

where θ = pressure angle

γ = cone angle

The forces on the larger bevel gear would be calculated in a similar manner.

The strength of the mounting shaft and the associated bearings needs to be corrected for this loading situation. In addition to the radial load, an axial load exists because of the taper. This axial load tends to create a thrust

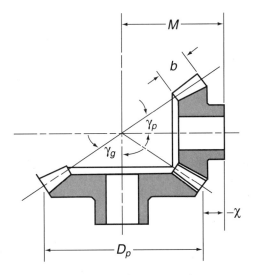

FIGURE 13.3 Bevel gear angles.

load on the bearings that support the two shafts. For the pinion, it would be calculated as follows:

$$F_a = F_t \tan \theta \sin \gamma \tag{13.8}$$

The following example shows the analysis of these loads.

Example Problem 13.3

If the bevel gear set shown in Figure 13.3 transmits 8 hp at 2,200 rpm, determine the transmitted, separating, and axial forces on the pinion and gear for the following parameters:

$$N_p = 24$$
$$N_g = 36$$
$$P_d = 8$$
$$\gamma_p = 33.7°$$
$$\gamma_g = 56.3°$$
$$\theta = 14½° \text{ (pressure angle)}$$
(Note: $\gamma_p + \gamma_g = 90°$)

First, find D_p.

$$D_p = \frac{N_p}{P_d} = \frac{24}{8} = 3 \text{ in.}$$

Transmitted force:

$$T = \frac{63,000P}{n} \tag{2.6}$$

$$T = \frac{63,000(8)}{2,200}$$

$$T = 230 \text{ in.-lb}$$

$$F_t = \frac{2T}{D_p} \tag{12.3}$$

$$F_t = \frac{2(230 \text{ in.-lb})}{3 \text{ in.}}$$

$$F_t = 153 \text{ lb}$$

Separating force:

Pinion:

$$F_n = F_t \tan \theta \cos \gamma \qquad (13.7)$$
$$F_n = 153 \text{ lb} \tan 14\tfrac{1}{2}° \cos 33.7°$$
$$F_n = 33 \text{ lb}$$

Gear:

$$F_n = 153 \text{ lb} \tan 14\tfrac{1}{2}° \cos 56.3°$$
$$F_n = 22 \text{ lb}$$

Axial force:

Pinion:

$$F_a = F_t \tan \theta \sin \gamma \qquad (13.8)$$
$$F_a = 153 \text{ lb} \tan 14\tfrac{1}{2}° \sin 33.7°$$
$$F_a = 22 \text{ lb}$$

Gear:

$$F_a = 153 \text{ lb} \tan 14\tfrac{1}{2}° \sin 56.3°$$
$$F_a = 33 \text{ lb}$$

Note that

$$F_n \text{ (pinion)} = F_a \text{ (gear)} = 33 \text{ lb}$$
and
$$F_a \text{ (pinion)} = F_n \text{ (gear)} = 22 \text{ lb}$$

13.5 WORM GEAR FORCES AND GEOMETRIES

Worms and worm gears are used to transmit motion and/or power between shafts mounted at right angles. Typically, worm-gear drives are the smoothest and quietest form of gearing when properly lubricated and

maintained. They offer very high-speed ratios in a relatively limited space and, furthermore, are often self-locking so that the output shaft is locked when input motion stops. The worm in a worm-gear set can best be thought of as a lead screw and the worm gear as an unfolded nut; therefore, the principles of lead screws are more applicable than most of the principles of gearing. Because this motion is developed through sliding action between the worm and the worm gear, friction and lubrication issues are even more important than with other forms of gearing. The thrust load on the worm is equal to the force transmitted on the worm-gear, making this analysis very straightforward. The output torque from a worm-gear set is the ratio of the system minus the **inefficiency**; thus, the output torque then becomes equal to

> **inefficiency**
> The opposite of efficiency or simply the measure of wasted energy.

$$T_{out} = e\, V_r\, T_{in} \qquad (13.9)$$

where

$$V_r = \frac{N_{t\,gear}}{N_{starts\,worm}} \qquad (13.10)$$

From Chapter 11, we learned that the ratio of a worm-gear set for a single-start or single-lead system is simply the number of teeth on the worm gear because worm gear progresses one tooth per revolution of the worm. The lead angle of the worm is important to know in order to determine the efficiency and self-locking ability of a worm-gear set. This can be found as follows:

$$\lambda = \tan^{-1}\left(\frac{L}{\pi D}\right) \qquad (13.11)$$

where λ = angle of worm

  L = lead of worm

Example Problem 13.4

A worm-gear set has a single-start worm. The worm has a pitch diameter of 2 inches. The gear has 60 teeth with a diametral pitch of 6. The input shaft is driven at 1,750 rpm from a 5-hp motor. The pressure angle is 20°. Find the pitch diameter of the gear, circular pitch, lead angle, centerline distance, velocity ratio, and output speed.

Pitch diameter:

$$D_p = \frac{N_g}{P_d} \qquad (11.4)$$

$$D_p = \frac{60}{6} = 10 \text{ in.}$$

Circular pitch:

$$P_c = \frac{\pi D_p}{N_g} \qquad (11.3)$$

$$P_c = \frac{\pi 10 \text{ in.}}{60}$$

$$P_c = .524 \text{ in.}$$

As this is a single-start worm,

$$L = P_c$$

Lead angle:

$$\lambda = \tan^{-1}\left(\frac{L}{\pi D}\right) \qquad (13.11)$$

$$\lambda = \tan^{-1}\left(\frac{.524 \text{ in.}}{\pi\, 2 \text{ in.}}\right)$$

$$\lambda = 4.77°$$

Centerline distance:

$$C\text{-}C = \frac{D_w + D_g}{2}$$

$$C\text{-}C = \left(\frac{2 \text{ in.} + 10 \text{ in.}}{2}\right)$$

$$C\text{-}C = 6 \text{ in.}$$

Velocity ratio:

$$V_r = \frac{N_{t\,gear}}{N_{starts\,worm}} \qquad (13.10)$$

$$V_r = \frac{60}{1} = 60 \text{ to } 1$$

Output speed:

$$n_{out} = \frac{n_{in}}{V_r}$$

$$n_{out} = \frac{1}{60}(1,750 \text{ rpm})$$

$$n_{out} = 29.2 \text{ rpm}$$

It is also necessary to use the lead angle to calculate a normal circular pitch, which would be the equivalent pitch at the angle of gearing. This can be determined as follows:

$$P_{cn} = P_c \cos \lambda \qquad (13.12)$$

where P_c = circular pitch of the worm gear

P_{cn} = normal circular pitch

λ = lead angle

See Figure 13.4.

The strength of the gear is typically lower than that of the worm. We can estimate the strength of the worm gear as follows:

$$F_s = \frac{S\,Y\,b\,P_{cn}}{\pi} \text{ or } \frac{S_n\,Y\,b\,P_{cn}}{\pi} \qquad (13.13)$$

An approximate Lewis form factor can be determined from Table 13.1. Note that φ_n is the pressure angle of the worm gear teeth.

The dynamic load can be estimated in the following manner if the speed is high:

$$F_d = \left(\frac{1,200 + V_{mg}}{1,200}\right) F_t \qquad (13.14)$$

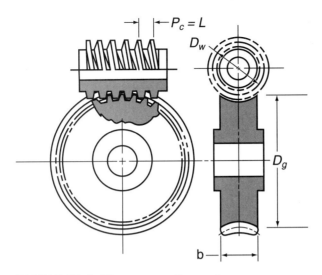

FIGURE 13.4 Worm gear dimensions.

TABLE 13.1

APPROXIMATE LEWIS FORM FACTOR FOR WORM GEAR TEETH

φ_n	Y
14½°	.314
20°	.392
25°	.470
30°	.550

where F_t is for the gear and can be found from

$$F_t = \frac{33,000 \cdot \text{hp}}{V_{mg}} \quad (13.15)$$

where V_{mg} = pitch-line speed of the gear in ft/min

As with spur gears, F_s should be greater than F_d and should include a service or safety factor as follows:

$$\frac{F_s}{N_{sf}} \geq F_d$$

Example Problem 13.5

In Example Problem 13.4, if both the gear and worm are made from annealed AISI 3140 steel and the gear is 1 inch wide, determine the normal circular pitch, dynamic load, force allowable on the gear, and effective service/safety factor.

Normal circular pitch:

$$P_{cn} = P_c \cos \lambda \qquad (13.12)$$

$$P_{cn} = .524 \cos 4.77°$$

$$P_{cn} = .522 \text{ in.}$$

Assume 100 percent efficiency for determining force transmitted.

$$T = \frac{63,000 \cdot hp}{n} \qquad (2.6)$$

$$T = \frac{63,000(5)}{29.2}$$

$$T = 10,800 \text{ in.-lb}$$

$$F_t = \frac{2T}{D_p} \qquad (12.3)$$

$$F_t = \frac{2(10,800 \text{ in.-lb})}{10 \text{ in.}}$$

$$F_t = 2,160 \text{ lb}$$

$$V_m = \frac{\pi D_g n}{12} \qquad (12.5)$$

$$V_m = \frac{\pi(10)29.2}{12}$$

$$V_m = 76.4 \text{ ft/min}$$

Dynamic load:

$$F_d = \left(\frac{1,200 + V_{mg}}{1,200}\right) F_t \qquad (13.14)$$

$$F_d = \left(\frac{1,200 + 76.4}{1,200}\right) 2,160 \text{ lb}$$

$$F_d = 2,300 \text{ lb}$$

Force allowable:

$$F_s = \frac{S_n\, Y\, b\, P_{cn}}{\pi} \qquad (13.13)$$

$$S_u = 95 \text{ ksi} \qquad \text{(Appendix 4)}$$

Lewis form factor:

$$Y = .392 \qquad \text{(Table 13.1)}$$

$$F_s = \frac{.5(95,000 \text{ lb/in.}^2)(.392)(1 \text{ in.})(.522 \text{ in.})}{\pi}$$

$$F_s = 3,100 \text{ lb}$$

$$\frac{F_s}{N_{sf}} \geq F_d$$

$$\frac{3,100}{N_{sf}} = 2,300$$

$$N_{sf} = 1.35$$

Efficiency

The **efficiency** of a worm-gear drive depends on the lead angle of the worm and the type of lubrication. The efficiency of the worm-gear system depends on the lead angle along with the overall ratio. A lower worm-gear lead angle typically reduces the overall efficiency. Due to the sliding action between the worm and the worm gear, the efficiency depends on this lead angle and the coefficient of friction of the contact surface. We can estimate the efficiency of a worm-gear set as follows:

$$e = \frac{\tan \lambda (1 - f \tan \lambda)}{f + \tan \lambda} \qquad (13.16)$$

where λ = worm lead angle

f = coefficient of friction

As will be shown later in this chapter, the inefficiency or energy lost due to friction in the worm-gear system becomes heat that needs to be dissipated when worm-gear sets are used for extensive periods of time. The

> **efficiency**
> The ratio of the power or torque into a system to the output.

> Worm efficiency (13.16)
> $$e = \frac{\tan \lambda (1 - f \tan \lambda)}{f + \tan \lambda}$$

coefficient of friction
A term used for the ratio of the force required to drag an object across a surface to its weight or normal force.

coefficient of friction would depend on the materials and the type of lubrication used. The coefficient of friction also changes with sliding speed, especially at very low speeds. For a bronze worm gear and a hardened steel worm, a coefficient of friction in the range of .03 to .05 is often used. Gear manufacturers should be consulted for the appropriate actual values for different combinations of materials and types of lubrication.

Example Problem 13.6

In Example Problems 13.4 and 13.5, if the coefficient of sliding friction is .03, find the efficiency of this worm-gear set and the actual torque output.

Efficiency:

$$e = \frac{\tan \lambda (1 - f \tan \lambda)}{f + \tan \lambda} \quad (13.16)$$

$$e = \frac{\tan 4.77 (1 - .03 \tan 4.77)}{.03 + \tan 4.77}$$

$$e = .73 \text{ or } 73\%$$

Torque input:

$$T = \frac{63{,}000 \cdot \text{hp}}{n} \quad (2.6)$$

$$T = \frac{63{,}000(5)}{1{,}750}$$

$$T = 180 \text{ in.-lb}$$

$$T_{out} = e\, V_r T_{in} \quad (13.9)$$

$$T_{out} = (.73)\frac{60}{1}(180 \text{ in.-lb})$$

$$T_{out} = 7{,}884 \text{ in.-lb}$$

Self-Locking

Most worm-gear sets are self-locking. That is, if torque is applied to the output shaft, the input worm will not turn. This locking is produced by the friction force between the worm threads and the worm gear and the relationship between this friction force and the lead angle. Typically, a lead angle of 6° or less would be self-locking, between 6° and 11° it would

depend on the type of lubrication present, and above 11° self-locking should not be expected to exist.

Wear

Wear of worm gears is a product of the hardness of the material and the type of lubrication used. Wear is fairly difficult to predict in some cases because of the accuracy and surface finish of the parts. Lubricating oil contamination and other factors can significantly affect the life of these types of gearing. Often a bronze worm gear will "wear in" and thus adjust for minor manufacturing discrepancies. In general, the worm is hardened steel and the gear may be steel or, in many cases, a bronze-type material that allows for some wear-in to be achieved. Lubrication is frequently done by mineral oil. However, in many cases, an SAE 80 or 90 gear oil with a splash lubrication system may be specified or, in other cases, a high-temperature synthetic oil or extreme pressure lubricant may be substituted in order to increase either lifetime or load ratings.

13.6 WORM GEAR THERMAL AND MECHANICAL RATINGS

An interesting property of worm-gear systems and, in particular, prepackaged worm-gear boxes is that the manufacturers often give these products two ratings: a mechanical rating, often called an intermittent rating, and a thermal or continuous rating based not on the ability of the gear sets to carry the load but on the ability of the gearboxes to dissipate heat. Because a worm-gear system relies on sliding friction, heat is generated that must be dissipated to the atmosphere in order to keep the lubricants at a temperature where they can perform properly. We calculate the heat dissipation of a gearbox system based on the efficiency of the gear sets. For continuous service, the heat gained has to equal the heat lost; otherwise, the system temperature will continue to rise. We can calculate this heat gain as follows:

$$Q = (1 - e)\, hp_{in} = (1 - e)\, hp_{in}\, (33{,}000) \qquad \text{ft-lb/min} \quad (13.17)$$

or

$$Q = (1 - e)\, hp_{in}\, (2{,}544) \qquad \text{Btu/hr} \quad (13.18)$$

Heat generated (13.18)

$Q = (1 - e)\, hp_{in}\, (2{,}544)$
Btu/hr

If the exterior of a gearbox system cannot dissipate heat at this rate, the lubricating properties of the oils will break down and the system will ultimately fail. Figure 13.5 is a page from a manufacturer's catalog showing different thermal and mechanical ratings for the same systems, as well as ratings based on the use of different lubricants. It is interesting to note that when the same gearbox systems use a synthetic oil capable of operating at higher temperatures, the ratings increase, not because less heat is generated, but because the higher oil temperature results in a higher external case temperature, which dissipates a greater amount of heat.

Example Problem 13.7

For the worm-gear set in Example Problems 13.4 through 13.6, determine the heat generated by this system.

$$Q = (1 - e)\, hp_{in}(2,544) \quad (13.18)$$

$$Q = (1 = .73)(5)(2,544)$$

$$Q = 3,430 \text{ Btu/hr}$$

13.7 SUMMARY

This chapter reviewed three different types of gearing that can be found in many types of systems. We also discussed the process of using preengineered products or selecting gears from manufacturers' catalogs. Especially in low-volume applications, the designer often specifies preengineered products because the cost of design can often be greater than the cost of the products. With preengineered products, time and money can be saved. We have also reviewed the loading principles that must be considered when designing shafts and bearings for these gears, and we have briefly reviewed the structural analysis necessary for the design of gears of these types. The problem set at the end of this chapter will emphasize the practice of selecting manufacturers' preengineered products. In some cases you will reanalyze these products' expected load-carrying capabilities, and you will analyze external factors in the design of shafts, bearings, and other features.

Helical, Bevel, and Worm Gears

SINGLE REDUCTION HORSEPOWER AND TORQUE RATINGS

REDUCER SIZE 924

HORSEPOWER AND TORQUE RATINGS (INCH POUNDS)

2.375 CENTER DISTANCE

RATIO[1]	INPUT RPM[2]	OUTPUT RPM	MECHANICAL[3] 1.00 SERVICE FACTOR				1.25 SF		1.50 SF		1.75 SF		THERMAL[4]		SYNTHETIC OIL MECHANICAL[5]			THERMAL	
			INPUT HP	OUTPUT HP	OUTPUT TORQUE	EFF[6]	INPUT HP	OUTPUT TORQUE	INPUT HP	OUTPUT TORQUE	INPUT HP	OUTPUT TORQUE	INPUT HP	OUTPUT TORQUE	INPUT HP	OUTPUT TORQUE	EFF[6]	INPUT HP	OUTPUT TORQUE
4 (4)	2500	625	5.79	5.44	548	94	4.65	439	3.89	365	3.35	313	3.72	350	5.68	548	96	5.68	548
	1750	438	4.85	4.53	653	94	3.89	522	3.25	435	2.79	373	3.77	506	4.74	653	96	4.74	653
	1160	290	4.17	3.86	840	93	3.34	672	2.79	560	2.39	480	3.48	700	4.05	840	95	4.05	840
	870	218	3.56	3.28	950	92	2.85	760	2.38	633	2.05	543	3.21	856	3.45	950	95	3.45	950
	600	150	2.79	2.54	1066	91	2.23	853	1.86	711	1.60	609	2.79	1066	2.69	1066	94	2.69	1066
	300	75	1.62	1.44	1212	89	1.30	969	1.08	808	0.93	692	1.62	1212	1.55	1212	93	1.55	1212
	100	25	0.61	0.52	1320	86	0.49	1056	0.41	880	0.35	754	0.61	1320	0.57	1320	92	0.57	1320
5 (5)	2500	500	5.08	4.70	593	93	4.08	474	3.41	395	2.94	339	3.79	440	4.96	593	95	4.96	593
	1750	350	4.38	4.03	726	92	3.52	581	2.94	484	2.53	415	3.72	616	4.26	726	95	4.26	726
	1160	232	3.68	3.35	911	91	2.95	729	2.46	607	2.12	521	3.38	835	3.56	911	94	3.56	911
	870	174	3.18	2.87	1039	90	2.55	831	2.13	693	1.82	594	3.10	1012	3.06	1039	94	3.06	1039
	600	120	2.51	2.24	1174	89	2.01	939	1.68	783	1.44	671	2.51	1174	2.40	1174	93	2.40	1174
	300	60	1.48	1.28	1345	87	1.19	1076	0.99	897	0.85	768	1.48	1345	1.40	1345	92	1.40	1345
	100	20	0.57	0.47	1472	83	0.45	1178	0.38	982	0.32	841	0.56	1472	0.52	1472	90	0.52	1472
7½ (7½)	2500	333	3.98	3.63	686	91	3.20	549	2.68	457	2.31	392	3.27	562	3.87	686	94	3.87	686
	1750	233	3.41	3.09	834	91	2.74	668	2.29	556	1.97	477	3.20	784	3.30	834	94	3.30	834
	1160	155	2.80	2.51	1022	90	2.24	818	1.88	681	1.61	584	2.80	1022	2.69	1022	93	2.69	1022
	870	116	2.34	2.08	1129	89	1.88	903	1.57	753	1.35	645	2.34	1129	2.24	1129	93	2.24	1129
	600	80	1.80	1.57	1239	87	1.44	991	1.20	826	1.03	708	1.80	1239	1.71	1239	92	1.71	1239
	300	40	1.03	0.87	1373	85	0.82	1099	0.69	916	0.59	785	1.03	1373	0.96	1373	91	0.96	1373
	100	13	0.38	0.31	1471	82	0.31	1177	0.26	981	0.22	841	0.38	1471	0.35	1471	89	0.35	1471
10 (10)	2500	250	3.18	2.85	719	90	2.56	575	2.15	479	1.85	411	2.89	652	3.08	719	93	3.08	719
	1750	175	2.77	2.48	892	89	2.23	714	1.87	595	1.61	510	2.77	892	2.67	892	93	2.67	892
	1160	116	2.20	1.95	1057	88	1.77	845	1.48	705	1.27	604	2.20	1057	2.11	1057	92	2.11	1057
	870	87	1.82	1.59	1149	87	1.46	919	1.22	766	1.05	656	1.82	1149	1.73	1149	92	1.73	1149
	600	60	1.37	1.18	1241	86	1.10	993	0.92	828	0.79	709	1.37	1241	1.30	1241	91	1.30	1241
	300	30	0.77	0.64	1353	84	0.62	1083	0.52	902	0.44	773	0.77	1353	0.72	1353	90	0.72	1353
	100	10	0.28	0.23	1433	80	0.23	1147	0.19	956	0.16	819	0.28	1433	0.26	1433	88	0.26	1433
15 (15)	2500	167	2.41	2.08	786	86	1.95	629	1.64	524	1.42	449	2.12	687	2.31	786	90	2.31	786
	1750	117	2.10	1.80	971	86	1.69	777	1.42	647	1.23	555	2.08	960	1.99	971	90	1.99	971
	1160	77	1.67	1.41	1147	84	1.35	917	1.13	764	0.97	655	1.67	1147	1.57	1147	90	1.57	1147
	870	58	1.38	1.14	1244	83	1.11	995	0.93	829	0.80	711	1.38	1244	1.29	1244	89	1.29	1244
	600	40	1.05	0.85	1342	81	0.84	1074	0.70	895	0.61	767	1.05	1342	0.97	1342	88	0.97	1342
	300	20	0.59	0.46	1461	78	0.48	1169	0.40	974	0.34	835	0.59	1461	0.54	1461	86	0.54	1461
	100	6.7	0.22	0.16	1545	74	0.18	1236	0.15	1030	0.13	883	0.22	1545	0.19	1545	84	0.19	1545
20 (20)	2500	125	1.94	1.61	813	84	1.57	650	1.32	542	1.14	465	1.76	738	1.83	813	88	1.83	813
	1750	88	1.66	1.37	988	83	1.34	790	1.13	658	0.97	564	1.66	988	1.56	988	88	1.56	988
	1160	58	1.31	1.06	1151	81	1.05	921	0.88	767	0.76	658	1.31	1151	1.21	1151	87	1.21	1151
	870	44	1.08	0.86	1241	80	0.87	993	0.72	827	0.62	709	1.08	1241	0.99	1241	87	0.99	1241
	600	30	0.81	0.63	1331	78	0.65	1065	0.55	887	0.47	760	0.81	1331	0.74	1331	86	0.74	1331
	300	15	0.46	0.34	1438	74	0.37	1151	0.31	959	0.27	822	0.46	1438	0.41	1438	84	0.41	1438
	100	5.0	0.17	0.12	1515	70	0.14	1212	0.15	1010	0.10	866	0.17	1515	0.15	1515	82	0.15	1515
25 (25)	2500	100	1.62	1.30	822	81	1.31	657	1.11	548	0.96	470	1.53	771	1.52	822	86	1.52	822
	1750	70	1.38	1.10	988	80	1.11	791	0.94	659	0.81	565	1.38	988	1.28	988	86	1.28	988
	1160	46	1.08	0.84	1143	78	0.87	914	0.73	762	0.63	653	1.08	1143	0.99	1143	85	0.99	1143
	870	35	0.89	0.68	1227	77	0.71	982	0.60	818	0.52	701	0.89	1227	0.80	1227	85	0.80	1227
	600	24	0.67	0.50	1311	75	0.54	1049	0.45	874	0.39	749	0.67	1311	0.60	1311	84	0.60	1311
	300	12	0.38	0.27	1412	71	0.30	1129	0.25	941	0.22	807	0.38	1412	0.33	1412	82	0.33	1412
	100	4.0	0.14	0.09	1483	67	0.11	1186	0.09	989	0.08	847	0.14	1483	0.12	1483	79	0.12	1483

1. Numbers shown in ( ) are exact ratios.
2. If input speed is below 1160 RPM, please specify speed and mounting position to insure proper lubrication.
3. See engineering section, pages 228-230, for further discussion regarding service factors.
4. Actual input HP must not exceed the thermal input HP capacity on a continuous basis.
5. 1.00 Service Factor.
6. See engineering section, page 225, for further discussion regarding gear reducer efficiencies.

FIGURE 13.5 Worm gear mechanical and thermal ratings. (Courtesy of Peerless-Winsmith, Inc.)

13.8 PROBLEMS

1. A helical gear with a D_p of 9 inches transmits a torque of 1,800 in.-lb at 2,400 rpm. There are 45 teeth on this gear. The axial plane pressure angle is 20°. The helix angle is 30°. Determine:
   a. normal plane pitch
   b. pitch-line velocity
   c. transmitted force at the pitch line
2. In Problem 1, find the:
   a. axial force
   b. separating force
3. In Problem 1, if this gear is commercial grade and made from hot-drawn AISI 1050 steel and is 2 inches wide, find the:
   a. allowable force
   b. dynamic force
   c. safety/service factor

From Figure 13.2, select a set of gears for Problems 4 and 5.
4. Input 1,800 rpm, 1½ hp. Output 600 rpm.
5. Same as Problem 4, except $N_{sf} = 2$.
6. A miter-gear set transmits 5 hp at 1,750 rpm. If the gears have 24 teeth, are 6 pitch, and have a 20° pressure angle, find the:
   a. transmitted force
   b. separating force
   c. axial force
   d. Compare the answers to parts b and c, and explain why they are the same.
7. A worm-gear set has a double-start worm with a pitch diameter of 1½ inches. The gear has 72 teeth. If the gear is an 8 pitch; the input is a 1 hp, 3,450 rpm electric motor; and the pressure angle is 20°, find the:
   a. lead angle
   b. output speed
   c. centerline distance
8. In Problem 7, if the gear is made from phosphor bronze and is 1 inch wide, determine the power rating for long life and a service factor of 1.
9. In Problems 7 and 8, if the coefficient of friction is .04, find the:
   a. efficiency
   b. output torque
10. In Problem 9, find the heat generated by this gear set based on the 1 hp input.

11. In Example Problems 13.4 through 13.6, assuming the 5-hp drive motor and worm-gear system is 60 percent efficient and electricity costs 10¢ per kwh, what is the annual electricity cost if this system will be used 40 hours per week?
12. In Problem 11, if a helical double-reduction system with an 85 percent efficiency could be substituted, but its purchase cost is $750 more, determine the payback period.

Problems 13 through 17 refer to Figures 13.6, 13.7, and 13.8.

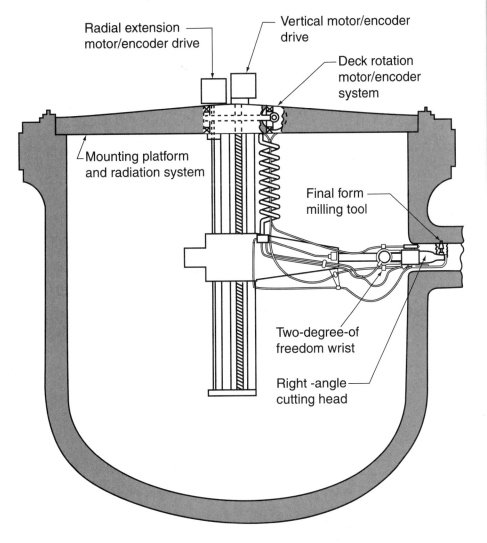

System shown in nozzle bore

FIGURE 13.6 Machine for underwater machining.

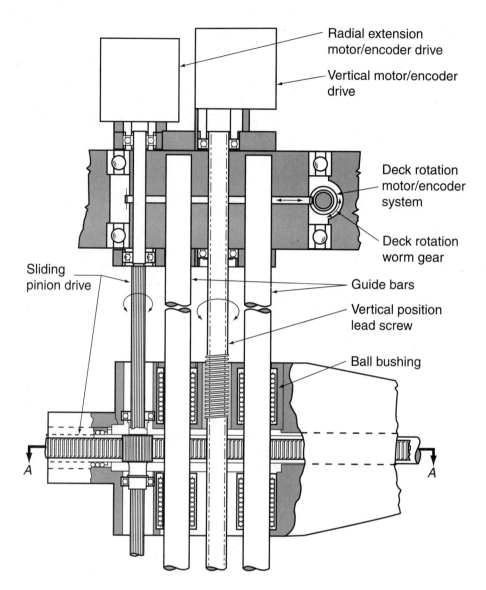

Details of vertical and radial extension system

FIGURE 13.7 Use of worm, rack-and-pinion, sliding spline, and lead screw systems.

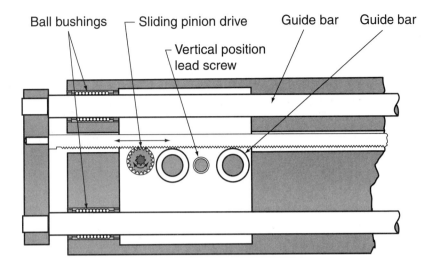

Section A-A
Details of vertical and radial extension system

FIGURE 13.8 Section view of system shown in Figure 13.7.

13. If the worm-gear deck rotation system gear has 150 teeth and needs to turn at 3 rpm, determine the worm rotational speed if it is a
    a. single-start
    b. double-start.
14. If the vessel in which this system operates is 12 feet in diameter and the force resisting rotation at the wall is 100 pounds, determine the worm torque required and the power needed for the single-start worm (assume $e = 100\%$).
15. Assume the gear is made from AISI 303 annealed stainless steel, has a 20° pressure angle, and is 8 pitch. The worm is 1¾ inches in diameter at the pitch line for an N_{sf} of 1.5 and is 1 inch wide. Is this acceptable for long life?
16. What is the efficiency of this system if the coefficient of sliding friction is $f = .05$?
17. What is the input torque and power for this efficiency?

13.9 CUMULATIVE PROBLEMS

18. The truck in previous chapters has a 1:2 ratio in the differential using a bevel gear system with 36 and 72 teeth, $P_d = 8$, and $\theta = 14\frac{1}{2}°$. For the conditions in Chapter 11, Problem 13, find the maximum separating and axial forces (see Figure 17.6).
19. From Problem 18, find the torque at the wheel for each gear.
20. From Problems 18 and 19, if the truck has a 42-inch-diameter tire, find the speed in each gear and the force between the tire and the road.

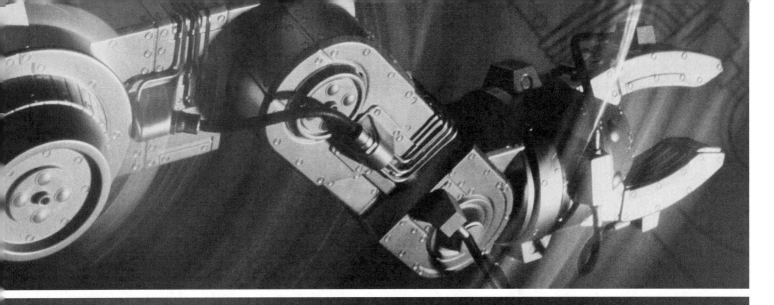

CHAPTER 14

Belt
and
Chain
Drives

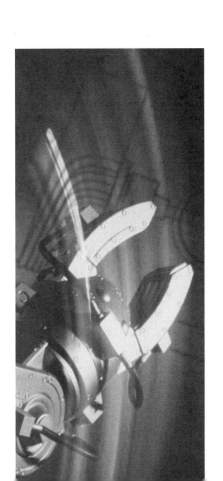

Chapter 14

> **flexible-drive systems**
> A term used to encompass belt and chain drives or other drives where a fixed distance between the shaft centerlines is not required.

The spectrum of available **flexible-drive systems** has expanded significantly over the years. The earliest flexible-drive systems were flat belts made from cowhide that were sewn together by seamstresses. With the introduction of rubber compounds, flexible-drive systems evolved into flat and V-belt rubber drives with reinforced fabric belting made from cotton or other natural fibers. Now these systems have evolved again due to the use of many synthetic high-strength materials that significantly increase their capacity. The introduction of timing belts allowed a flexible-drive system to have a fixed control of ratios similar to gears or chain drives. Many new types of drive systems, such as cable chains that can even turn corners, offer many alternatives to the designer. When higher loads or exact timing is required, chains may often be substituted for belts. Chain drives can consist of miniature systems, small systems such as those used on bicycles, medium systems such as those on motorcycles, large systems such as on large construction or industrial equipment, and very large systems such as those that open locks on a canal. A good example of the evolution of a flexible-drive system would be a motorcycle. Most of the early motorcycles used a chain-drive system, whereas now many use belts that offer lighter weight, quieter operation, and lower maintenance.

14.1 OBJECTIVES

In this chapter, we will learn about many of the different types of flexible-drive systems available and the basic principles of their operation. After completing this chapter, you should be able to:

- Understand the principles of operation of flexible-drive systems.
- Determine allowable forces and torques for flexible-drive systems, along with the necessary sprockets or sheaves.
- Identify some of the newer types of flexible-drive systems.
- Describe the basic features of belt-drive systems.
- Describe the basic features of chain-drive systems.
- Understand the principles of operation of different types of chain drives.
- Specify types and sizes of chain drives and their associated sprockets for different applications.
- Recognize applications where flexible-drive systems, including both belt and chain drives, are advantageous.
- Recognize the advantages and disadvantages of different types of flexible-drive systems.

14.2 BELT DRIVES

All belt drives require some tension in the belt system over and above the driving force. Flat belts, for instance, require a significant force separating the pulleys in order to create the necessary levels of friction. V-belts require less separating force because of their wedging action. Timing belts, because of the physical grooves in the belt that act like cogs, require even less separating force. In order to understand the principles of operation, we will label the two forces in the belt as a front force and a back force, and the difference between them becomes the net driving force. We do this because there is still some tension on the slack side or return side of the belt that provides the appropriate friction. Hence, we can envision this back force as subtracting from the front force. The resultant torque on the system becomes the difference between them (see Figure 14.1):

$$F_d = F_f - F_b \tag{14.1}$$

Because of this differential force acting on the two pulleys, we can calculate the net resultant torque as follows:

$$T = (F_f - F_b)\, r \tag{14.2}$$

or

$$T = F_d\, r \tag{14.3}$$

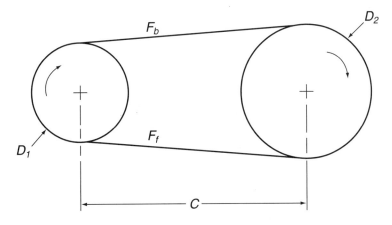

FIGURE 14.1 Typical drive belt geometry: left pulley is driven.

sheave
The term for a pulley used in a belt system.

The ratio of a belt drive is determined in a manner similar to a gearing ratio. The rotational speed is the ratio of the diameter or circumferences of the two **sheaves**. This ratio can be determined as follows:

$$m_w = \frac{D_2}{D_1} = \frac{C_2}{C_1} \qquad (14.4)$$

Note that in most types of belt-drive systems, in particular flat belts and to a lesser extent with V-belts, minor slippage occurs due to the deflection of the belt. The ratio expressed in Equation 14.4, unlike gear or timing belt drives, may not be absolute, and a slight reduction in the ratio is normally expected.

The surface velocity of belt-drive systems is often used in selecting the appropriate type of belting. Potential belt whipping, as well as centrifugal forces in a chain system due to the weight of the chain, can be of concern at high speeds. The surface velocity of a belt or chain system can be determined as follows:

$$V_m = \frac{\pi D n}{12} \qquad \text{ft/min} \qquad (14.5)$$

where D is in inches

n is in rpm

V_m = surface speed in feet/minute

The following example illustrates this.

Example Problem 14.1

A flat-belt system has a small pulley with a 6-inch diameter and a large pulley with a 10-inch diameter. The small pulley transmits 2 hp at 2,450 rpm, and a back force of 10 pounds in the belt is required for frictional purposes. Determine the front force, net driving force, and force on the shafts. Also, find the belt surface speed and the output speed.

First, find the torque on the small pulley.

$$T = \frac{63,000 \text{ hp}}{n} \qquad (2.6)$$

$$T = \frac{63,000(2)}{2,450}$$

$$T = 51.4 \text{ in.-lb}$$

If

$$T = F_d\, r \quad (14.3)$$

$$F_d = \frac{T}{r}$$

$$F_d = \frac{51.4 \text{ in.-lb}}{\frac{6 \text{ in.}}{2}}$$

$$F_d = 17 \text{ lb}$$

then, the front force would be

$$F_d = F_f - F_b \quad (14.1)$$
$$F_f = F_d + F_b$$
$$F_f = 17 \text{ lb} + 10 \text{ lb}$$
$$F_f = 27 \text{ lb}$$

The force pulling the shafts together would be

$$F_t = F_f + F_b$$
$$F_t = 27 \text{ lb} + 10 \text{ lb}$$
$$F_t = 37 \text{ lb}$$

Find the surface speed.

$$V_m = \frac{\pi D n}{12} \quad (14.5)$$

$$V_m = \frac{\pi\, 6 \text{ in.}\, 2{,}450 \text{ rpm}}{12}$$

$$V_m = 3{,}848 \text{ ft/min}$$

Find the ratio.

$$m_w = \frac{D_2}{D_1} \quad (14.4)$$

$$m_w = \frac{10}{6} = \frac{1.67}{1}$$

The output speed is

$$n = \frac{2,450 \text{ rpm}}{\frac{1.67}{1}} = 1,467 \text{ rpm}$$

Note: This is approximate as minor slippage may occur.

In general, most belt-drive systems are capable of fairly high surface velocities, whereas chain-drive systems are best used in lower-velocity applications. However, some of the lighter chain-drive systems, such as those made of plastic, may be capable of fairly high velocities, depending on the application, tension, and other factors.

In order to size belt or chain systems, we must determine the length of the belt and/or chain. For chains or timing belts, it is necessary to work in complete link distances, whereas other belts can be special-ordered in any length needed. Most styles of belts are available in standard sizes. However, some can be "field spliced" to make any desired length. The length of a belt or chain would be determined as follows:

$$L = 2C + 1.57(D_1 + D_2) + \frac{(D_2 - D_1)^2}{4C} \quad (14.6)$$

The centerline distance, which is the centerline spacing between the mounting shafts, would be determined as follows:

$$C = \frac{B + \sqrt{B^2 - 32(D_2 - D_1)^2}}{16} \quad (14.7)$$

where B is calculated as

$$B = 4L - 6.28(D_2 + D_1) \quad (14.8)$$

It is often necessary to calculate the **angle of contact** for flexible-drive systems. In the case of belt systems, the angle of contact is needed to obtain the appropriate friction, and for chain systems, it is necessary to provide appropriate interaction between the sprocket and the chain. The angle of contact can be calculated as follows:

$$\theta = 180° \pm 2 \sin^{-1}\left(\frac{D_2 - D_1}{2C}\right) \quad (14.9)$$

Belt length (14.6)

$$L = 2C + 1.57(D_1 + D_2) + \frac{(D_2 - D_1)^2}{4C}$$

Pulley spacing (14.7)

$$C = \frac{B + \sqrt{B^2 - 32(D_2 - D_1)^2}}{16}$$

(14.8)

$$B = 4L - 6.28(D_2 + D_1)$$

Wrap angle (14.9)

$$\theta = 180° \pm 2 \sin^{-1}\left(\frac{D_2 - D_1}{2C}\right)$$

---

angle of contact

Sometimes referred to as "angle of wrap" for flexible-drive systems; the angle at which the belt or chain is in engagement with the drive pulley or sprocket.

where the positive value would be for the larger sprocket and the negative value for the small sprocket or sheave. Different kinds of flexible-drive systems require different minimum angles of contact, but in general, the greater the angle of contact, the better for both maintaining contact and friction. In addition, nearly all belt systems must be adjustable, both for installing the belt and for compensating for stretch over time. Therefore, either the center distance must be adjustable or the system must include an idler pulley. If possible, this idler is placed on the slack side of the small pulley to increase the contact angle. However, this is not always the case: for example, a complicated **serpentine** drive system in an automobile engine may not contain an idler because some components, such as alternators, allow for adjustments to be made. An example of a typical layout using an idler is shown in Figure 14.2. Example Problem 14.2 demonstrates sizing of belt-drive systems.

serpentine
A term used for a complicated belt or chain drive system that may go around many components and in some cases uses both sides of the belt or chain.

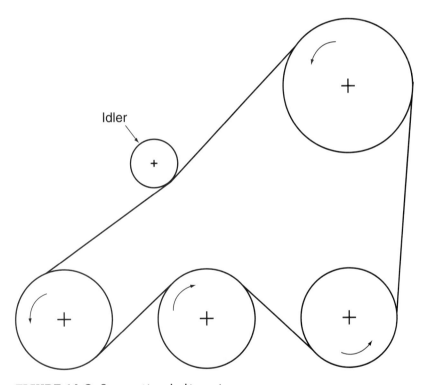

FIGURE 14.2 Serpentine belt system.

Example Problem 14.2

We wish to mate a 4-inch-diameter sheave with a 6-inch-diameter one, and their centerlines can range from 18 to 20 inches apart. Select a length of belt and determine the actual centerline distance; if the small sheave rotates at 1,800 rpm, determine the ratio and surface speed. Also, find the angle of contact for the small sheave.

For determining the length of belt, try a centerline spacing of 19 inches:

$$L = 2C + 1.57(D_1 + D_2) + \frac{(D_2 - D_1)^2}{4C} \tag{14.6}$$

$$L = 2(19) + 1.57(4 + 6) + \frac{(6-4)^2}{4(19)}$$

$$L = 53.75 \text{ in.}$$

Assuming a 54-inch belt is available.

Find the centerline distance.

$$C = \frac{B + \sqrt{B^2 - 32(D_2 - D_1)^2}}{16} \tag{14.7}$$

$$B = 4L - 6.28(D_2 + D_1) \tag{14.8}$$

$$B = 4(54) - 6.28(6 + 4)$$

$$B = 153.2$$

$$C = \frac{153.2 + \sqrt{153.2^2 - 32(6-4)^2}}{16}$$

$$C = 19.12 \text{ in.}$$

Find the ratio.

$$m_w = \frac{D_2}{D_1} = \frac{6}{4} = 1.5 \tag{14.4}$$

Find the surface speed.

$$V_m = \frac{\pi D n}{12} \tag{14.5}$$

$$V_m = \frac{\pi \; 4 \text{ in. } 1,800 \text{ rpm}}{12}$$

$$V_m = 1,885 \text{ ft/min}$$

Find the angle of contact.

$$\theta = 180° - 2\sin^{-1}\left(\frac{D_2 - D_1}{2C}\right) \quad (14.9)$$

$$\theta = 180° - 2\sin^{-1}\left(\frac{6 - 4}{2(19.12)}\right)$$

$$\theta = 174°$$

14.3 V-BELTS

V-belts are the most common type of flexible-drive system, and they have been in existence for many years. One of the greatest advantages of the V-belt system is that, because of its shape, it creates a wedging action in the sheaves so that the frictional driving force is increased significantly. Because of this, less back force is needed than for a flat-belt system. This results in a higher torque in relation to the shaft bending loads, which results in a smaller shaft, lower bearing loads, and smaller overall size. V-belts can be grouped in multiple-belt systems fairly easily and, due to the variety of backing materials available, can transmit a reasonably large load, depending on their size and the strength of this backing. Generally, the strength of a V-belt is almost entirely in the high-strength cords that are molded into the belt, whereas the rubber V-shape section is only used to prevent slipping. The cords can be made from natural fibers, synthetic strands, steel, or even newer composites such as **Kevlar.** Many standard sizes are available, including heavy-duty industrial V-belts, industrial narrow-section V-belts, light-duty fractional horsepower belts, and automotive series belts. Most of these sizes have an equivalent metric size, and the manufacturer's literature should be consulted for the dimensions of these standard shapes. V-belts are also made in standard lengths but, as there are many series of belts, manufacturers' literature should be consulted for availability. Modifications to the basic V-belt system are continually evolving to meet new and changing applications, such as load-carrying cords, and to take advantage of different materials such as rubber compounds that increase friction and life.

Power ratings for V-belt drives are best found from manufacturers' catalogs. Factors such as reinforcing materials and rubber compounds can significantly affect the rated capacity. Other factors such as minimum sheave size, temperature, and expected life will vary significantly based on these material properties. Hence, the ratings can best be found from the manufacturer's independent analysis and testing. Figure 14.3 shows the V-belt wedging action, along with the basic dimensions of a V-belt sheave groove, and Figure 14.4 shows some of the standard V-belt sizes.

Kevlar
A term used for reinforced strands of materials that are made from a carbon fiber possessing high strength and flexibility.

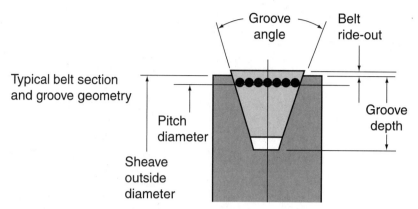

FIGURE 14.3 Cross-section of V-belt in sheave groove.

TABLE 14.1

STANDARD BELT LENGTHS FOR 3V, 5V, AND 8V BELTS (in.)				
3V Only	3V and 5V	3V, 5V, and 8V	5V and 8V	8V Only
25	50	100	150	375
26.6	53	106	160	400
28	56	112	170	425
30	60	118	180	450
31.5	63	125	190	475
33.5	67	132	200	500
35.5	71	140	212	
37.5	75		224	
40	80		236	
42.5	85		250	
45	90		265	
47.5	95		280	
			300	
			315	
			335	
			355	

Table 14.1 includes standard belt lengths for series 3, 5, and 8 V-belt systems. Figure 14.5 is a page from a manufacturer's catalog delineating the horsepower ratings for their series 5 and 8 Aramide cord belt systems. Note that this is a specialized material which results in a very high power rating. Example Problem 14.3 reviews the selection process for selecting a V-belt drive system.

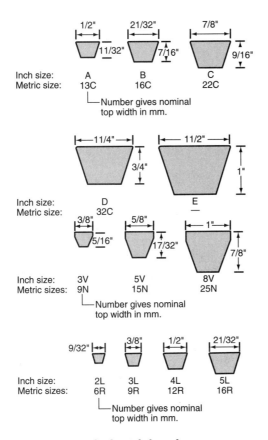

Industrial series

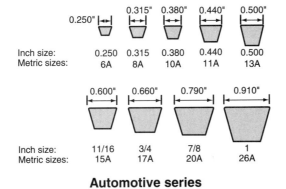

Automotive series

FIGURE 14.4 Standard V-belt sizes.

5VF BASIC HORSEPOWER RATINGS
Aramide Cord Belt

SEE CAUTION BELOW

Faster Shaft RPM	Rated HP per Belt for Small Sheave O.D. of:														Additional HP per Belt for Speed Ratio of: †				
	7.1	8.0	8.5	9.0	9.75	10.3	10.9	11.3	11.8	12.5	13.2	14.0	15.0	16.0	1.02 to 1.20	1.21 to 1.50	1.51 to 2.19	2.20 to 3.32	3.33 and up
200	3.55	4.42	4.91	5.39	6.11	6.63	7.2	7.58	8.05	8.71	9.36	10.1	11.0	11.9	.10	.24	.33	.37	.38
300	5.01	6.29	6.99	7.89	8.74	9.5	10.3	10.9	11.5	12.5	13.5	14.5	15.9	17.2	.15	.36	.50	.55	.57
400	6.39	8.05	8.97	9.88	11.2	12.2	13.3	14.0	14.9	16.1	17.4	18.8	20.5	22.2	.19	.47	.66	.74	.76
500	7.71	9.74	10.9	12.0	13.6	14.8	16.2	17.0	18.1	19.6	21.1	22.8	24.9	27.0	.24	.59	.83	.92	.94
600	8.96	11.4	12.7	14.0	15.9	17.4	18.9	19.9	21.2	23.0	24.8	26.7	29.2	31.7	.29	.71	.99	1.11	1.13
700	10.2	12.9	14.4	15.9	18.2	19.8	21.6	22.8	24.2	26.3	28.3	30.5	33.3	36.1	.34	.83	1.16	1.29	1.32
800	11.3	14.4	16.1	17.8	20.4	22.2	24.2	25.5	27.1	29.4	31.6	34.2	37.3	40.4	.39	.95	1.32	1.48	1.51
900	12.4	15.9	17.8	19.7	22.5	24.5	26.7	28.2	29.9	32.4	34.9	37.7	41.1	44.5	.44	1.07	1.49	1.66	1.70
1000	13.5	17.3	19.4	21.5	24.5	26.7	29.1	30.7	32.7	35.4	38.0	41.1	44.8	48.4	.49	1.18	1.65	1.85	1.89
1100	14.6	18.7	20.9	23.2	26.5	28.9	31.5	33.2	35.3	38.2	41.1	44.3	48.2	52.1	.53	1.30	1.82	2.03	2.08
1200	15.6	20.0	22.4	24.8	28.4	30.9	33.7	35.5	37.8	40.9	43.9	47.4	51.5	55.6	.58	1.42	1.99	2.22	2.27
1300	16.6	21.3	23.9	26.4	30.2	32.9	35.9	37.8	40.2	43.5	46.7	50.3	54.7	58.9	.63	1.54	2.15	2.40	2.45
1400	17.5	22.5	25.3	28.0	32.0	34.9	37.9	40.0	42.5	45.9	49.3	53.0	57.6	61.9	.68	1.66	2.32	2.59	2.64
1600	19.3	24.9	27.9	30.9	35.3	38.5	41.8	44.1	46.8	50.5	54.1	58.1	62.8	67.3	.78	1.89	2.65	2.96	3.02
1800	20.9	27.0	30.3	33.6	38.4	41.7	45.4	47.7	50.6	54.5	58.2	62.3	67.2	71.6	.88	2.13	2.98	3.32	3.40
2000	22.4	29.0	32.6	36.0	41.1	44.7	48.5	50.9	53.9	57.9	61.7	65.8	70.5	74.8	.97	2.37	3.31	3.69	3.78
2200	23.8	30.8	34.5	38.2	43.5	47.2	51.2	53.7	56.7	60.7	64.5	68.5	72.9	76.7	1.07	2.60	3.64	4.06	4.15
2400	24.9	32.4	36.3	40.1	45.6	49.4	53.4	55.9	58.9	62.8	65.5	70.2	74.1	77.3	1.17	2.84	3.97	4.43	4.53
2600	26.0	33.7	37.8	41.7	47.3	51.1	55.1	57.6	60.5	64.3	67.6	70.9	...	...	1.26	3.08	4.30	4.80	4.91
2800	26.9	34.8	39.0	43.0	48.6	52.4	56.3	58.7	61.4	64.9	67.8	...	...	...	1.36	3.31	4.63	5.17	5.29
3000	27.6	35.8	40.0	44.0	49.6	53.3	56.9	59.2	61.7	64.7	...	...	...	...	1.46	3.55	4.96	5.54	5.66
3200	28.1	36.4	40.7	44.6	50.1	53.6	57.0	59.0	61.2	...	...	...	...	...	1.56	3.79	5.30	5.91	6.04
3400	28.4	36.8	41.0	44.9	50.1	53.4	56.5	56.2	...	...	...	...	...	...	1.65	4.02	5.63	6.28	6.42
3600	28.6	36.9	41.1	44.8	49.7	52.7	...	...	...	...	...	...	...	...	1.75	4.26	5.96	6.65	6.80

8VF

SEE CAUTION BELOW

Faster Shaft RPM	Rated HP per Belt for Small Sheave O.D. of:												Additional HP per Belt for Speed Ratio of: †				
	12.5	13.2	14.0	15.0	16.0	17.0	18.0	19.0	20.0	21.2	22.4	24.8	1.02 to 1.20	1.21 to 1.50	1.51 to 2.19	2.20 to 3.32	3.33 and up
200	12.6	14.5	16.6	19.3	21.9	24.6	27.2	29.8	32.433	35.5	38.6	44.7	.59	1.43	2.00	2.24	2.29
250	15.0	17.4	20.0	23.3	26.5	29.7	33.0	36.2	39.4	43.2	46.9	54.5	.74	1.79	2.51	2.80	2.86
300	17.4	20.1	23.2	27.1	30.9	34.7	38.5	42.3	46.1	50.5	55.0	63.8	.88	2.15	3.01	3.36	3.43
350	19.6	22.7	26.3	30.7	35.2	39.6	43.9	48.3	52.6	57.7	62.8	72.9	1.03	2.51	3.51	3.91	4.00
400	21.7	25.2	29.3	34.3	39.3	44.2	49.1	54.0	58.9	64.6	70.4	81.7	1.18	2.87	4.01	4.47	4.57
450	23.7	27.7	32.2	37.7	43.2	48.7	54.2	59.6	65.0	71.4	77.7	90.2	1.33	3.23	4.51	5.03	5.15
500	25.6	30.0	34.9	41.0	47.1	53.1	59.1	65.0	70.9	77.8	84.8	98.4	1.47	3.58	5.01	5.59	5.72
600	29.3	34.4	40.1	47.3	54.4	61.4	68.4	75.3	82.2	90.2	98.2	113.9	1.77	4.30	6.01	6.71	6.86
700	32.6	38.4	45.0	51.2	61.2	69.2	77.1	84.9	92.6	101.7	110.7	128.1	2.06	5.02	7.01	7.83	8.00
800	35.6	42.1	49.5	58.6	67.6	76.5	85.2	93.8	102.3	112.3	122.1	141.1	2.36	5.73	8.02	8.95	9.15
900	38.3	45.5	53.6	63.6	73.5	83.1	92.7	102.0	111.2	121.9	132.4	152.5	2.65	6.45	9.02	10.1	10.3
1000	40.7	48.5	57.4	68.2	78.8	89.2	99.4	109.4	119.1	130.5	141.5	162.4	2.95	7.17	10.0	11.2	11.4
1100	42.9	51.3	60.7	72.3	83.6	94.7	105.5	116.0	126.2	138.0	149.4	170.7	3.24	7.89	11.0	12.3	12.6
1200	44.7	53.7	63.7	75.9	87.9	99.5	110.8	121.7	132.2	144.4	155.9	177.3	3.53	8.60	12.0	13.4	13.7
1300	46.2	55.7	66.3	79.2	91.6	103.7	115.3	126.5	137.3	149.5	161.1	181.8	3.83	9.32	13.1	14.5	14.8
1400	47.4	57.3	68.4	81.8	94.7	107.1	119.0	130.4	141.2	153.4	164.5	184.5	4.12	10.1	14.1	15.7	16.0
1500	48.3	58.6	70.1	83.9	97.1	109.8	121.8	133.2	143.9	155.8	165.7	184.9	4.42	10.7	15.0	16.8	17.1
1600	48.8	59.5	71.3	85.4	98.9	111.7	123.7	135.0	145.4	156.3	165.0	...	4.71	11.5	16.0	17.9	18.3
1700	49.0	59.0	72.0	86.4	100.0	112.8	124.7	135.7	145.8	156.3	165.5	...	5.01	12.2	17.0	19.0	19.4
1800	48.8	60.0	72.2	86.8	100.4	113.0	124.6	135.1	144.5	154.2	...	...	5.30	12.9	18.0	20.1	20.6
1900	48.2	59.5	71.9	86.5	100.0	112.4	123.5	133.4	141.9	...	...	...	5.60	13.6	19.0	21.2	21.7
2000	47.2	58.7	71.1	85.6	98.8	110.7	121.3	130.3	...	...	...	...	5.89	14.3	20.0	22.3	22.8
2100	45.8	57.3	69.7	84.0	96.8	108.1	117.9	...	...	...	...	...	6.19	15.0	21.0	23.5	24.0
2200	43.9	55.5	67.7	81.7	94.0	104.5	...	...	...	...	...	...	6.48	15.8	22.0	24.6	25.1

Shaded Areas indicate rim speeds exceeding 6500 FPM which require higher strength sheaves.
▲ Subject to Arc and Length Correction Factors

TOTAL RATING = Rated HP + "additional HP" from right hand column.
† Additional HP below 1.02 ratio equals zero.

CAUTION: Belt horsepower ratings may exceed design capacity of stock sheaves. Consult factory for recommendations.

FIGURE 14.5 Typical power ratings for standard V-belts. (Courtesy of Rockwell Automation, Inc.)

Example Problem 14.3

From Figure 14.5, determine the power rating for a 5 VF Aramide cord belt for an input speed of 1,800 rpm, an output speed of 900 rpm, with an input sheave OD of 8 inches and a service factor of 1.5. Determine a standard belt length if the input and output shafts need to be about 20 inches apart.

(From Figure 14.5, the basic rating from the 1,800-rpm row and the 8-inch column is 27 hp.)

In addition, add the horsepower for a 2:1 ratio of 2.98:

$$P = 27 + 2.98 = 29.98 \text{ hp}$$

Adding in the service factor,

$$P = \frac{29.98}{1.5} = 20 \text{ hp}$$

The output sheave would need to have a 16-inch diameter to obtain the 900 rpm speed.

Find the approximate length of belt needed.

$$L = 2C + 1.57(D_1 + D_2) + \frac{(D_2 - D_1)^2}{4C} \tag{14.6}$$

$$L = 2(20) + 1.57(8 + 16) + \frac{(16 - 8)^2}{4(20)}$$

$$L = 78.5 \text{ in.}$$

Use an 80-inch belt. **(Table 14.1)**

14.4 FLAT-BELT SYSTEMS

Flat-belt systems offer an interesting perspective into history. In the early days of power transmission, flat belts were used almost exclusively. Later, with the introduction of V-belts and timing belts, flat belt usage decreased almost to extinction. Later, with the introduction of serpentine-belt systems, they returned to frequent use. Many modern automobiles use flat belts to a significant extent in order to save space. It is interesting to note the following section on the strength of leather flat belts from a textbook written originally in the 1930s, revised in the 1950s, and used extensively into the 1970s:

Since the hides of no two steers are likely to be of the same quality and since the strength of leather also depends upon the method of tanning, the variability is expected to be large. The breaking strength of oak tanned belting varies from 3 to more than 6 ksi. Mineral tanned leather is stronger, say 7 to 12 ksi. To make the belt endless, the ends must be joined by cementing. . . .

Obviously very few flat-belt systems are now made of leather. Modern polymers, along with different cord systems, have replaced leather. With the advent of modern materials has come the extensive application of serpentine-belt systems. If you open the hood of your car, you will probably see that a two-sided serpentine-belt system drives the engine compartment auxiliary systems such as the alternator, power steering, and air conditioner. These serpentine-belt systems are flat belts, but they may incorporate features of a V-belt system (such as small V sections) or, in some cases, timing belt cogs. The operating principles of flat-belt systems are applicable to these modified flat-belt systems, and the forces discussed in the earlier section of this chapter still apply.

14.5 TIMING BELTS

If we envision a flat belt, especially in its modern adaptations, as a serpentine system that uses both sides of the belt, then add to that the advantages of a V-belt system, which reduces the tension required for friction, and then add the exact timing associated with gear systems, we would have a timing belt. Because of the timing belt's built-in gear or cog system, exact timing can be obtained (as is needed for opening and closing valves in an automobile engine) along with the space advantages of flat-belt systems. Figure 14.6 shows a timing-belt system. Forces and torques are found in the same manner as for flat belts. As with other types of belt systems, the strength of a timing belt is derived from the strength of the reinforcing cords. Manufacturers' catalogs should be consulted for the torque and power ratings.

14.6 CHAIN DRIVES

The distinction between a chain drive and a timing belt is significant only due to the material from which they are made. Most chains are made from high-strength steel and can transmit high loads at low and moderate speeds, as opposed to timing belts which are typically used to transmit lower loads at higher speeds. Their principles of operation are very similar: both use a type of pitch that is the equivalent of the circular pitch for gears. There are many types of chain systems, the most common of which is a

FIGURE 14.6 Timing belts and pulleys. (Courtesy of W. M. Berg, Inc.)

roller chain. However, many applications have spawned the development of many types of specialty chains. Chains can be used for reducing ratios in a power-drive system and for transporting parts. Many adaptations and modifications of chain systems are available. There are two somewhat similar, yet different, ways of specifying the load capacities of chain systems. In one approach, the tensile strength of a chain is specified (as shown in Table 14.2) for different series of chain sizes, along with the associated pitch. When the average tensile strength is used to select a chain, a factor of safety is incorporated. In the other approach, we consult the manufacturers' catalogs for a horsepower chart based on the size of the chain, speed, and number of teeth in the smaller sprocket. Figure 14.7 shows an example of such a chart. Either method is a suitable way of selecting an appropriate chain system. In the second method, a service factor is often included.

Probably the greatest difference between a chain and a belt-drive system is that the roller chain typically needs some form of regular lubrication.

TABLE 14.2
PROPERTIES OF STANDARD ROLLER CHAINS (REGULAR)

Chain No.	Pitch	Average Ultimate Strength (lb)	Approximate Weight (lb/ft)	Limiting Speed (ft/min)[†]
25*	¼	875	.09	3,500
35*	⅜	2,100	.21	2,800
41	½	2,000	.26	2,300
40	½	3,700	.42	2,300
50	⅝	6,100	.68	2,000
60	¾	8,500	1.00	1,800
80	1	14,500	1.73	1,500
100	1¼	24,000	2.50	1,300
120	1½	34,000	3.69	1,200
140	1¾	46,000	5.00	1,100
160	2	58,000	6.50	1,000
180	2¼	76,000	9.06	950
200	2½	95,000	10.65	900

*Rollerless
[†]Guideline for lubricated chain

Lubrication is important because there is motion within the chain, specifically in the roller that goes around the central pin and the flexure of the links on the sides of the chain. The internals of a typical roller-chain system are shown in Figure 14.8. There are many different ways to lubricate roller chains. A splash-type system where the chain dips into an oil reservoir as it rotates is the most common, but drip-type systems, or even occasionally manual oiling, are used in some applications that require only intermittent service.

Another factor to consider with chains is they have significant weight. Hence they operate better at lower speeds because of the inertia effects from the moving chain. Unlike any of the belt-drive systems, chains are almost never used with a back force because a slight amount of slack is required to prevent binding. Lastly, roller chains must be used in whole-pitch lengths. However, chains can be field-cut and joined so that, unlike most belt systems, they do not need to be purchased from the manufacturer in a fixed length. There is also relatively little stretch in a chain system, and for most systems, significant take-up or adjustment capabilities are not needed. This is also dissimilar from belt drive systems. Many standard adapters can be purchased for chain systems, especially for applications where the chain is being used to transport a part. These adapters include such items as oversized rollers so that the chain can roll on a flat surface. Brackets, adapters, and crowned tops can be added for applications that transport parts or materials, such as those that move logs into place in a sawmill or transport parts for heat treatment in an industrial oven. See Figure 14.9.

Small Sprocket		HP RATINGS—STANDARD SINGLE * STRAND ROLLER CHAIN—NO. 40-1/2" PITCH																		
RPM → Teeth	P.D.	10	20	30	50	75	100	125	150	200	250	300	400	500	600	900	1200	1500	1800	2400	3000
11	1.77	.054	.10	.15	.23	.33	.43	.53	.62	.80	.98	1.16	1.50	1.84	2.16	3.11	4.03	4.93	4.66	3.03	2.17
13	2.09	.065	.12	.17	.28	.40	.52	.63	.74	.96	1.18	1.39	1.80	2.20	2.59	3.73	4.83	5.91	5.99	3.89	2.79
15	2.40	.076	.14	.20	.32	.46	.60	.74	.87	1.12	1.37	1.62	2.10	2.56	3.02	4.35	5.64	6.89	7.43	4.82	3.45
17	2.72	.087	.16	.23	.37	.53	.69	.84	.99	1.29	1.57	1.85	2.40	2.94	3.45	4.98	6.45	7.89	8.96	5.82	4.17
19	3.04	.098	.18	.26	.42	.60	.78	.95	1.12	1.45	1.77	2.09	2.71	3.31	3.90	5.62	7.27	8.89	10.5	6.88	4.92
21	3.35	.109	.20	.29	.46	.67	.87	1.06	1.25	1.62	1.98	2.33	3.02	3.69	4.34	6.26	8.11	9.91	11.7	7.99	5.72
23	3.67	.120	.22	.32	.51	.74	.96	1.17	1.38	1.78	2.18	2.57	3.33	4.07	4.79	6.90	8.94	10.9	12.9	9.16	6.55
25	3.99	.132	.25	.35	.56	.81	1.05	1.28	1.51	1.95	2.38	2.81	3.64	4.45	5.24	7.55	9.78	12.0	14.1	10.4	7.43
Lubrication #		Type I								Type II						Type III		Type IV			

TYPE I—MANUAL LUBRICATION
Manual lubrication is accomplished by applying oil with a brush or spout can to the inside of the chain at the edges of the side plates. Volume and frequency should be determined by periodic inspection.

TYPE II—DRIP LUBRICATION
Oil is directed between link plate edges to a drip lubricator. Only enough oil to keep the chain moist is necessary and a light metal splash guard will keep the floor and surroundings clean.

TYPE III—BATH OR DISC LUBRICATION
With bath lubrication, the lower strand of the chain runs through a sump of oil. The oil level should reach the pitch line of the chain at its lowest point while operating. With disc lubrication, the chain operates above the oil level. The disc picks up oil from the sump and deposits it on the chain, usually by means of a trough. The disc diameter should be such as to produce rim speeds from 600 minimum to 8000 maximum FPM. This type of lubrication requires that the drive be enclosed in an oil tight chain case.

TYPE IV—OIL STREAM LUBRICATION
The lubricant is usually supplied by a circulating pump capable of supplying the chain drive with a continuous stream of oil. The oil should be applied inside the chain loop evenly across the chain width, and directed at the lower strand. This type of lubrication requires that the drive be enclosed in an oil tight chain case.

FIGURE 14.7 Number 40 roller chain horsepower ratings. (Courtesy of Boston Gear)

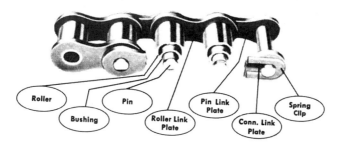

FIGURE 14.8 Cutaway view of roller chain. (Courtesy of Boston Gear)

FIGURE 14.9 Roller conveyor chain attachments. (Courtesy of Teleflex Morse Pte Ltd.)

Example Problem 14.4

From Figure 14.7, determine the horsepower rating for a number 40 roller chain if the input is through a 12-tooth sprocket turning at 1,800 rpm and the output turns at 900 rpm.

1. $P = 5.31$ hp
2. Select the output sprocket for this application.

$$12 \text{ tooth } \frac{1,800}{900} = 24 \text{ teeth}$$

3. Determine the surface speed for this application.

$$D = \frac{P_c N_t}{\pi} \quad (11.3)$$

$$D = \frac{.5\,(12)}{\pi} \approx 1.91 \text{ in.}$$

Note that on small sprockets, this is approximate.

$$V_m = \frac{\pi D n}{12} \quad (14.5)$$

$$V_m = \frac{\pi\, 1.91 \bullet 1,800}{12}$$

$$V_m = 900 \text{ ft/min}$$

4. What type of lubrication is recommended for this application? (Bath or disc lubrication from the bottom of Figure 14.7)
5. If the sprocket centerlines are to be 10 inches apart, how long a chain will be needed in full links?

First, find the large sprocket diameter.

$$D = \frac{P_c N_t}{\pi}$$

$$D = \frac{.5(24)}{\pi} = 3.82 \text{ in.}$$

$$L \approx 2C + 1.57(D_1 + D_2) + \frac{(D_2 - D_1)^2}{4C} \quad (14.6)$$

$$L \approx 2(10) + 1.57(1.91 + 3.82) + \frac{(3.82 - 1.91)^2}{4(10)}$$

$$L \approx 29.08$$

Use a 29- or 30-inch chain. Note that this formula is only approximate for chains.

6. What is the force in the chain if the power transmitted is 5.31 hp?

$$T = \frac{63,000 \text{ hp}}{n} \quad (2.6)$$

$$T = \frac{63,000 \cdot 5.31}{1,900}$$

$$T = 186 \text{ in.-lb}$$

7. How does this force compare to the ultimate strength?

$$F = \frac{2T}{D} \quad (12.3)$$

$$F = \frac{(2)186 \text{ in.-lb}}{1.91}$$

$$F = 195 \text{ lb}$$

8. Compare this result to the ultimate strength of 3,700 lb (not a valid comparison because of speed, etc.).

14.7 CABLE CHAINS AND OTHER SPECIALTY CHAIN DRIVES

In Section 14.6, we saw that many adaptations to chains are possible. However, over the years, due to the increased use of plastics and other polymers, many specialized types of chains have been created. A good example of this is a cable chain (as shown in Figure 14.10), which uses formed, plastic knobs on a cable so that, with special sprockets, this chain can be used in many angles and orientations, can turn corners easily, and can create large ratios if needed. Many chains of this type are now used in specialized applications such as vending machines and small part positioners. Load capacities for special chain drives should be determined from data supplied by the manufacturers.

14.8 SUMMARY

After reading this chapter, you should be gaining familiarity with the many different types of flexible-drive systems and their advantages and disadvantages. You should also understand where V-belts may be advantageous as opposed to flat belts, serpentine belts, timing belts, or chains. You should now be aware of the unique and different applications for many of the specialized chain products and how they can be used. Also, you should now understand the basics of pitch, length, separating distances, back forces, and driving forces and their associated torques. You should now begin to recognize where these types of products are currently used and how, when, and where they could be used.

14.9 PROBLEMS

1. Determine the approximate flat-belt length for an application that uses a 3-inch input pulley with a 9-inch output, if the shafts are 12 inches apart.
2. In Problem 1, if the input power source is a 5-hp electric motor turning at 1,800 rpm and a back force of 20 pounds is needed, determine the driving force, front force, force pulling the shafts together, surface speed, and output speed.
3. In Problem 1, determine the angle of wrap for both pulleys.

FIGURE 14.10 Cable chain. (Courtesy of W.M. Berg, Inc.)

4. We need to reduce the speed of a 10-hp electric motor from 1,750 rpm to 500 rpm using a V-belt system. Determine possible pulley sizes if a belt surface speed of about 1,800 ft/min is desired.
5. For the same conditions as in Problem 4, if you selected a 4-inch-drive pulley and the back force needs to be 10 percent of the driving force, find the front force and the force on the mounting shafts.
6. If the desired force in the belt should be below 50 pounds, how many V-belts should be used?
7. An existing 30-hp motor-driven compressor system uses four series 5 V-belts. If these V-belts were replaced by Aramide cord belts from Figure 14.5, how many belts would now be required, assuming a service factor of 1.8? The motor is an 1,150-rpm motor with an 8½-inch sheave, and the compressor uses a 12½-inch sheave.
8. In Problem 7, for the same approximate conditions, select a series 8 V-belt system. Determine the pulley sizes to maintain a similar compressor speed and the number of belts needed.
9. From Figure 14.7, for a number 40 chain, what is the power rating for a 20-tooth sprocket turning at 1,800 rpm? What is the tensile load on this chain, and how does this compare to the ultimate strength?
10. If the chain in Problem 9 drives a 45-tooth sprocket and the shafts are 20 inches apart, find the output torque and speed and the length of chain needed.
11. The drive system in Figure 14.11 uses a V-belt from the motor to the intermediate shaft and a number 40 chain from the intermediate shaft to the output shaft. If the input drive is a 1,750 rpm, 1-hp electric motor and a service factor for this system is 1.5, find the:
    a. speed in each shaft
    b. torque in each shaft
    c. surface speed for the chain
12. In Problem 11, is a number 40 chain acceptable, and how should it be lubricated?
13. In Problem 11, if a back force of 20 percent of the belt's net driving force is required and the chain has no back force, determine the two loads on the intermediate shaft.
14. If the intermediate shaft has a ¾-inch diameter, what is the torsional stress?

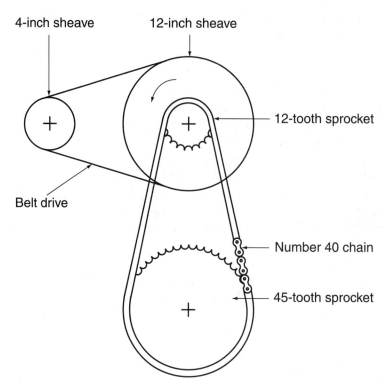

FIGURE 14.11 Belt and chain systems for Problems 9 through 12.

14.10 CUMULATIVE PROBLEMS

15. In Problem 5, if the output shaft pulley is centrally mounted between the two self-aligning bearings that are 12 inches apart and the shaft is ¾ inch in diameter, find the combined stress.
16. If a safety factor of 2 based on yield is desired in Problem 15 and the shafting is cold-drawn AISI 1040 steel, is this acceptable?

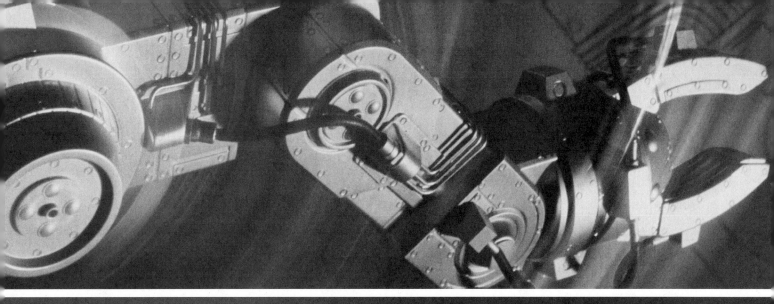

CHAPTER 15

Keys and Couplings

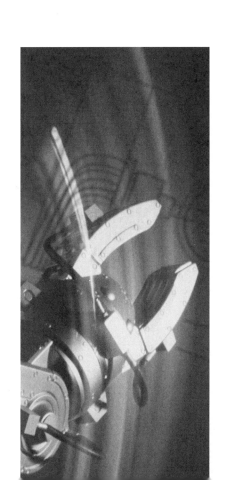

couplings
Any method used for attaching two shafts; may allow for angular misalignment.

clutch
Any method used to engage or disengage one rotating shaft from another; for instance, for attaching the rotating parts of a transmission to an automobile engine.

Often parts have to be attached to rotating shafts in order to transmit power. One of the more common ways of accomplishing this is to use a key to transmit torque between a shaft and the associated hub. In this chapter, we will review the basic principles of keys, how to size keys, and how to determine whether or not they are suitable for the transmitted torque.

We also often need to attach two shafts. This is typically done with **couplings,** splines, or **clutches.** The basic principles of couplings and splines will be reviewed in this chapter, and Chapter 16 will discuss clutches and how they are designed and used.

15.1 OBJECTIVES

In addition to learning about keys and shaft attachment mechanisms, you will also review the structural analysis of the more common ones. After completing this chapter, you should be able to:

- Recognize the different types of keys and their standard sizes.
- Size keys for appropriate structural loads.
- Recognize many types of couplings and their advantages and disadvantages.
- Understand the principles of splines and be able to analyze appropriate loads.
- Understand the basic types of universal joints and how and when they may be used.
- Recognize and understand principles of other miscellaneous shaft attachment mechanisms such as setscrews, clamps, and cross pins.

15.2 TYPES OF KEYS

The most common types of key systems are square and rectangular or flat keys. Figure 15.1 shows a typical gear attached to a shaft with a square key and the basic dimensions for keys. Table 15.1 shows standard key sizes for different shaft diameters. Standard key sizes are needed because standard items such as gears, couplings, and clutches are often manufactured and supplied with a keyway. The size of this keyway is based on the shaft size. Thus, if we have a standard-size key for each shaft size, we can adjust the length of the key to allow for the necessary torque-carrying capabilities.

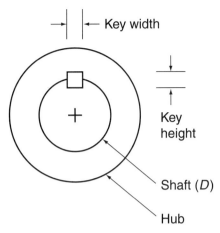

FIGURE 15.1 Key dimensions: Note that height and width are the same for a square key.

TABLE 15.1

Shaft Diameter (Inclusive)	Width* b	Height* Rectangular t
½–9/16	1/8	3/32
5/8–7/8	3/16	1/8
15/16–1¼	1/4	3/16
1 5/16–1 3/8	5/16	1/4
1 7/16–1 3/4	3/8	1/4
1 13/16–2¼	1/2	3/8
2 5/16–2 3/4	5/8	7/16
2 7/8–3¼	3/4	1/2
3 3/8–3 3/4	7/8	5/8
3 7/8–4½	1	3/4
4 3/4–5½	1¼	7/8
5 3/4–6	1½	1

*Width and height are the same for a square key, i.e., ($b \times b$).

Other types of keys in common usage are shown in Figure 15.2 and are somewhat self-explanatory. Tapered keys can be driven into position in cases where the shaft and hub may not be accurately manufactured or wear may be present. A gib key has, in essence, a handle on the end to aid in removal. With a woodruff key, concerns about the key falling out are eliminated as a similar-radius cutout is made in the shaft. The pin key shown has certain advantages in that it can be installed after the assembly of the hub and the shaft, which may be important in some cases where timing or angular alignment are critical and must be set at assembly. There are many other kinds of key systems, but these are some of the most common.

One of the concerns with many types of keys is holding them in position. That is why the woodruff key and, in some cases, the gib key have advantages. Often with rectangular or square keys, a setscrew is used on top of the key to hold it in position if vibration might cause it to move axially. Most keys are made from low-carbon, cold-drawn steel. However, in certain applications, higher-strength keys may be required. On rare occasions, a key may be heat treated to further enhance its material properties.

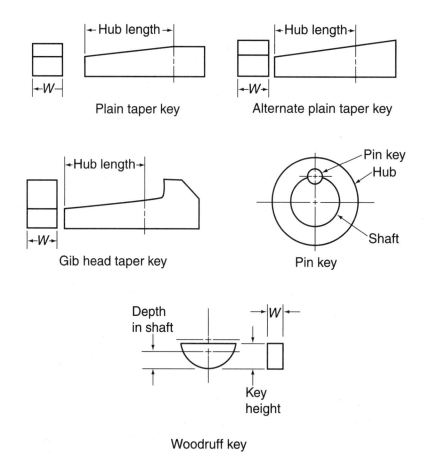

FIGURE 15.2 Common key types.

If, however, repeated loading or reverse loading is present, the higher ductility of non–heat-treated materials may be advantageous.

15.3 DESIGN OF KEYS

There are two primary considerations in the design of typical square or rectangular keys. These are the two modes of failure, shear and compression failure. In the shear mode, the key would fail at the hub-shaft intersection as shown in Figure 15.3. The key could fail in the compression mode when the key is deformed, causing clearances to increase. This would result in unsuitable service due to impact loading when the shaft was subjected to repeated loads or reversing directions.

The shear failure in a key can be calculated as follows:

$$T = \frac{S_s bLD}{2} \tag{15.1}$$

where S_s = shear stress allowable
b = key width
L = length of key
D = shaft diameter

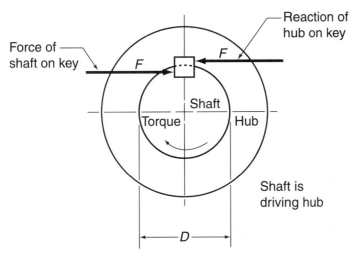

FIGURE 15.3 Shear plane for key failure.

Because standard-size keys are typically used, it is convenient to solve for a minimum length of key:

Key shear

$$L = \frac{2T}{S_s bD} \tag{15.2}$$

Note that we would typically include a factor of safety ranging from 1.5 to 4.5 for severe shock loading, depending on the type of loading or other considerations. In situations where loading is repeated or direction is reversed, it would also be necessary to verify that the compression loading is suitable. The appropriate torque for compression loading can be found as follows:

$$T = \frac{S_c t LD}{4} \tag{15.3}$$

where S_c = compression strength, typically S_y

t = height of key

or, to rearrange this to solve for the minimum length necessary,

Key compression

$$L = \frac{4T}{S_c t D} \tag{15.4}$$

In addition to solving for the appropriate key length, we should review the compression loading on the shaft and hub. This can be done by using the lowest value of compressive stress allowable for the key, shaft, or hub, which in some cases may require the key length to be increased. Typically, the yield stress allowable (S_y) is used for this value. In some cases, where a shaft and a hub assembly are subject to high loads, multiple keys may be used. When this is necessary, the length solved for in Equations 15.2 and 15.4 would simply be the total length of the keys. The following example problem demonstrates this process.

Example Problem 15.1

A ½-inch shaft transmits 5 hp at 1,750 rpm. The shaft is made from cold-drawn 1040 steel, the hub is made from hot-rolled 1213 steel, and a cold-drawn 1020 steel key is to be used. If the direction is regularly reversed, determine the length of the square key required for a SF = 2.5.

Material properties:

$$CD\ 1040\quad S_y = 71\ \text{ksi} \quad\quad\quad\quad \textbf{(Appendix 4)}$$
$$HR\ 1213\quad S_y = 58\ \text{ksi}$$
$$CD\ 1020\quad S_y = 51\ \text{ksi} \quad S_u = 61\ \text{ksi}$$

First, determine the torque.

$$T = \frac{63,000\ P}{n} \quad\quad (2.6)$$

$$T = \frac{63,000(5)}{1,750}$$

$$T = 180\ \text{in.-lb}$$

(From Table 15.1, for ½-inch shafting, the key is ⅛ × ⅛ in.)
Determine the length of the key for shear.

$$\text{Use } S_s = .5(S_u) = .5(61\ \text{ksi})$$

$$S_s = 30.5\ \text{ksi}$$

$$L = \frac{2T}{S_s b D} \quad\quad (15.2)$$

$$L = \frac{2(180\ \text{in.-lb})}{30,500\ \text{lb/in.}^2\ .125\ \text{in.}\ .5\ \text{in.}}$$

$$L = .188\ \text{in.}$$

With a SF = 2.5,

$$L = 2.5(.188)$$

$$L = \tfrac{1}{2}\ \text{in.}$$

Find the length required for compression, using the lowest value of yield, which is the key value $S_y = 51$ ksi.

$$L = \frac{4T}{S_c t D} \quad (15.4)$$

$$L = \frac{4(180 \text{ in.-lb})}{51,000 \text{ lb/in.}^2 \; .125 \text{ in.} \; .5 \text{ in.}}$$

$$L = .226$$

With a SF = 2.5:

$$L = 2.5(.226)$$

$$L = .565 \text{ in.}$$

This would be the minimum length. However, a longer key may be useful if the hub length is longer.

15.4 SPLINES

Splines are somewhat similar in appearance to keys, but they serve a very different purpose. Whereas a key transmits rotational torque, a spline does that while simultaneously allowing a change in the length of the drive shaft. Many different spline geometries are used. The most common is a straight-sided spline, but other profiles, such as the involute gear-tooth shape, are often used. Even rectangular bars and tubes could be considered splines. Figure 15.4 shows a straight-sided spline configuration, and Figure 15.5 shows an involute gear-shaped spline. We can determine the torque-carrying capability of a spline system as follows:

$$T = \left(\frac{S_s \pi D L}{8}\right)\left(\frac{D}{2}\right) = \frac{S_s \pi D^2 L}{16} \quad (15.5)$$

In most all cases, however, splines are not designed based on their stress-carrying capability but on sliding surface loading and lubrication concerns. Because a spline must often slide under load, when this is an important consideration, we often significantly de-rate the capability of a spline system based on the appropriate surface contact pressures of the sliding surface.

If sliding under load is expected, the torque capacity for this condition can be estimated by

$$T = SAr_m \tag{15.6}$$

where S = desired contact stress, typically 1,000 psi for lubricated steel, but can be increased with special coatings

A = surface area subject to sliding, approximately equal to

$\left(\dfrac{D-d}{2}\right) L$ (number of splines)

r_m = average radius of surface area, approximately $(D + d)/4$ from Figure 15.4

A straight-spline selection is shown in Example Problem 15.2.

Example Problem 15.2

A straight-sided spline like the one shown in Figure 15.4 has the following dimensions: $D = 1$ inch; 6 splines; $d = .810$ inch. Determine the torque capacity if this system is made from the same 1020 steel as in Example Problem 15.1. Assume SF = 2 and the spline has an engagement length of 2 inches.

$$T = \frac{S_s \pi D^2 L}{16} \tag{15.5}$$

$$T = \frac{30,500 \text{ lb/in.}^2 \pi (1 \text{ in.})^2 \, 2 \text{ in.}}{16}$$

$$T = 11,977 \text{ in.-lb}$$

With a safety factor of 2,

$$T = 5,988 \text{ in.-lb}$$

If sliding under load is needed, assume 1,000-psi contact pressure.
Find the area.

$$A = \frac{(D-d)L}{2} \text{ (number of splines)}$$

$$A = \frac{(1-.810)}{2} 2(6)$$

$$A = 1.14 \text{ in.}^2$$

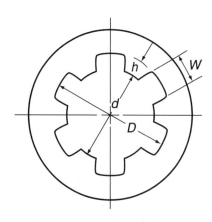

FIGURE 15.4 Straight-sided spline.

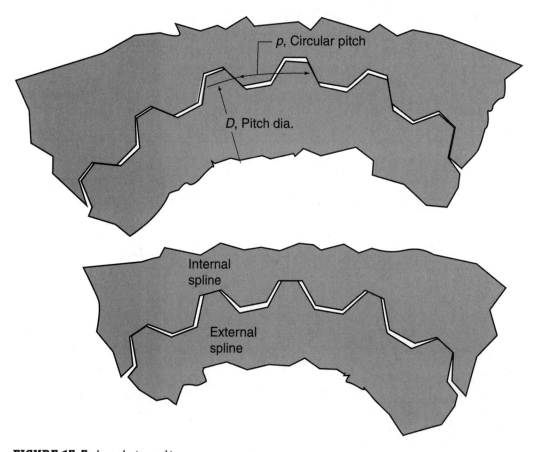

FIGURE 15.5 Involute spline.

$$T = SAr_m \tag{15.6}$$

$$T = 1{,}000 \text{ lb/in.}^2 \; 1.14 \text{ in.}^2 \left(\frac{1+.810}{4}\right)$$

$$T = 516 \text{ in.-lb}$$

Note that this is far less than for the strength of the spline.

15.5 COUPLINGS

Couplings are used in cases where torque needs to be transmitted through shafts where there is a potential for misalignment. In most cases, this would happen when separately mounted devices are connected, such as a motor to a pump, where two mating shafts need to be joined. Many types of couplings are available. The most common is the star coupling, which can be used with many types of inserts for different applications. These inserts include rubber for absorbing shock, bronze, high-strength plastics, or even steel. A star coupling is useful when minor amounts of misalignment need to be accommodated. Many other types of couplings are available such as bellows, gear, chain, or rigid couplings that do not compensate for misalignment. Generally, the manufacturers should be consulted for the appropriate torque ratings for their couplings. Figures 15.6 through 15.8 show common types of couplings.

FIGURE 15.6 Three-jaw (star) coupling. (Courtesy of Boston Gear)

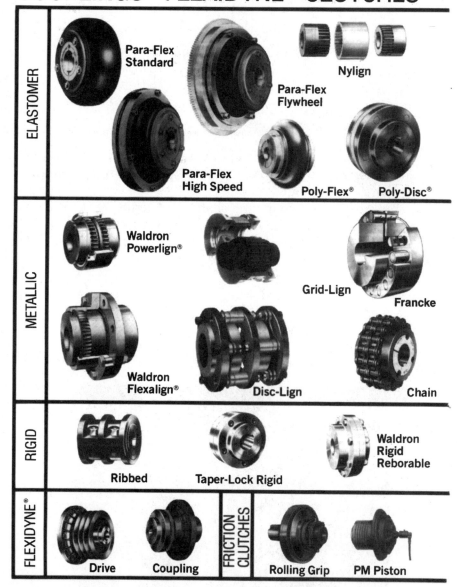

FIGURE 15.7 Types of couplings. (Courtesy of Rockwell Automation, Inc.)

Waldron Francke®

Taper-Lock Chain

Waldron Reborable

Taper-Lock Rigid

Ribbed Rigid

FIGURE 15.8 Additional coupling types. (Courtesy of Rockwell Automation, Inc.)

15.6 UNIVERSAL JOINTS

Universal joints differ from couplings in that they allow for greater angles of misalignment and are typically used in pairs, often with a spline. This allows significant misalignment to exist between shafts. It also allows the misalignment to change, as in a rear-wheel-drive automobile, where the differential and wheel assemblies move up and down. A pair of two universal joints are typically used in these applications. Figures 15.9 through 15.11 show some types of universal joints. These include a simple pin joint, a bearing-type spider joint, which is used in a rear-wheel-drive automobile, and a constant-velocity joint that is used in front-wheel drives where space often does not allow for two sets of universal joints. It should be noted that universal joints, especially the spider-type used with an automotive differential, require that the two sets of yokes be aligned because the shaft between the two does not rotate at a constant velocity. Hence, it is important in order to minimize vibration that the yokes be aligned. This phenomenon of the center part of the shaft not rotating at a constant velocity is why it is necessary to use these types of universal joints in pairs. This is the reason why the constant-velocity joint used in front-wheel-drive cars was developed.

universal joints
A flexible-shaft attachment method that allows for angular misalignment between shafts.

FIGURE 15.9 Pin-type universal joint. (Courtesy of Boston Gear)

FIGURE 15.10 Needle bearing spider-type universal joint. (Courtesy of Boston Gear)

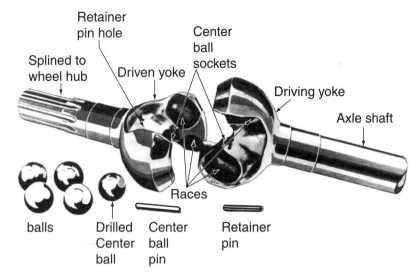

FIGURE 15.11 Constant-velocity joint.

15.7 OTHER SHAFT ATTACHMENT METHODS

There are far too many different types of shaft attachment methods to be discussed in the scope of this text. However, a few more warrant mention. **Snap rings** are often used to keep components, especially small components, from sliding along shafts. Figure 15.12 shows some of the types of snap rings available, and Figure 15.13 shows snap rings mounted on a shaft to hold a typical bearing in position. Other types of shaft attachment methods include **setscrews**, which simply clamp a hub to a shaft; collars, which may be slid over a shaft and may be used along with, for example, set screws; lock nuts; or spring nuts. These are only a few of the types of systems available for attaching shafts. If you look around at many machines in operation, you will see multiple types of systems used to couple shafts, to lock shafts, and to keep parts from sliding along shafts.

snap rings
Commonly referred to as "spring rings." Small components that can be used to keep a component from sliding along a shaft. They require a groove or other recess.

setscrew
A cap screw–type mechanism mounted perpendicular to a shaft that, due to the friction in the tip of the screw, can hold a hub in position on a shaft.

Internal	Basic N5000 For housings and bores Size .250—10.0 in. Range 6.4—254.0 mm.	External	Bowed 5101* For shafts and pins Size .188—1.750 in. Range 4.8—44.4 mm.	External	Reinforced 5115 For shafts and pins Size .094—1.0 in. Range •	External	Heavy-duty 5160 For shafts and pins Size .394—2.0 in. Range 10.0—50.8 mm.	External
Internal	Bowed N5001* For housings and bores Size .250—1.750 in. Range 6.4—44.4 mm.	External	Beveled 5102 For shafts and pins Size 1.0—10.0 in. Range 25.4—254.0 mm.	External	Bowed e-ring 5131 For shafts and pins Size .110—1.375 in. Range 2.8—34.9 mm.	External	Klipping® 5304 T-5304 For shafts and pins Size .156—1.000 in. Range 4.0—25.4 mm.	External
Internal	Beveled °N5002/°N5003 For housings and bores 1.0Z—10.0 in. 25.4—254.0 mm. 1.56—2.81 in. 39.7—71.4 mm.	External	Crescent® 5103 For shafts and pins Size .125—2.0 in. Range 3.2—50.8 mm.	External	E-ring 5133 For shafts and pins Size .040—1.375 in. Range 1.0—34.9 mm.	External	Gripping® 5555 For shafts and pins Size .079—.750 in. Range 2.0—19.0 mm.	External
Internal	Circular 5005 For housings and bores Size .312—2.0 in. Range •	External	Circular 5105 For shafts and pins Size .094—1.0 in. Range •	External	Radial gripping® 5135 For shafts and pins Size .094—.375 in. Range 2.4—9.5 mm.	External	Radial gripping® 5560* For shafts and pins Size .101—.328 in. Range •	External
Internal	Inverted 5008 For housings and bores Size .750—4.0 in. Range 19.0—101.6 mm.	External	Interlocking 5107* For shafts and pins Size .469—3.375 in. Range 11.9—85.7 mm.	External	Prong-lock® 5139* For shafts and pins Size .092—.438 in. Range •	External	Permanent shoulder 5590* For shafts and pins Size .250—.750 in. Range 6.4—19.0 mm.	External
External	Basic 5100 For shafts and pins Size .125—10.0 in. Range 3.2—254.0 mm.	External	Inverted 5108 For shafts and pins Size .500—4.0 in. Range 12.7—10.6 mm.	External	Reinforced e-ring 5144 For shafts and pins Size .094—.562 in. Range 2.4—14.3 mm.	External	* Non-stocking ring type: available on special order only	

FIGURE 15.12 Types of retaining rings. (Courtesy of Truarc Company LLC)

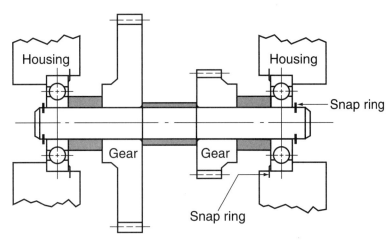

FIGURE 15.13 Snap rings for resisting axial loads.

15.8 SUMMARY

You should now be familiar with the types of keys available, some of the principles of keys, and how to analyze the appropriate torque or length required for flat and square keys. You should also understand the basic principles of splines and couplings, how they are used in design, and what factors should be considered in the use of these types of products.

Universal joints, both spider-type and constant-velocity joints, should now be familiar. You should understand where these joints may be useful, especially in moving equipment or in applications where shafts need to be rotated while still being able to move with respect to each other. Other shaft attachment methods such as snap rings and set screws should now be recognized along with their potential applications.

15.9 PROBLEMS

1. A 6061-T6 aluminum pulley transmits 75 hp at 1,750 rpm. The shaft is 1¾ inches in diameter and is made from cold-drawn 1040 steel. The key is to be made from cold-drawn 1020 steel. For a safety factor of 3, compute the required square key length for both shear and compression.
2. In Problem 1, solve for a flat key.
3. Determine the shear and compressive stress induced in a standard square key 2 inches long in a 1½-inch shaft, transmitting 200 hp at 875 rpm.
4. In Problem 3, select a key material if a safety factor of 2 is desired.
5. A 25-mm shaft has a 6-mm square key 40 mm long. Find the shear and compressive stresses in the key and the shaft when transmitting 5 kw at 2,450 rpm.
6. Specify a key for a gear mounted on a shaft of 1¾-inch diameter. The transmitted torque is 20,000 in.-lb. The hub is 3½ inches long and the system changes direction. Specify the key length and material.
7. An electric motor drives a machine through a V-belt pulley. The motor is rated at 6,000 in.-lb. of torque and has a 1½-inch diameter output shaft. This system does not reverse direction, and it stops and starts infrequently. Design the key for this application.
8. In Problem 7, if the drive is a chain system, the direction reverses, and the sprocket is made of class 20 cast iron, design the key for this application.
9. Determine the torque and horsepower rating for a straight-sided spline with the following parameters (see Figure 15.4). Sliding under load is required, and the spline rotates at 1,800 rpm.

$$D = 2 \text{ inches}$$
$$d = 1.62 \text{ inches}$$
$$L = 2 \text{ inches}$$
$$10 \text{ splines}$$

Material is 1040 cold-rolled steel and lubricated with oil.

10. Determine the torque and horsepower for the spline in Problem 9 that can be transmitted at 1,800 rpm if sliding under load is not needed. Assume a safety factor of 2.
11. A gear is mounted to a 2-inch-diameter shaft with a cross bolt that has a ⅜-inch diameter and a shear strength of 35,000 psi. Determine the torque in the shaft that would cause this pin to fail (assume double shear).
12. In Problem 11, describe what may happen when this bolt fails and why this may be a poor design.

15.10 CUMULATIVE PROBLEMS

13. The wheels on the locomotive in Problems 12 and 13 of Chapter 4 are keyed to the shaft. Design a key for this application.
14. Design a straight-sided spline for the transmission output driveshaft for the conditions in Chapter 11, Problems 13 and 14. This spline needs to slide under load.

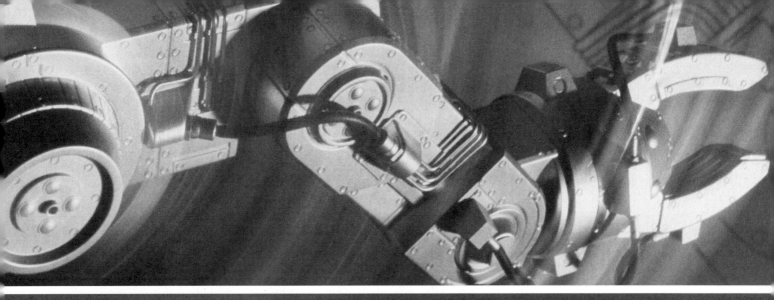

CHAPTER 16

Clutches and Brakes

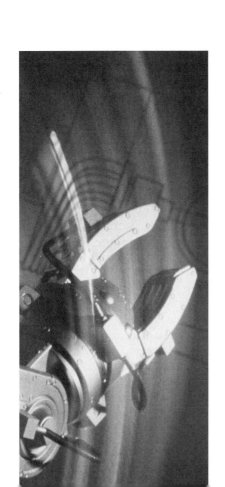

The operating principles of brakes and clutches are almost identical. Both systems use a friction surface to either accelerate or decelerate a rotating shaft system. For example, in an automobile where the engine is turning continuously, the clutch is used to bring the transmission and drive wheels up to speed. When you shift gears, the clutch is disengaged to allow the gears to be changed, and then it is reengaged to match the input speed of the transmission to the engine speed. A brake also uses a friction surface, but the difference from a clutch is that its purpose is only to slow the rotation of a shaft or to stop it entirely.

In this chapter, we will learn how to analyze the forces between brake and clutch surfaces and their associated torques in order to correctly size these systems. We will also learn about many of the different types of brakes and clutches available and how they can be used in a design. Depending on the application, the types of clutches available range from a fairly conventional plate clutch, as used in an automobile, to many kinds of multiple-disc, cone, and other configurations used for specialized purposes. Similarly, brakes are made in different configurations depending on the type of system, such as the squeeze braking or caliper braking system on a bicycle and the drum or disc brakes in an automobile. Elevators, cranes, and construction equipment may also use their own types of braking systems.

In the past, brake systems for most cars were drum style because, based on their geometry, they could be made partially **self-activating.** This meant that the pedal force would not need to be as great, an inherent advantage with this type of system. More recent braking systems use a **power accumulator,** more commonly called power brakes; hence, disc brakes have become more common because greater hydraulic force is available from the accumulator and the self-energizing feature of drum brakes is no longer needed. Disc brakes also lend themselves to antilock systems, which minimize some of the concerns about automobiles skidding when the brakes are applied. They may also have more heat-dissipating ability than drum brakes.

The disposal of heat is probably the area in which a typical clutch and a brake assembly differ most. Because a clutch is used rather infrequently, the heat generated from its friction surfaces is usually insignificant. In braking systems, by contrast, the dissipation of this heat can be of great concern. In an automobile, we often refer to **brake fade,** which is the decrease in the coefficient of friction of the braking material as it is heated by extensive use. The potential energy in a car going down a hill can result in a significant amount of energy that must be absorbed by the braking system. This can result in a significant increase in temperature. For example, on a typical caliper braking system on a bicycle, the rims can become quite hot after a

self-activating
A property of certain styles of brakes whereby a slight force causes the brake system to pull itself into harder engagement.

power accumulator
A term used for multiplying the power, typically for braking systems, in order to minimize the amount of force necessary to activate the brakes.

brake fade
A property of brakes whereby the braking power is reduced when the brake pad material heats up with use, either because of glazing of the pad or other temperature-related property changes.

long period of braking. If you ride your bicycle down a mountain pass, or drive a car for that matter, it may be necessary to stop occasionally to let the braking systems cool in order to dissipate this heat. Long-haul trucks are often seen stopped on steep hills for the same reason. Therefore, unlike clutches, which are typically rated in terms of torque or horsepower capability, a brake system may be rated in terms of its ability to dissipate heat. This rating might be in units of brake horsepower, which is the ability of the brake system to dissipate heat.

In the design of a clutch or brake system, it is often necessary to compensate for both linear and rotational inertia. In an automobile, for example, the braking system stops the rotating parts as well as the forward motion. The clutch system brings both the rotating parts and the automobile up to speed. These aspects will be discussed as we go through this chapter.

16.1 OBJECTIVES

This chapter is intended to give you an understanding of the different types of clutch and brake systems and to help you learn to analyze the forces, torque, and power in these systems. After completing this chapter, you should be able to:

- Recognize the basic geometries of clutch and brake systems.
- Calculate the frictional forces in different types of clutches.
- Calculate the torque transfer available from different clutch systems.
- Calculate the frictional forces in brake systems.
- Determine the torque capabilities of braking systems.
- Understand the principles of heat generation and heat removal from brake systems.
- Calculate frictional brake horsepower and recognize how to use it.
- Learn some of the principles of self-energizing and self-locking brake systems.

16.2 TYPES OF CLUTCHES AND BRAKES

Clutches and brakes that use friction materials to transmit a torque to start a mechanism, or to create a torque to stop a mechanism, can be classified according to their general shape and the location of the friction surfaces

along with their means of activation. In many cases, the same geometry can be used to create either a clutch or a brake. The following are some of the more common types of clutch and brake systems.

Plate Clutch or Brake

In this type of system, the friction surfaces are arranged on a flat surface and the two surfaces are either pushed together or spring-loaded together and pulled apart to disengage. This system is often used, for example, as a clutch in a manual-shift automobile. Some machine brakes are also made in a similar manner. Figure 16.1 shows this basic geometry.

Cone Clutch or Brake

A cone clutch or brake system is very similar to a plate, with the exception that because the friction surface is tapered, a wedging action occurs that increases the frictional force. Despite being more difficult to make because of the complicated geometry, this system has an advantage where limited force is available for the engagement of the clutch or brake. Figure 16.2 shows a schematic of a cone clutch or brake system.

Caliper Disc Brake

A caliper disc brake is used in many modern automobiles; a form of it is used on bicycles as well as many motorcycles. In the case of the bicycle, the disc is the actual rim of the bicycle wheel, whereas in an automobile or motorcycle it is typically a separate disc assembly. In either case, the friction materials are pressed against both sides of the disc in order to create an appropriate friction surface. This can be done by mechanical leverage,

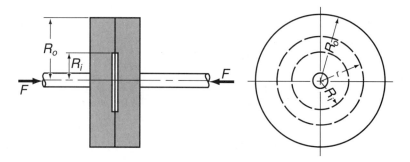

FIGURE 16.1 Plate brake or clutch geometry.

air pressure, or hydraulic pressure. This configuration is primarily used for braking systems. See Figure 16.3.

Drum Brakes

Drum brakes can have either an internal or an external friction surface. The most common is the type of drum brake used on an automobile, where the friction surfaces are pressed outward in order to create a braking system for each wheel. This was the primary braking system in automobiles for many years because of its unique ability to self-lock or partially self-lock. This

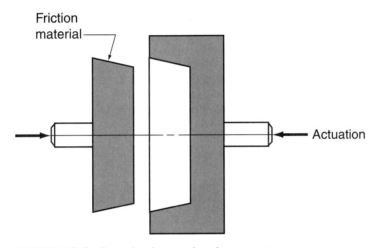

FIGURE 16.2 Cone brake or clutch geometry.

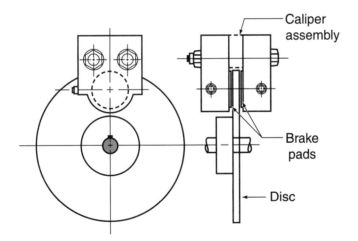

FIGURE 16.3 Caliper-style disc brake system.

means that the force of friction tends to pull the brake into harder engagement, reducing the amount of force required from the operator. However, an inherent disadvantage of this system is that it can lock up too quickly if too much force is applied. Because of this, and because of the advent of power disc brake systems, drum brake systems have become less common. Power disc brakes have become the primary type of braking mechanism for newer automobiles.

A drum configuration can have its friction surfaces on the outside of the drum; they can either encircle the drum or partially encircle it. When the friction surfaces partially encircle the drum, this is referred to as a block brake system. Figure 16.4 shows an internal drum system, and Figure 16.5 shows an external block brake system along with a circular shoe system.

Band Brake

A band brake system uses a belt that wraps around a drum, often spring-loaded but not necessarily. It is often used in applications where a different braking force is required in one direction than the other. It can be self- or partially self-activating in one direction because the friction force can pull the band into tighter engagement if designed properly. Figure 16.6 shows a schematic of a band brake system.

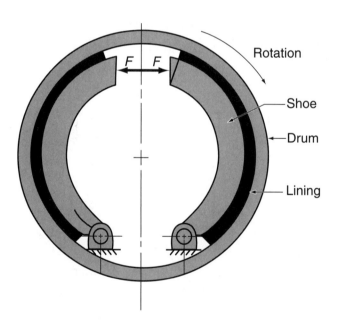

FIGURE 16.4 Internal-shoe drum brake.

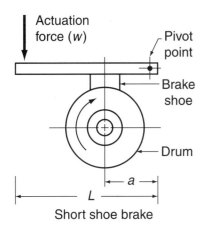

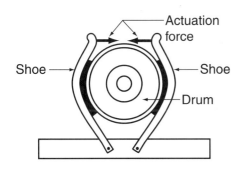

FIGURE 16.5 Short- and long-shoe external brake assemblies.

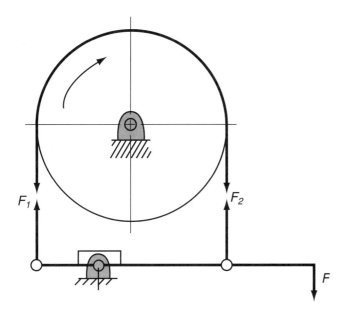

FIGURE 16.6 Band brake schematic.

fail-safe brake
A term used for brakes that require power to unlock the brake system so that they are automatically set in the event of a power failure.

centrifugal clutch
A type of clutch in which a rotating weight is spun more and more rapidly, causing a friction material to be pushed into engagement and causing the clutch to engage.

electromagnetic
By imparting an electrical field, a force is created to engage a clutch or brake system. This often may be referred to as a "solenoid" brake or clutch system.

Brakes and clutches can be activated in many ways. They may be hydraulically activated, as in mobile equipment such as construction equipment and automobiles. They may be mechanically activated or, solenoid or electrically activated. They may be either force activated or force released. What is meant by all this? Envision a clutch in an automobile. This clutch is always engaged, held by the spring system, unless it is physically pulled out of engagement. By contrast, the typical brake in a car is not engaged unless it is pushed into engagement. This can be useful if we would like a brake system to be automatically applied or released in the absence of a power supply. A spring-activated system that needs force to release it is often referred to as a **fail-safe brake.** Elevators, for instance, may use this type of system in the event of a power failure. Another type of clutch system that is often used in some simpler machines is a **centrifugal clutch,** in which the rotation of a weight causes friction between the outer surface and, usually, a drum that brings it into engagement. Good examples of centrifugal clutches appear in portable chain saws, some small minibikes, and many of the early snowmobiles. The advantage of this type of system is that only the throttle needs to be varied to bring the system into engagement. These types of systems tend to have a higher slippage during engagement, but their simplicity can be advantageous.

Another activation system for clutches and brakes, in addition to mechanical, hydraulic, and pneumatic, is an **electromagnetic** system in which solenoids are used to create motion in an electrical coil. In some cases, the armature could be part of the central shaft. Figure 16.7 shows some of these types of clutch and brake systems.

16.3 FRICTION MATERIALS AND COEFFICIENTS OF FRICTION

In order for clutch or brake systems to operate effectively, their mating surfaces must either be coated or have a separate disc or plate made of material that transfers the frictional forces in a reasonable manner. For years, many friction materials for clutches and brakes were made from asbestos fibers embedded in an epoxy-type material, which provided good thermal properties as well as reasonably high coefficients of friction in the range of .35 to .50. However, environmental concerns have eliminated almost all use of asbestos fiber, and now many alternate materials are available. Most modern friction plate materials are made of epoxy or polymer compounds with some form of impregnated material such as small metal shavings, graphite, sintered iron, or other materials that are generally capable of

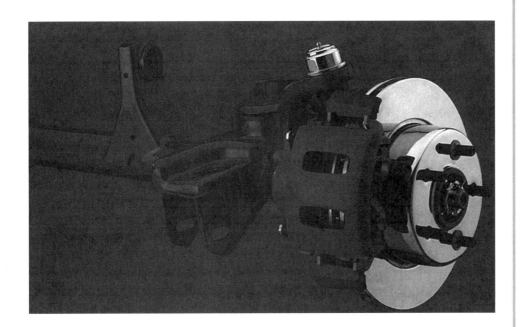

FIGURE 16.7 Disc brake and limited-slip differential with integral clutch plate. (Courtesy of Dana Torque & Traction Technologies)

glazing

One of the properties that causes brake fade. The surface of the friction material becomes either smoothed over or, under extreme temperature, becomes semiliquid, thereby reducing the coefficient of friction.

good performance under different temperature conditions and have reasonable resistance to wear.

In many applications, a high coefficient of friction is not the primary consideration. Too high a coefficient of friction can prohibit smooth engagement of the friction surfaces. Consistent values for the coefficient of friction over a range of operating conditions are often more important than a high coefficient of friction. Chemical compatibility in many applications is critical, as is the ability to resist wear under long-term loading conditions. A good example of the latter would be the brakes in an automobile because they are used extensively and need to last for a relatively long time. Another important consideration for friction pad materials is their ability to resist phenomena such as **glazing** under high-temperature conditions. Glazing can cause the coefficient of friction to change significantly or break down under high temperature. Many friction materials require a short break-in period to achieve their optimum friction values or to shape themselves to the mating surfaces. Table 16.1 shows some coefficients of friction under dynamic load conditions for some common friction pad materials.

TABLE 16.1
COEFFICIENTS OF FRICTION

Friction Material	Dynamic Friction Coefficient		Pressure Range	
	Dry	In Oil	(psi)	(kPa)
Molded components	.25–.45	.06–.10	150–300	1035–2070
Woven materials	.25–.45	.08–.10	50–100	345–690
Sintered metal	.15–.45	.05–.08	150–300	1035–2070
Cork	.30–.50	.15–.25	8–15	55–100
Wood	.20–.45	.12–.16	50–90	345–620
Cast iron	.15–.25	.03–.06	100–250	690–1725
Paper-based		.10–.15		
Graphite/resin		.10–.14		
Rigid molded asbestos	.35–.41	.06	<300	<2070

Compiled from a variety of sources.

16.4 TORQUES AND FORCES ON CLUTCHES AND BRAKES

For a plate-type clutch or brake system, the sliding friction can be calculated by the relationship from Newtonian physics as follows:

$$F_f = fN \qquad (16.1)$$

where f = coefficient of friction

N = normal force

This would give us the same frictional force that we would have if we were to slide the friction material across a flat surface. The normal force would be the force pushing the surfaces together or simply the weight of the object if an external force was not applied.

The rotating torque, if we had a plate-type clutch or brake system as shown in Figure 16.1, could be found as follows:

$$T_f = fN\left(\frac{r_o + r_i}{2}\right) \qquad (16.2)$$

where r_o = outside radius

r_i = inside radius

Note that the average radius becomes the moment arm for the typical torque equation. It is interesting to note that increasing the inside radius has the net effect of moving this moment arm further out and increasing the frictional torque available. Obviously, however, this also reduces the area of the friction surface, which can affect the wear life. Frictional power can then be determined from the relationship

$$P_f = \frac{T_f n}{63,000} \qquad (16.3)$$

where T_f, inch-pounds

n, rpm

P_f, hp

Clutch/brake torque (16.2)

$$T_f = fN\left(\frac{r_o + r_i}{2}\right)$$

Frictional power (16.3)

$$P_f = \frac{T_f n}{63,000}$$

From this, we can determine the power that can be transmitted at different rotational speeds in a clutch or the power to stop in a brake system. The following example illustrates this technique.

Example Problem 16.1

A plate-type clutch like the one in Figure 16.1 has the following properties r_o = 12 in., r_i = 9 in., engagement force of 120 lb (normal force), and turns at 2,000 rpm. The friction disc has a coefficient of friction of .3. Determine the torque and power that can be transmitted by this system.

Torque capacity:

$$T_f = fN\frac{(r_o + r_i)}{2} \qquad (16.2)$$

$$T_f = .3(120 \text{ lb})\left(\frac{12 \text{ in.} + 9 \text{ in.}}{2}\right)$$

$$T_f = 378 \text{ in.-lb}$$

Power:

$$P_f = \frac{T_f n}{63,000} \qquad (16.3)$$

$$P_f = \frac{378(2,000)}{63,000}$$

$$P_f = 12 \text{ hp}$$

For a caliper disc brake system, the analysis would be done in a similar manner: the outside radius and the inside radius become the distance from the inner edge and the corresponding distance to the outer edge of the disc pad.

For a drum brake, it would merely be the inside drum diameter for an internal brake or the outside drum diameter for an external brake. However, in these cases, the calculation of the normal force may be more difficult to determine. In some cases, due to the mechanism for engaging brakes of this type, self-energizing features may make this analysis fairly difficult. The following example illustrates a block drum brake system. Note that for longer-length drum brakes and band brakes, determining contact force is complicated because the force is not necessarily uniform over the surface area.

Example Problem 16.2

For the short-shoe drum brake shown in Figure 16.5, determine the braking torque for the following dimensions: $a = 4$ in., $L = 20$ in., $D = 12$ in., $f = .4$, and $W = 100$ lb.

First, find moments to determine normal force.

$$\Sigma M_p = WL - aN$$

$$\Sigma M_p = 100 \text{ lb } 20 \text{ in.} - 4 \text{ in. } N$$

$$N = 500 \text{ lb}$$

Torque friction:

$$T_f = fN\frac{D}{2}$$

$$T_f = .4(500 \text{ lb})\frac{12 \text{ in.}}{2}$$

$$T_f = 1,200 \text{ in.-lb}$$

This analysis assumes that the lever arms stay approximately horizontal.

For a cone clutch or brake system, the analysis is complicated by the fact that there is a force-compounding effect due to the angle, and the normal force needs to be calculated based on that friction angle. Our frictional force relationship can then be found by

$$T_f = F_f r_m = fNr_m \tag{16.4}$$

but our normal force must be modified by the relationship

$$N = \frac{F_a}{\sin \alpha + f \cos \alpha}$$

where $F_a =$ axial force

$\alpha =$ cone angle

If we combine these two, we get the frictional torque on a cone brake or clutch system.

$$T_f = \frac{fr_m F_a}{\sin \alpha + f \cos \alpha} \tag{16.5}$$

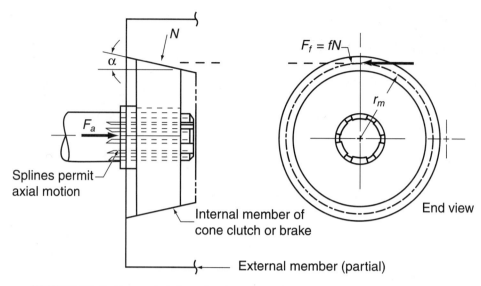

FIGURE 16.8 Cone clutch or brake geometry.

Example Problem 16.3 illustrates this.

Example Problem 16.3

For the cone clutch shown in Figure 16.8, determine the torque-transmitting capacity based on the following parameters: D_{mean} = 12 inches; F_a = 75 lb; f = .35; α = 20°. Also solve if α = 10°, and compare the results.

$$T_f = \frac{f r_m F_a}{\sin \alpha + f \cos \alpha} \tag{16.5}$$

$$T_f = \frac{.35 \left(\dfrac{12 \text{ in.}}{2}\right) 75 \text{ lb}}{\sin 20° + .35 \cos 20°}$$

$$T_f = 235 \text{ in.-lb}$$

For α = 10°,

$$T_f = \frac{.35 \left(\dfrac{12 \text{ in.}}{2}\right) 75 \text{ lb}}{\sin 10° + .35 \cos 10°}$$

$$T_f = 304 \text{ in.-lb}$$

Note that the smaller angle creates a greater wedging force and, correspondingly, larger torque capacity.

Often in clutch or brake design, the forces required to accelerate or decelerate the object are primarily linear. For example, if we were designing a braking system for a truck of significant mass, the rotational inertia of the wheels and the drive system would be relatively small compared to the overall mass of the truck. In this case, depending on the rate at which we would like to stop the truck, we can determine the force necessary to decelerate it and then, by analyzing the effective radius of our brake system, determine the necessary contact forces. The following example problem illustrates this approach. Note that we have ignored any rotational inertia of the truck system.

Example Problem 16.4

A truck has a total weight of 40,000 lb and is traveling 60 mph. The brake design calls for it to be able to stop in 400 feet. Determine the stopping force required. Determine the stopping torque required if the wheels are 36 inches in diameter. Determine the torque per brake, assuming there are 10 sets of brakes. Assuming that each brake is a disc brake with a mean radius of 10 inches, determine the normal brake force if $f = .4$.

First, find the rate of deceleration. Converting 60 mph to ft/sec,

$$60 \text{ mph} \; \frac{5,280 \text{ ft}}{\text{mile}} \; \frac{\text{hr}}{3,600 \text{ sec}} = 88 \text{ ft/sec}$$

Find the stopping rate.

$$D = V_a t$$

$$t = \frac{D}{V_a}$$

$$t = \frac{400 \text{ ft}}{\frac{88}{2} \text{ ft/sec}}$$

$$t = 9 \text{ sec}$$

$$V = at$$

$$a = \frac{V}{t}$$

$$a = \frac{88 \text{ ft/sec}}{9 \text{ sec}}$$

$$a = 9.8 \text{ ft/sec}^2$$

Find the stopping force.

$$F = \frac{W}{g} a$$

$$F = \frac{40,000 \text{ lb}}{32.2 \text{ ft/sec}^2} \; 9.8 \text{ ft/sec}^2$$

$$F = 12,100 \text{ lb}$$

Find the torque, if the wheels are 36 inches in diameter.

$$T = Fr$$

$$T = 12,100 \text{ lb} \; \frac{36 \text{ in.}}{2}$$

$$T = 217,800 \text{ in.-lb}$$

For each wheel,

$$T = 21,780 \text{ in.-lb}$$

Braking normal force:

$$T_f = fN r_m \quad \quad (16.2)$$

$$N = \frac{T_f}{f r_m}$$

$$N = \frac{21,780 \text{ in.-lb}}{.4 \bullet 10 \text{ in.}}$$

$$N = 5,450 \text{ lb}$$

This is a significant normal force, especially for a disc brake system.

16.5 ROTATIONAL INERTIA AND BRAKE POWER

In many cases, unlike the truck example just described, the rotational inertia can be significant. A large **flywheel**, for instance, can store a significant amount of energy, and its inertia can add to the required braking power or the torque needed in a clutch system to bring the flywheel or other mass up to speed. We will not determine the rotational inertia for complex systems in this chapter; for these calculations you must do further study. In most cases, unless the mass of the rotating objects is significant with regard to the overall mass, a small factor can be included to compensate for that additional inertia effect.

flywheel
A large, rotating weight in which energy could be stored as rotational inertia.

Brake systems are often classified in terms of frictional horsepower because the energy contained is a product of the work done in stopping or slowing an object. The energy from a rotating torque can be found as follows:

$$U_f = F \pi D N_t = T_f 2\pi N_t \quad (16.6)$$

where U_f = frictional work

D = effective diameter

N_t = number of turns

The power associated with stopping the object or changing its speed can be found as follows:

$$P_f = \frac{T_f n}{63,000} = \frac{U_f}{550t} \quad (16.7)$$

Frictional energy (16.7)

$$P_f = \frac{T_f n}{63,000} = \frac{U_f}{550t}$$

where T_f, inch-pounds

n, rpm (average)

t, seconds

Note that the energy to be absorbed can be either in terms of potential energy (for example, an automobile starting down a mountain pass) or the kinetic energy that is related to the velocity and mass of an object. The associated potential energy can be found as follows:

$$\Delta PE = W(h_1 - h_2) \quad (16.8)$$

The similar kinetic energy from the velocity and mass of the object can be found as follows:

$$\Delta KE = \frac{W}{2g}\left(V_1^2 - V_2^2\right) \qquad (16.9)$$

or, for a rotating object,

$$\Delta KE = \frac{I}{2}\left(\omega_1^2 - \omega_2^2\right) \qquad (16.10)$$

where I = moment of inertia
ω = radians per second

or, for both horizontal and rotary motion,

$$\Delta KE = \frac{W}{2g}\left(V_1^2 - V_2^2\right) + \frac{I}{2}\left(\omega_1^2 - \omega_2^2\right) \qquad (16.11)$$

This, for example, could apply to a vehicle or any rolling object.

Note that for short time periods it is generally assumed that the frictional energy would be absorbed by the adjacent metal in a brake system. The total energy to be absorbed would be the sum of the changes in potential and kinetic energy. The expected temperature rise can be estimated as follows:

$$\Delta T = \frac{U_f}{W_m c} \qquad (16.12)$$

where c = specific heat
W_m = weight of brake system that can absorb heat

specific heat
A property of material that is the amount of energy required to raise its temperature. It is usually expressed in terms of Btu's or calories per unit weight or mass.

Note that c is the **specific heat** of the material from which the system is made. For estimation purposes, we could use a value of c for cast iron of approximately 101 foot-pounds per pound per degree F; for steel, approximately 93 foot-pounds per pound per degree F; and for aluminum, approximately 195 foot-pounds per pound per degree F. Note that this analysis will only estimate an approximate temperature change because invariably some of the energy will be dissipated during this process. The following example shows the technique for determining the estimated temperature rise of the brakes when stopping an automobile.

Example Problem 16.5

A 3,500-pound automobile is traveling 50 mph and decelerates on flat ground at a rate of 20 ft/sec². Each of the four steel brake drums weighs 10 pounds. Assuming that all heat is absorbed by the drums during this period, find the energy absorbed, the average frictional power, and the temperature rise of the drums.

Converting 50 mph to ft/sec,

$$50 \text{ mph} \frac{\text{hr}}{3,600 \text{ sec}} \frac{5,280 \text{ ft}}{\text{mile}} = 73 \text{ ft/sec}$$

Kinetic energy to be absorbed:

$$KE = \frac{WV^2}{2g} \qquad (16.9)$$

$$KE = \frac{3,500 \text{ lb } (73 \text{ ft})^2}{2(32.2 \text{ ft/sec}^2) \text{ sec}^2}$$

$$KE = 289,620 \text{ ft-lb}$$

(Energy gain U = KE lost.)

$$U_f = Wc\Delta T \qquad (16.12)$$

$$\Delta T = \frac{U_f}{Wc}$$

$$\Delta T = \frac{289,620 \text{ ft-lb}}{40 \text{ lb } 93 \frac{\text{ft-lb}}{\text{lb°F}}}$$

$$\Delta T = 78°$$

Finding the stopping time,

$$\Delta V = at$$

$$t = \frac{\Delta V}{a}$$

$$t = \frac{73 \text{ ft/sec}}{20 \text{ ft/sec}^2}$$

$$t = 3.7 \text{ sec}$$

Frictional power could then be found.

$$f_{hp} = \frac{U_f}{550t} = \frac{KE}{550t} \tag{16.7}$$

$$f_{hp} = \frac{289,620}{550(3.7)}$$

$$f_{hp} = 142$$

16.6 DESIGN OF BRAKE AND CLUTCH SYSTEMS

One aspect that has not been covered in this chapter, especially as it relates to brake systems, is the energy absorption that can degrade the performance of a braking system. As discussed in the introduction to this chapter, when an automobile goes down a long incline and the brakes are used continuously, considerable heat is generated. The brake systems need to be able to dissipate that heat in order to prevent significant fade and even warping of the assembly. If we look at the disc brakes in an automobile, we will often see "ventilation" that consists of a very small fan between two plates. This is designed to force air through the disc system in order to dissipate a greater amount of heat. If heat is not adequately dissipated, the disc itself can become so hot as to permanently warp. Mountain bicycle racers can create enough heat in the tire rims with their caliper-style brakes that there have been cases of the tires popping from this heat, leading to rider injuries. Obviously, this energy can be considerable. However, the analysis of the generation of that heat, or more specifically the dissipation of that heat, is beyond the scope of this course. You would need a course in heat transfer to better understand those aspects of clutch and brake design.

Other types of brake and clutch systems have their own specific design constraints or advantages, and many systems exist that do not appear to be a conventional form of brake or clutch system. For example, the tensioner assembly in a flat or V-belt system could allow the belt to slip or to be tightened, acting as a clutch. Many hydraulic clutch or brake systems exist, such as those used in the automatic transmission of an automobile, in many hydrostatic lawn tractors, or in some cases heavy construction equipment where hydraulic fluid pump systems or closely rotating vanes create the ability to connect or disconnect mating shafts. For years, flat belts have also been used as a form of clutch system. Especially in cases where overload protection is needed, a flat belt may be an ideal solution, depending on the

desired lifetime. Some assemblies, such as small inboard/outboard boat motors, do not use a conventional clutch. Instead, they use a mating-gear system or jaw clutch so that they are actually pulled into hard engagement. Roller or cam clutches can be used to limit motion to one direction. Thus, clutches and brakes can include many components that may not normally be recognized as clutch or brake systems.

Lastly, in many brake systems the most difficult part of the analysis is determining the activating force. Often a lever arm, over-center mechanism, or other force-magnification system is used to produce the required normal force.

16.7 SUMMARY

You should now be familiar with the basic types of clutches and brakes and have gained an understanding of how the coefficient of friction of the friction surfaces affects their performance. You should also be able to analyze the available torques of conventional brake and clutch systems and to determine the necessary clutch or braking power required.

16.8 PROBLEMS

1. A clutch like the one shown in Figure 16.1 turns at 2,000 rpm when engaged and has the following properties: OD = 10 inches; ID = 6 inches; axial force on the clutch disc (F_a) = 300 pounds; and the friction disc material is sintered metal $f = .3$. Find the:
   a. torque that can be transmitted
   b. frictional power of this clutch
2. In Problem 1, substitute the following properties: OD = 300 mm; ID = 150 mm; F_a = 1,000 N.
3. A disc brake system like the one shown in Figure 16.3 needs to stop a car going 50 mph in 300 feet. The car weighs 3,000 pounds; the brakes have a mean effective radius of 6 inches; the tires have an effective diameter of 24 inches; and $f = .35$. Find the compressive force needed on each brake.
4. In Problem 3, if the brake is activated by a hydraulic cylinder with a diameter of 2½ inches, what hydraulic pressure is required?
5. A short-shoe brake is activated by a lever arm that multiplies the activating force by a 3:1 ratio. If the drum is 500 mm in diameter and turns at 10 revolutions per second for an activating force of 250 N, find the torque and frictional power if $f = .25$.

6. In Problem 5, solve for a drum size of 10 inches and an activating force of 60 pounds.
7. A cone clutch like the one in Figure 16.2 has the following properties: a mean radius of 8 inches, a cone angle of 15°, and an axial force of 150 pounds. If the coefficient of friction is .3, find the torque and power rating at 2,400 rpm.
8. In Problems 3 and 4, if each brake is made of cast iron and weighs 10 pounds, determine the energy gained, the expected temperature rise if no heat is lost to the environment, and the frictional power rating using the energy lost method.
9. If the cone clutch in Problem 7 drives a go-kart through a 4:1 reduction in the gear system to a 12-inch wheel and the go-kart weighs 550 pounds, find the rate of acceleration possible.
10. An old factory used to use large grindstones (as shown in Figure 16.9) for grinding wood to make paper using a hydraulic impulse turbine. These grindstones are now used only as a brake for this turbine system, which now drives an electric generator. If the grindstones are 3 feet in diameter and a large wooden wedge with a 20° taper is driven into a cavity tangent to the stone with a 500-pound force, estimate the stopping torque. Give an expected range for your answer and explain your technique.

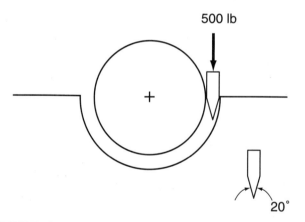

FIGURE 16.9 Grindstone and wedge for Problem 10.

16.9 CUMULATIVE PROBLEMS

11. For the truck designed in the cumulative problems earlier in this text, design a caliper disc brake system. From Chapter 2, Problem 12, the truck weighs 60,000 pounds, has 10 brake systems, the tires are 42 inches in diameter, and the truck needs to stop in 200 feet from 60 mph. Determine the disc diameters, friction material, normal forces, piston size, and pressure needed. Also, determine the energy given up and the temperature rise if each disc absorbs the heat, weighs 20 pounds, and is made of steel. What is the frictional power of each brake?

12. The clutch for the truck in Problem 11 uses 24 springs: spring number C0720-105-1500 (Appendix 12). The clutch springs are 1.105 inches long when the clutch is engaged and 1.0 inch when disengaged. If the clutch material has an $f = .3$, an OD = 24 inches, and an ID = 16 inches, what torque can be transmitted? If the engine puts out 220 hp at 2,200 rpm, is this design suitable? Also, what force is required to fully disengage this clutch?

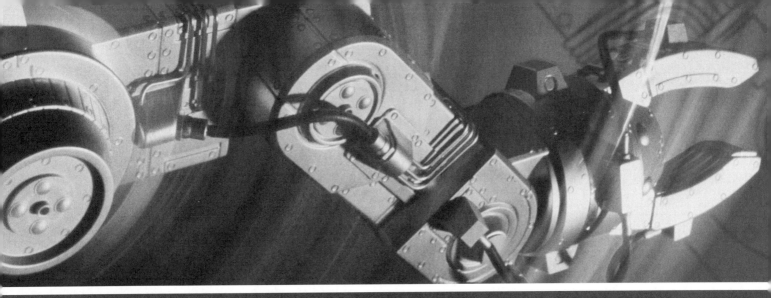

CHAPTER 17

Shaft Design

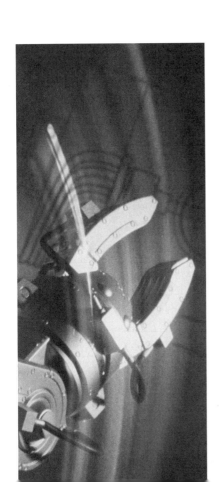

In many ways, this chapter is a review of material presented in earlier chapters. The design of shafting, where loads are transmitted in torsion, uses the principles of combined stresses discussed in Chapter 4 and then incorporates the repeated loading or fatigue analysis that was covered in Chapter 5. In this chapter we will also review how the loads are transmitted to a shaft or from the shaft by gears, as was discussed in Chapters 11, 12, and 13. We will review the loads and torques that could be transmitted by belt and chain drives, first discussed in Chapter 14. Hence, while this chapter presents little new material, it will serve as a review of previous material and will explain how to incorporate it into the design of a power-transmitting shaft.

This topic is often considered difficult because extreme care is needed to determine what the area of interest should be in the shaft system—that is, what are the points of highest stress that should be analyzed in order to size the shafting. The loads and moments are often in different directions, and combining them can be complicated. However, in this chapter, once we identify an area of interest (i.e., maximum stress), we will simply identify the torques and moments or shear stresses acting at that location. We will not try to identify all the different angles at which they may be applied. In general, in a rotating shaft, not identifying the angles will result in a safe result, albeit slightly conservative.

When we are reviewing the design of a power-transmitting shaft, the torque and associated power may be added and removed at many different points. There may be many areas at which analysis is appropriate. In some shafts, the size of the shaft may change at different locations due to different-sized pulleys, sprockets, or gears are placed on it. Stress risers such as shoulders, keys, and other factors discussed in Chapter 5 may require additional points of analysis. For example, envision a shaft supported by two bearings driven by a gear system (whether it be a spur, bevel, or worm gear) that transmits a torque into the shaft. There is a bending load from the driving force on the gear and a separation load from the gear where the shaft transmits this power to another device by means of a belt or chain drive system in addition to the torsional stress. This belt or chain drive system has an associated force that is also trying to bend the shaft in a different location. In the case of a belt, for example, there is a driving force and a back force that both put a load on the shaft. Once we determine the torque in the shaft, the easiest approach is often to create a shear and bending diagram to determine the location of the maximum moments. As discussed previously, if there is more than one moment from different bending loads, we will usually take the simplistic approach of adding them together regardless of angles. Then, if the shaft is subject to fatigue, we compare the stresses, using the combined stress analysis from Chapter 4, with the stress allowable, including stress concentration factors that were developed in our study of Chapter 5. Many examples appear in this chapter to illustrate this technique.

17.1 OBJECTIVES

The design of power-transmitting shafting includes the determination of the loads that are acting on the shafts, whether from a coupling or clutch or from loads from gears, pulleys, or sprockets, and the determination of the associated stresses. These stresses would then need to be compared to the endurance limits, including any stress concentration factors necessary for the design of a shaft system. After completing this chapter, you should be able to:

- Compute the forces acting on shafts from gears, pulleys, and sprockets.
- Find the bending moments from the gears, pulleys, or sprockets that are transmitting loads to or from other devices.
- Add any moments or forces as appropriate to determine the combined moments acting on the shafts.
- Determine the torque in the shafts from the gears, pulleys, sprockets, clutches, and couplings.
- Calculate the stresses from bending moments, shear loads, and torsional loads.
- Combine the stresses as needed, using appropriate combination techniques.
- Compare the combined stresses to the suitable allowable stresses, including any required stress reduction factors such as stress concentration factors and factors of safety.
- Determine the suitability of the shaft design and/or the necessary size of the shafting.

17.2 SOURCES OF LOADS ON SHAFTS

From Chapter 2, the torque in the shaft can be calculated from the relationship

$$T = Fd \qquad (2.5)$$

and torque can also be found, if we know the power input, from the relationship

$$P = \frac{F2\pi r}{t} \quad (2.4)$$

Since

$$T = Fr = Fd$$

then

$$hp = \frac{2\pi Tn}{396,000} = \frac{Tn}{63,000} \quad (2.6)$$

or

$$T = \frac{63,000 \text{ hp}}{n} \quad (17.1)$$

Torque (T, in.-lb; n, rpm) (17.1)

$$T = \frac{63,000 \text{ hp}}{n}$$

From this, the stress in the shaft for a round shaft can be determined as follows using the moment of inertia from Appendix 3:

For a round shaft,

$$Z' = \frac{\pi D^3}{16}$$

The stress is

$$S_s = \frac{Tc}{J} = \frac{T}{Z'} \quad (3.6)$$

Substituting and rearranging,

$$S_s = \frac{16T}{\pi D^3} \quad (17.2)$$

and solving for diameter,

$$D = \sqrt[3]{\frac{16T}{\pi S_s}} \quad (17.3)$$

Shaft sizing (17.3)

$$D = \sqrt[3]{\frac{16T}{\pi S_s}}$$

In this formula, safety factors can be applied to the stress allowable or torque as appropriate.

Spur Gear Loads

From Chapter 12, the loading on the shaft from the combination of the transmitted and separating forces can be found as follows:

$$F_r = \frac{F_t}{\cos \theta} \tag{12.2}$$

This can be seen from Figure 17.1. From this force, the bending moment can be calculated. If this gear is very close to a shaft support such as a bearing, the shear stress in the shaft may be more significant than the bending moment.

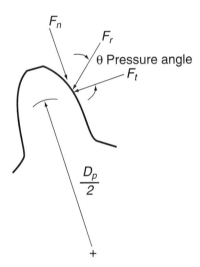

FIGURE 17.1 Transmitted and separating force from a spur gear.

Loads from Bevel Gears

Bevel gears have three forces acting on them because of the angle of bevel: a transmitted force similar to that on spur gears, a separating force that is also similar to that on spur gears, and an axial force due to the angle of bevel. For the pinion or the small gear, the transmitted force and the separating force are usually much larger than the axial force. For the larger gear, however, the axial force becomes larger than the radial separating force. On the smaller pinion gear, the axial force is relatively insignificant because on many shafts this may be the only axial force, and the bending component is small because of the pinion gear's small diameter. When this is true, in the case of the pinion, we can ignore the axial force and often the separating force for the large gear. We can calculate these values as follows:

The force transmitted is

$$F_t = \frac{T}{r} = \frac{T}{d} \qquad (2.5)$$

Note: d is the distance, not diameter (or $D_p/2$).
Then, the separating radial force can be found as

$$F_n = F_t \tan \theta \cos \gamma \qquad (13.7)$$

and the corresponding axial force can then be found as

$$F_a = F_t \tan \theta \sin \gamma \qquad (13.7)$$

The directions of these forces are shown in Figure 17.2. Then we can add the two larger forces using the Pythagorean theorem because they are perpendicular.

$$F_r = \sqrt{F_t^2 + F^2} \quad \text{(add larger of } F_n \text{ or } F_a \text{ to } F_t\text{)} \qquad (17.4)$$

From these forces we then calculate the appropriate bending moments. If the separating and axial forces are similar in magnitude, they can be added

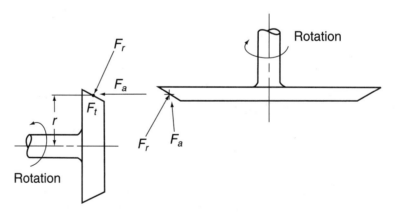

FIGURE 17.2 Forces on shafts from bevel gears.

using the Pythagorean theorem before inserting into Formula (17.4) and finding the bending moment. Note that the axial load, if significant, would still create an axial stress.

Worm Gear Loads

The forces on worm gears can be found in a somewhat similar manner. The difference, however, is that the separating force between the worm and the worm gear is equal, and the axial load on the worm is equal to the driving force on the worm gear, which can be found as follows:

$$F_t = \frac{T_o}{r_{wg}} \quad (17.5)$$

where F_t = thrust on worm

T_o = output torque

and

$$F_s = \frac{F_t \sin \phi}{\cos \phi \cos \lambda - f \sin \lambda} \quad (17.6)$$

where λ = lead angle

ϕ = normal pressure angle

f = coefficient of friction

Figure 17.3 shows the worm gear forces.

Note also that the force-resisting rotation of the worm is equal to the axial force on the worm gear and can be found as follows:

$$F_{a(gear)} = F_{t(gear)} \left(\frac{\cos \phi \sin \lambda + f \cos \lambda}{\cos \phi \cos \lambda - f \sin \lambda} \right) \quad (17.7)$$

Note that the coefficient of friction typically changes with rotational "sliding" speed.

Belt and Chain Drive Loads

From Chapter 14, we can determine the forces from belt and chain drives as being equal to the driving force in the case of chains and the front force plus the back force in the case of a belt.

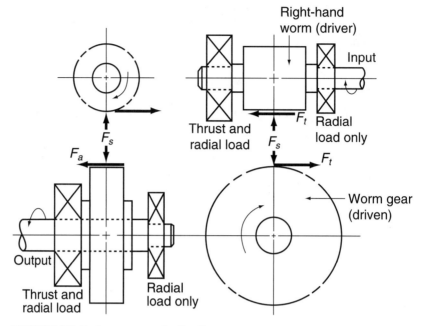

FIGURE 17.3 Forces on shafts from worm gears.

For a belt, the total load on the shaft would be equal to

$$F_t = F_f + F_b \tag{17.8}$$

as

$$F_d = F_f - F_b \tag{14.1}$$

and

$$T = F_d r \tag{14.3}$$

where F_d = net driving force
F_f = front force
F_b = back force
r = effective radius of the pulley or sprocket

For a chain, the back force is typically zero, so the load on the shaft is simply the driving force.

17.3 DESIGN STRESSES IN SHAFTS

Now that the loads in the shaft have been determined, the next step is to simplify where possible. This involves taking any loads that are acting at the same axial location on the shaft and adding vectorially to obtain a resultant force. In addition, any torques that could be added should be done at this time. For instance, the driving force and separating force for a spur or worm gear system, since they are perpendicular to each other, can be added using the Pythagorean theorem, which results in a single force. Then, taking into account the placement of the bearings and other forces acting on the shaft, moment diagrams can be created. Remember from the introduction that, if the moments are not in the same plane, they are not added directly. However, doing so will result in an acceptable conservative solution. Therefore we will take that approach in this text.

There are now two distinctly different approaches that are often taken in shaft design. If the design has a proposed shaft size, we can solve for the stresses at the areas of interest in the shaft. This would involve determining the bending stress from the moments determined previously and the shearing stresses from either shear loads or torque, as appropriate. Example Problem 17.1 demonstrates this technique.

Example Problem 17.1

The shaft shown in Figure 17.4 drives a gear set that is transmitting 5 hp at 1,750 rpm. The shaft is supported in self-aligning ball bearings, and the gears are both 10 pitch, 40 tooth, 20° spur gears. Find the torsional and bending stresses in the shaft.

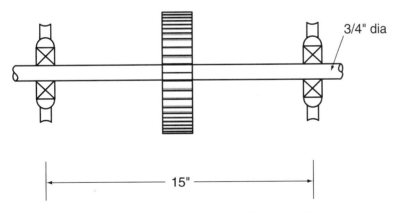

FIGURE 17.4 Shaft with spur gear for Example Problems 17.1 through 17.3.

Find the torsion in the shaft.

$$\text{hp} = \frac{Tn}{63,000} \qquad (2.6)$$

Then

$$T = \frac{63,000 \text{ hp}}{n} \qquad (17.1)$$

$$T = \frac{63,000(5)}{1,750}$$

$$T = 180 \text{ in.-lb}$$

Find the torsional stress in the shaft.

First find Z'.

$$Z' = \frac{\pi D^3}{16} \qquad \text{(Appendix 3)}$$

$$Z' = \frac{\pi(.75 \text{ in.})^3}{16}$$

$$Z' = .083 \text{ in.}^3$$

$$S_s = \frac{T}{Z'} \qquad (3.6)$$

$$S_s = \frac{180 \text{ in.-lb}}{.083 \text{ in.}^3}$$

$$S_s = 2,170 \text{ lb/in.}^2$$

Find the load at the gear pitch circle.

$$D_p = \frac{N_t}{P_d} \qquad (11.4)$$

$$D_p = \frac{40}{10}$$

$$D_p = 4 \text{ inches}$$

$$F_t = \frac{2T}{D_p} \quad \text{(12.3)}$$

$$F_t = \frac{2(180 \text{ in.-lb})}{4 \text{ in.}}$$

$$F_t = 90 \text{ lb}$$

Find the resultant force on the shaft.

$$F_r = \frac{F_t}{\cos \theta} \quad \text{(12.2)}$$

$$F_r = \frac{90 \text{ lb}}{\cos 20°}$$

$$F_r = 96 \text{ lb}$$

Find the maximum moment.

$$M_m = \frac{FL}{4} \quad \textbf{(Appendix 2)}$$

$$M_m = \frac{96 \text{ lb } (15 \text{ in.})}{4}$$

$$M_m = 360 \text{ in.-lb}$$

Find the stress.

$$S = \frac{M}{Z}$$

$$Z = \frac{\pi D^3}{32} \quad \textbf{(Appendix 3)}$$

$$Z = \frac{\pi(.75 \text{ in.})^3}{32}$$

$$Z = .041 \text{ in.}^3$$

$$S = \frac{M}{Z}$$

$$S = \frac{360 \text{ in.-lb}}{.041 \text{ in.}^3}$$

$$S = 8,780 \text{ lb/in.}^2$$

The second approach involves trying to solve for the shaft size using the maximum shear stress theory or normal shear stress theory as discussed in Chapter 4. This approach has some inherent simplicities, and estimates of shaft sizes can be obtained. However, as discussed later in this chapter, this approach does not easily allow for other effects, such as stress concentrations. It may be most useful as a preliminary sizing approach; once the size is determined, the actual stress suitability can be determined by other methods.

17.4 COMBINED STRESSES IN SHAFTS

In Chapter 4, we studied the principles of combined stresses, which are applicable here as well. The most common ways to combine stresses in shaft design problems would be to combine normal and shear stresses using the maximum normal stress theory or to combine stresses using the maximum shear stress theory. The maximum shear stress theory formula from Chapter 4 is

$$\tau = \left(S_s^2 + \left(\frac{S}{2} \right)^2 \right)^{1/2} \qquad (4.5)$$

We can find an equivalent shear stress to be compared to the shear stress allowable, or this formula can be rearranged in terms of diameter for a solid shaft, subject to a torque and bending moment, for estimated sizing purposes as follows:

$$\tau = \frac{5.1}{D^3} \left(T^2 + M^2 \right)^{1/2} \qquad (17.9)$$

Shaft shear stress (17.9)

$$\tau = \frac{5.1}{D^3} \left(T^2 + M^2 \right)^{1/2}$$

where τ = allowable shear stress (S_s)

T = torque in shaft

M = maximum moment

Note that in the maximum shear stress theory, the stress allowable would be the shear stress allowable that, as discussed earlier in the text, would typically be 50 to 60 percent of the normal stress allowable values. The following example problem demonstrates this technique using information from Example Problem 17.1.

Example Problem 17.2

From Example Problem 17.1, find the combined stress using the maximum shear stress theorem:

$$\tau = \left(S_s^2 + \left(\frac{S}{2}\right)^2 \right)^{1/2} \quad (4.5)$$

Substituting stresses from Example Problem 17.1:

$$\tau = \left((2{,}170 \text{ lb/in.}^2)^2 + \left(\frac{8{,}780}{2} \text{ lb/in.}^2\right)^2 \right)^{1/2}$$

$$\tau = 4{,}900 \text{ lb/in.}^2$$

Note that this should be compared to shear stress allowables.

If we were to use the maximum normal stress theory from Chapter 4, we would obtain a similar result, which in this case should be compared to the normal stress limit. The maximum normal stress theory is

$$\sigma = \frac{S}{2} \pm \left(S_s^2 + \left(\frac{S}{2}\right)^2 \right)^{1/2} \quad (4.4)$$

Note that, for estimated sizing purposes, we can do a similar substitution in this case to solve for the diameter of a solid shaft as follows:

$$\sigma = \frac{5.1}{D^3} \left[M + (T^2 + M^2)^{1/2} \right] \quad (17.10)$$

Shaft normal stress (17.10)

$$\sigma = \frac{5.1}{D^3}\left[M + (T^2 + M^2)^{1/2}\right]$$

The following example problem demonstrates this combination from Example Problem 17.1.

Example Problem 17.3

From Example Problem 17.1, find the combined stress using the maximum normal stress theory.

$$\sigma = \frac{S}{2} \pm \left(S_s^2 + \left(\frac{S}{2}\right)^2 \right)^{1/2} \quad (4.4)$$

Substituting stresses from Example Problem 17.1:

$$\sigma = \frac{8,780 \text{ lb/in.}^2}{2} + \left((2,170 \text{ lb/in.}^2)^2 + \left(\frac{8,780 \text{ in.}^2}{2}\right)^2\right)^{1/2}$$

$$\sigma = 9,300 \text{ lb/in.}^2$$

Note that this should be compared to the normal stress allowable.

17.5 COMPARISON OF STRESSES TO ALLOWABLE VALUES AND ENDURANCE LIMITS

In Chapter 5 we examined the analysis of fatigue loading or endurance limits. In shaft design problems, the load is usually repeated and reversed for the bending parts of the load. As the stresses have been combined in the previous section, it is conservative to then substitute into the Soderberg equation, assuming fully reversed and repeated loads. This simplifies the Soderberg equation to

$$\frac{1}{N} = \frac{\sigma \text{ or } \tau}{S_n} \qquad (17.11)$$

(Note that, if combined shear stress [τ] is used, S_n should include a factor for shear [.5 or .6].)

Where the mean stress is zero, we then compare the alternating stress, which is equal to the maximum stress, to the endurance limit, modifying the endurance limit as necessary for surface, size, and type as was done in Chapter 5. Therefore, we can combine this analysis with our combined stress analysis in Section 17.4 and substitute the revised endurance limit, along with an appropriate safety factor for the allowable stresses. Note in this technique we are not factoring in shape discontinuities or other factors. If this was necessary, it could also be done using the Chapter 5 techniques. This would then further modify the Soderberg equation as follows:

$$\frac{1}{N} = \frac{K_t (\sigma \text{ or } \tau)}{S_n} \qquad (17.12)$$

where these factors for K_t would further reduce the endurance limit by dividing the endurance limit by the values obtained for K_t. Note that this is a rough technique because many of the modifying factors had specialized cases depending on whether the part was loaded, for example, in torsion as opposed to bending. However, if we take the most conservative values, which would be the largest values for K_t, and use these values in our analysis, our shaft sizing will almost always be conservative. The following example problem demonstrates this technique using the shaft and gear set from previous example problems, where we have now modified the shaft to include a shoulder at the highest-stress area.

Example Problem 17.4

For the shaft in Example Problems 17.1 through 17.3, assume the shaft is made from OQT AISI 1040 steel with a machined surface and a safety factor of 2. The step has a radius of .030 inch, which results in an estimated K_t value of 2.3. Is this a suitable design for long-term use based on the Soderberg equation?

First, find the modifying factors for S_n.

$$S_u = 88 \text{ ksi} \qquad \text{(Appendix 4)}$$
$$C_{size} = .85 \qquad (\tfrac{1}{2} < D < 2)$$
$$C_{surface} = .88 \qquad \text{(Figure 5.7b)}$$
$$C_{type} = 1 \qquad \text{(Bending)}$$

Note that, as the bending stress is four times as large as the shear stress, shear factor is probably not important.

$$S_n = C_{size} \, C_{surface} \, C_{type} \, .5 S_u \qquad (5.1)$$
$$S_n = .85 \, (.88) \, 1 \, (.5) \, 88 \text{ ksi}$$
$$S_n = 32.9 \text{ ksi}$$
$$\frac{1}{N} = \frac{K_t \sigma}{S_n} \qquad (17.12)$$

Solve for N. If $N > 2$, it would be an acceptable design.

$$\frac{1}{N} = \frac{2.3 \, (9,300 \text{ lb/in.}^2)}{32,900 \text{ lb/in.}^2}$$
$$N = 1.54$$

This does not meet the criteria. Possible options could be:

1. Increase the shaft size.
2. Change to a higher-strength material.
3. Increase the step radius or reduce the step size.

17.6 STANDARD SHAFT SIZES

Because many standard pulleys, sprockets, gear hubs, couplings, and clutches are used with shafting materials, standard sizes are generally used. Between ½ inch and 2½ inches in diameter, shafts are usually available in increments of 1/16 inch. From 2½ to 4 inches in diameter, 1/8-inch increments are usually available. Above that, shafting is generally available in ¼-inch increments. However, in many cases the smaller sizes may only be obtained in 1/8-inch increments and the middle range in ¼-inch increments for mating with standard components. Actual manufacturers' parts should be reviewed when selecting shaft sizes. Generally, the shaft size found from the analysis would be rounded up to the next available standard size to eliminate the need for custom parts. Obviously, in many applications where weight is critical, nonstandard parts may be specified. In high-volume mass production, special sizes may actually reduce costs. However, for most standard design problems, a standard-size shafting is selected in order to utilize standard mating parts.

17.7 CRITICAL SPEED

> **critical speed**
> Refers to a speed at which a part would vibrate due to its natural frequency; usually extreme care is taken to avoid operating the part close to or at its critical speed.

> **natural frequency**
> The same as critical speed; the speed at which the part has an inherent tendency to vibrate.

Critical speed or **natural frequency** occurrences can be a significant problem in many aspects of machine design. Envision for a moment the antenna on your car, which vibrates in a sine-wave fashion at certain speeds. What is happening is that the antenna is being "excited" at its natural frequency. This is similar to the phenomenon of pushing someone in a swing when they are going away from us in order to increase the amplitude of the swing at its natural frequency. Generally, natural frequencies multiply. For example, if our antenna vibrated at 30 mph, it would likely vibrate with twice the fixed points at 60 mph. If we had a tire out of balance on our car that vibrated at 30 mph, it would likely vibrate again at 60 mph. Because of this phenomenon, the destructive forces on a part when it operates at its natural frequency can be significant in cases where forces are added at this frequency. Water hammer in pipes is another example of this phenomenon that you may have witnessed.

The analysis of natural frequency is difficult to do exactly. In order to effectively determine the natural frequency of a part, we need to analyze

the deflection under load, which can be difficult. However, it is often useful to try to estimate the natural frequency of a part. Because this is only an estimate, it is generally recommended that the normal operating speed should be no more than approximately 80 percent of the calculated critical speed and, if it is necessary to operate above the first critical speed or a multiple thereof, a machine should pass through this critical speed or natural frequency fairly quickly because operating within this range can be very detrimental.

The critical speed can be estimated as follows:

$$N_c = \frac{30}{\pi}\left(\frac{g_o \Sigma F \delta}{\Sigma F \delta^2}\right)^{1/2} \quad (17.13)$$

where F = each force acting on the shaft

δ = associated deflection from each force

g_o = 386 inches/sec^2

N_c = critical speed in rpm

This formula should be applied carefully because each load and its deflection must be calculated. We can estimate the critical speed of a shaft with a concentrated load located along a single span as follows:

$$N_c = \frac{188}{\sqrt{\delta}} \quad (17.14)$$

where N_c = critical speed in rpm

δ = deflection using standard beam formulas

Note that the deflection in this analysis would be found using standard methods. For a steel shaft supported rigidly in self-aligning bearings, again with a single load, we can estimate its approximate natural frequency as follows:

$$N_c = 387,000 \frac{D^2}{ab}\sqrt{\frac{L}{F}} \quad (17.15)$$

where D = diameter of shaft

L = distance between supports

a and b = distance from load to each of the bearings

> **Natural frequency (17.13)**
>
> $N_c = \frac{30}{\pi}\left(\frac{g_o \Sigma F \delta}{\Sigma F \delta^2}\right)^{1/2}$

If we assume the shaft is rigidly supported, an approximate estimate of the natural frequency, again for a single load in a span of a shaft, is

$$N_c = 387,000 \frac{D^2 L}{ab} \left(\frac{L}{Fab} \right)^{1/2} \quad (17.16)$$

To use Equations (17.15) and (17.16), the shaft needs to be of uniform diameter. If the weight of the shaft is significant, adding one-half of the weight to the load will increase the accuracy of this analysis. An example problem showing this technique follows.

Example Problem 17.5

Find the estimated critical speed for the shaft in Example Problem 17.1 (assume the entire shaft diameter is 3/4 inch).

First, find deflection.

$$\delta = \frac{FL^3}{48\,EI} \quad \text{(Appendix 2)}$$

$$I = -\frac{\pi D^4}{64} \quad \text{(Appendix 3)}$$

$$I = \frac{\pi(.75 \text{ in.})^4}{64}$$

$$I = .016 \text{ in.}^4$$

$$\delta = -\frac{96 \text{ lb } (15 \text{ in.})^4}{48\,(30 \times 10^6 \text{ lb/in.}^2)(.016 \text{ in.}^4)}$$

$$\delta = .21 \text{ inch}$$

$$N_c = \frac{188}{\sqrt{.21}}$$

$$N_c = 410 \text{ rpm} \quad (17.14)$$

Note that this is approximate, and additional multiples would exist at 820, 1,230, and 1,640 rpm.

The analysis presented here for determining critical speed is suitable for estimating purposes only. If you are concerned about the natural frequency in a part based on its expected use, either greater study should go into the analysis of natural frequency or testing should be done to determine these points.

17.8 SUMMARY

Upon completing this chapter, you should have obtained a basic understanding of how to identify the loads that act on shafts from standard components such as gears, belts, and chains. You should understand how to determine the stresses and, when appropriate, how to combine them. You should also be able to compare these stresses to suitable allowable stresses by using the methods presented in Chapters 4 and 5. You should also have obtained some basic understanding of how to roughly estimate the critical speed of a part, and you should understand that the analysis of natural frequency or the harmonics of a critical speed is very complicated. Exact analysis is beyond the scope of this text, and further study and research would be necessary to accurately determine these values.

17.9 PROBLEMS

(For Problems 1 through 15, refer to Figure 17.5.)

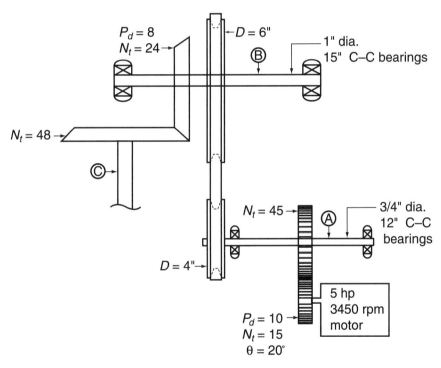

FIGURE 17.5 Figure for Problems 1 through 15.

1. If the 5-hp motor in Figure 17.5 has an output speed of 3,450 rpm, find the rotational speed of shafts A, B, and C.
2. In Problem 1, find the torque in each shaft.
3. Find the torsional stress in shaft A.
4. Find the resultant force from the gear in shaft A.
5. Find the bending stress in shaft A from the gear.
6. Find the combined stress at the gear in shaft A using the maximum normal stress theory.
7. Repeat Problem 6 using Equation 17.10. Compare this result with the results found in Problem 6.
8. Find the net force from the pulley on shafts A and B.
9. Find the combined shear and torsional stress from the 4-inch sheave. How does this compare to the stress found in Problems 6 and 7 at the gear location?

For Problems 10 through 15, assume both the pulley and bevel gear are approximately in the center of the 15-inch span in shaft B.

10. From observation only, where do you think the maximum stress occurs in shaft B? List the stresses at that location.
11. Determine the torsional stress in shaft B.
12. Find the load on the shaft from the bevel gear.
13. Find the combined stress in shaft B.
14. If this shaft is made from cold-drawn AISI 1050 steel and is subject to long rotating life, determine the factor of safety for shaft B. (Assume a stress concentration factor of 1.5 and a machined surface. Also assume both the pulley and bevel gear are approximately in the center of the 15-inch span in shaft B.)
15. Estimate the critical speed for shaft B. (Assume both the pulley and bevel gear are in the center of the 15-inch span.)

17.10 CUMULATIVE PROBLEM

16. The driveshaft of the truck described in previous chapters (as shown in Figure 17.6) has the following conditions: bevel gears from Chapter 13, Problem 18 and torque from Chapter 13, Problem 19. Assume a load on the end of the shaft to be 6,000 pounds (60,000 lb/10). Use an AISI 4140 OQT 1000 steel for this shaft. For infinite life, determine the shaft size at both points of likely high stress (both bearings). Use a safety factor of 2.

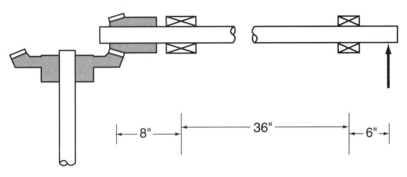

FIGURE 17.6 Truck drive shaft.

CHAPTER 18

Power Screws and Ball Screws

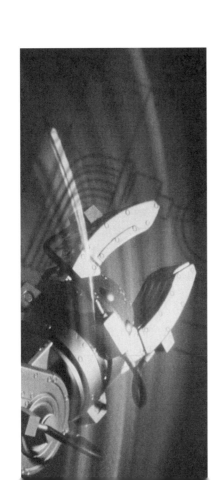

power screw
Often referred to as a "lead screw": a threaded fastener-type assembly used to convert rotary motion to linear motion.

ball screw
A power screw that, instead of using sliding friction, has rotating balls that act similar to a bearing to transform rotary to linear motion.

Power screws and ball screws are used to convert rotary to linear motion. In a power screw, which is also commonly called a lead screw, the screw or the nut could translate depending on the mounting. Many different types of power and ball screws are available. A good example of the use of a power screw is the mechanism used to advance the cutting carriage on a lathe or the table on a milling machine. The accuracy of movement from power screws can be quite exact, for example, in the case of machine tools. Other applications require less accuracy, such as a lead-screw-driven garage door opener or an automobile jack. Another application of power screws appears on airplane wings, where power screws are used to change the angle of flaps, to retract or extend the flaps, or to raise or lower the landing gear. Linear motion is often needed in design and, typically, the power source is an electric motor or internal combustion engine. Therefore, power screws are a vital part of the designer's inventory of products.

18.1 OBJECTIVES

This chapter is intended to introduce you to the principles of power screws, including how to calculate the required forces and necessary torques. Using the principles previously learned in other chapters, you will also calculate the structural capabilities of these products. After completing this chapter, you should be able to:

- Recognize and understand the advantages and disadvantages of the different types of power screws.
- Determine the power necessary for driving power screws at different speeds and torques.
- Understand the principles of operation of ball screws and how they differ from friction-type power screws.
- Understand and calculate the torque and efficiencies of power screws and ball screws.
- Begin to understand and envision how power screws and ball screws can be used in different designs.

18.2 TYPES OF POWER SCREWS

Figure 18.1 shows three types of power screw threads: square thread, acme thread, and buttress thread. Each has different advantages and disadvantages depending on the application. The square thread is generally consid-

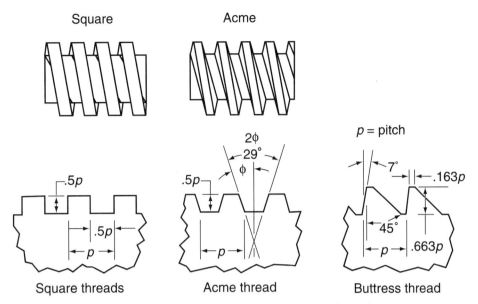

FIGURE 18.1 Profile of square, acme, and buttress threads.

ered to be the most efficient for transferring torque to linear motion. The acme thread is easier to make, may have advantages in applications where cleanliness may not be as high, and, if well lubricated, can achieve efficiencies that are only slightly less than those of a square thread. The buttress thread is more efficient than the acme thread, due to the fact that the forward angle is closer to a square thread, and is generally used in applications where the force is only being transmitted in one direction such as in an automobile jack.

A standard cut or rolled thread such as a Unified National (UNC) profile is sometimes used for lead screws as well, but generally only in applications where they are used intermittently. Some other modified forms of threads have also been developed for specialized applications.

Lead and pitch can be determined in the same manner as was done in Chapter 6. For a lead screw assembly, the nut would progress the distance of the pitch or lead in one revolution and therefore is equal to the reciprocal of the number of threads per unit length. The tensile area for a lead screw can be conservatively approximated by finding the area of the minor diameter. Some tables will give a value that is slightly higher, because technically the shear area would include the area as if the part had broken through one thread, which adds slightly to the area. Table 18.1 lists the standards for square threads, acme threads, and ball screws.

TABLE 18.1
STANDARDS FOR SQUARE THREADS, ACME THREADS, AND BALL SCREWS

Size, Inches	Square Threads		Acme Threads		Ball Screws
	Threads per Inch	Minor Diameter	Threads per Inch	Minor Diameter	Threads per Inch
¼	10	.163	16	.188	
⁵⁄₁₆			14	.241	
⅜	8	.266	12	.292	8
⁷⁄₁₆			12	.354	
½	6½	.366	10	.400	5
⅝	5½	.466	8	.500	
¾	5	.575	6	.583	2
⅞	4½	.681	6	.708	
1	4	.781	5	.800	
1⅛			5	.925	
1¼	3½	1.000	5	1.050	
1⅜			4	1.125	
1½	3	1.208	4	1.250	2
1¾	2½	1.400	4	1.500	
2	2¼	1.612	4	1.750	2
2¼	2¼	1.862	3	1.917	
2½	2	2.063	3	2.167	1
2¾	2	2.313	3	2.417	
3	1¾	2.500	2	2.500	1½
3½	1⅝	2.962	2	3.000	
4	1½	3.418	2	3.500	
4½			2	4.000	
5			2	4.500	

In the next section, we will analyze the required torque and the associated power for a lead screw, along with the associated efficiency. It is important to recognize that, particularly in an open lubricated system, the introduction of dirt or other contaminants can significantly affect the efficiency of power screws. Plated-on lubricant is often used to minimize the potential for dirt or other contaminants to stick to the lead screw systems. Extreme care should be taken when using lead screws to minimize the potential introduction of dirt if it is necessary to have accurate values for torque and power.

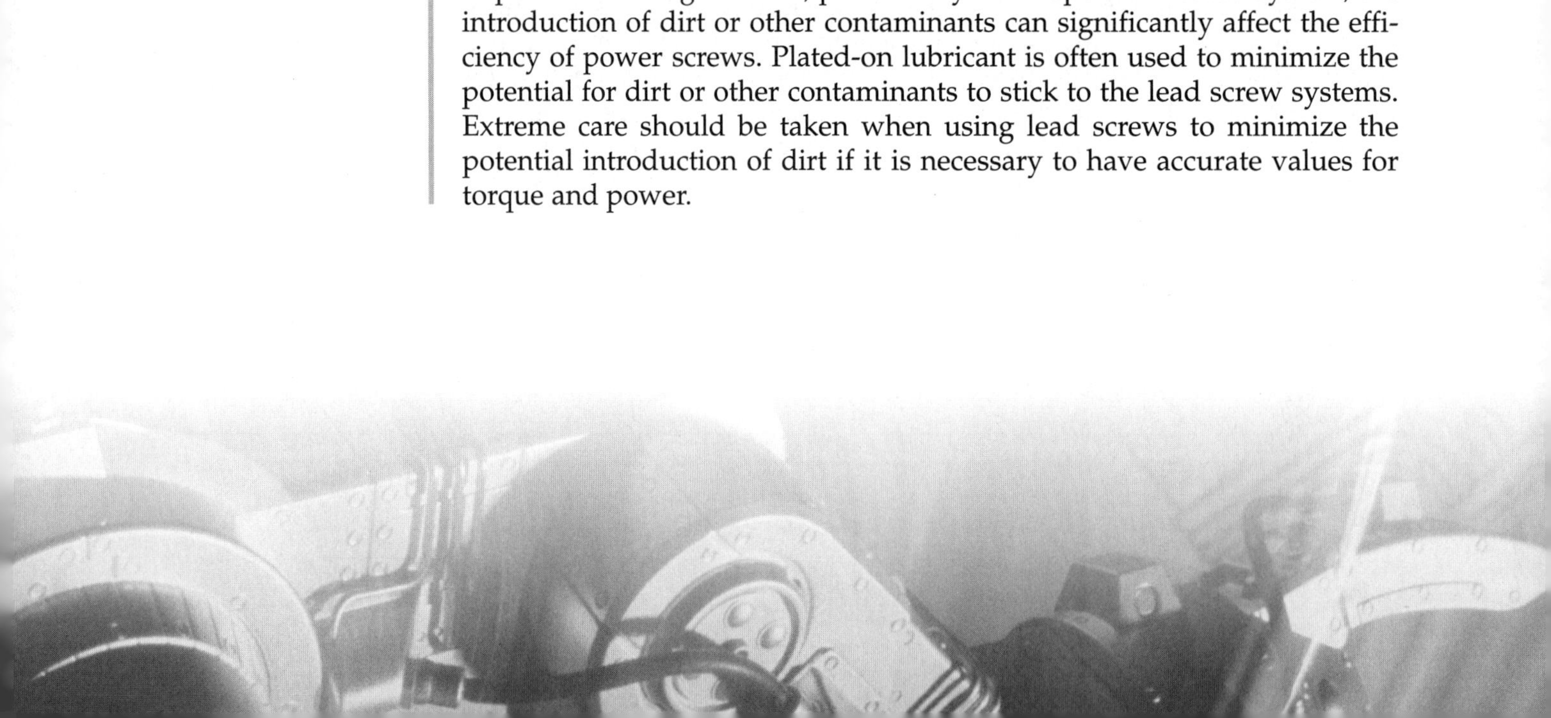

18.3 TORQUE, POWER, AND EFFICIENCY IN POWER SCREWS

A lead screw can be envisioned as an inclined plane problem; the lead screw in essence pushes the load around its circumference similar to the way a box is pushed up an inclined plane. Figure 18.2 shows the forces acting on a lead screw assembly, both for pushing the box up the plane and for pushing it down. A lead screw needs a different torque to raise than to lower a load. In order to calculate the torque for a lead screw, we need to find the angle of lead, which becomes the equivalent of the inclined plane angle.

$$\tan \lambda = \frac{L}{\pi D_p} \quad (18.1)$$

where λ = angle of incline
L = lead
D_p = pitch diameter

Lead angle (18.1)

$$\tan \lambda = \frac{L}{\pi D_p}$$

P = Force required to move the load
F_f = Friction force
N = Normal force
λ = Lead angle
D_p = Pitch diameter

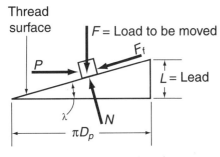

(a) Force exerted *up* the plane

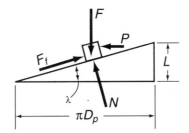

(b) Force exerted *down* the plane

FIGURE 18.2 Force required to push a box up or down an incline.

Note that the lead is the same as the pitch and is equal to $1/n$. As in gears, the pitch diameter is a property of the shape of the thread. The pitch diameter is often considered to be the same as the mean diameter (D_m), which is the average of the major and minor diameters.

We also need to know the coefficient of friction (f), which is a product of the materials from which the lead screw and nut are constructed and the method and type of lubrication. If values are not known, we can approximate this value as .15 for lubricated, well-manufactured lead screw assemblies. However, this value can vary significantly depending on surface finish, lubrication, and, especially, the potential for the introduction of dirt or other contaminants. We can find the torque needed to move a load up or horizontally against a force as follows:

$$T_{up} = \frac{FD_p}{2}\left(\frac{L + \pi f D_p}{\pi D_p - fL}\right) \quad (18.2)$$

where L = lead or pitch

f = coefficient of friction

D_p = pitch diameter

Correspondingly, the torque needed to lower a load can be determined as follows:

$$T_{down} = \frac{FD_p}{2}\left(\frac{\pi f D_p - L}{\pi D_p + fL}\right) \quad (18.3)$$

Formulas (18.2) and (18.3) are for a square thread, but they can be used to approximate torque for other types of threads.

An alternate form of Equations (18.2) and (18.3) is often useful. We can also determine the torque needed to move a load up using the following equation:

$$T_{up} = \frac{FD_p}{2}\left(\frac{(\tan \lambda + f)}{1 - f \tan \lambda}\right) \quad (18.4)$$

The torque required to lower the load would be

$$T_{down} = \frac{FD_p}{2}\left(\frac{(f - \tan \lambda)}{1 + f \tan \lambda}\right) \quad (18.5)$$

Power screw torque up (18.2)

$$T_{up} = \frac{FD_p}{2}\left(\frac{L + \pi f D_p}{\pi D_p - fL}\right)$$

Power screw torque down (18.3)

$$T_{down} = \frac{FD_p}{2}\left(\frac{\pi f D_p - L}{\pi D_p + fL}\right)$$

Example Problem 18.1

An elevator-type lift is being designed to raise a load of up to 5,000 pounds between floors. Two square-thread lead screws, 1½ inches in diameter, are used for this application. Determine the torque required to raise this lift if the moving part of the elevator weighs 800 pounds and the coefficient of friction is assumed to be .15.

From Table 18.1, the 1½-inch-square thread has three threads per inch:

$$D_m = \frac{1.5 + 1.208}{2} = 1.354 \text{ in.}$$

Determining the torque up,

$$T_{up} = \frac{FD_p}{2}\left(\frac{L + \pi f D_p}{\pi D_p - fL}\right) \quad (18.2)$$

$$T_{up} = \left(\frac{5,800 \text{ lb}}{2}\right)\left(\frac{1.354 \text{ in.}}{2}\right)\left(\frac{\tfrac{1}{3} \text{ in.} + \pi .15(1.354 \text{ in.})}{\pi(1.354 \text{ in.}) - .15(\tfrac{1}{3} \text{ in.})}\right)$$

$$T_{up} = 454 \text{ in.-lb (for each lead screw)}$$

Power Required to Drive a Lead Screw

The power necessary for driving a lead screw is a product of the velocity and the efficiency. It can be determined from the standard power formula from Chapter 2. That torque can be substituted if you wish to know the power for lowering a load.

$$P = \frac{Tn}{63,000} \quad (2.6)$$

Self-Locking

Lead screws are somewhat similar in their principles of operation to the worm gears studied earlier in this text. They can also be self-locking. With a self-locking lead screw, a load on the nut would not force rotation of the lead screw. Self-locking would certainly be desirable for jacks or other similar products where the load needs to stay in place without locking the lead. Self-locking is obtained when the following property is true:

$$f > \tan \lambda \quad (18.6)$$

It can be seen from this formula that, when the coefficient of friction drops, the necessary angle for self-locking would decrease as well. Depending on lubrication, most standard leads tend to be self-locking. Care should be taken, however, in applications where vibration is present as this may cause minor movement of the screw.

Efficiency of a Power Screw

The efficiency of a lead screw or power screw assembly is a product of the relationship of the friction and the energy lost to friction. We can determine the necessary torque for a friction-free environment as follows:

$$T' = \frac{FD_p}{2} \tan \lambda \quad (18.7)$$

A friction-free environment, of course, would never exist, but if we compare this to the actual friction, we can determine the efficiency of our lead screw assembly. The actual efficiency can be found as follows:

$$e = \frac{T'}{T} = \frac{\tan \lambda (1 - f \tan \lambda)}{\tan \lambda + f} \quad (18.8)$$

Power screw efficiency (18.8)

$$e = \frac{T'}{T} = \frac{\tan \lambda (1 - f \tan \lambda)}{\tan \lambda + f}$$

This formula, like Formulas (18.2) and (18.3), is for a square thread but can be used to approximate efficiency for other types of threads.

The following example problem demonstrates the technique used to determine the power required for a square thread and the efficiency of the system.

Example Problem 18.2

For the elevator described in Example Problem 18.1, find the efficiency and power required if each lead screw is driven at 175 rpm.

The lead angle is

$$\tan \lambda = \frac{L}{\pi D_p} \quad (18.1)$$

$$\tan \lambda = \frac{\frac{1}{3} \text{ in.}}{\pi \, 1.354 \text{ in.}}$$

$$\lambda = 4.5°$$

The efficiency would be

$$e = \frac{\tan \lambda (1 - f \tan \lambda)}{\tan \lambda + f} \quad (18.8)$$

$$e = \frac{\tan 4.5° (1 - .15 \tan 4.5°)}{\tan 4.5 + .15}$$

$$e = 34\%$$

Power would then be

$$P = \frac{Tn}{63,000} \quad (2.6)$$

$$P = \frac{454 \,(175)}{63,000}$$

$$P = 1.26 \text{ hp (per lead screw)}$$

or

$$P = 2.52 \text{ hp (total)}$$

Verifying that this is self-locking:

$$f > \tan \lambda \quad (18.6)$$

$$f > \tan 4.5°$$

$$.15 > .079$$

Yes, it is self-locking.

Acme Threads

The face of an acme thread, unlike a square thread, has an angle of 14½°; therefore, a correction needs to be made for this type of thread. (See Figure 18.3.) We can calculate the torque up and the torque down with an adjustment for this additional angle as follows:

The torque up would be

$$T_{up} = \frac{FD_p}{2} \left(\frac{\cos \phi \tan \lambda + f}{\cos \phi - f \tan \lambda} \right) \quad (18.9)$$

Chapter 18

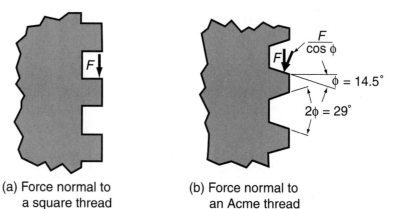

(a) Force normal to a square thread

(b) Force normal to an Acme thread

FIGURE 18.3 Forces on acme and square threads.

and the torque to move the load down would be

$$T_{down} = \frac{FD_p}{2}\left(\frac{(f - \cos\phi \tan\lambda)}{(\cos\phi + f\tan\lambda)}\right) \quad (18.10)$$

where ϕ = face angle of $14\tfrac{1}{2}°$

Example Problem 18.3 demonstrates the use of this modified formula for acme threads.

Example Problem 18.3

For the elevator in Example Problem 18.1, find the torque and power if a 1½-inch diameter acme thread was substituted:

$$D_p = D_m = \frac{1.5 + 1.25}{2} = 1.375 \quad \textbf{(from Table 18.1)}$$

$$\tan\lambda = \frac{L}{\pi D_p} \quad (18.1)$$

$$\tan\lambda = \frac{\tfrac{1}{4}}{\pi\, 1.375}$$

$$\lambda = 3.3°$$

$$T_{up} = \frac{FD_p}{2}\left(\frac{(\cos\phi \tan\lambda + f)}{(\cos\phi - f\tan\lambda)}\right) \quad (18.9)$$

$$T_{up} = \left(\frac{5,800 \text{ lb}}{2}\right)\left(\frac{1.375 \text{ in.}}{2}\right)\left(\frac{[\cos 14.5° \tan 3.3° + .15]}{[\cos 14.5° - .15(\tan 3.3°)]}\right)$$

$$T_{up} = 428 \text{ in.-lb}$$

Note that this result is lower than that of Example Problem 18.1 because the lead is lower, not because an acme thread is more efficient.

Find the power. For obtaining the same rate, however, the drive speed would need to be changed by the ratio of the leads.

$$n = 175 \cdot \frac{4}{3} = 233 \text{ rpm}$$

$$P = \frac{Tn}{63,000} \quad (2.6)$$

$$P = \frac{428 \text{ in.-lb } 233 \text{ rpm}}{63,000}$$

$$P = 1.59 \text{ hp (per lead screw)}$$

Note that this value reflects the lower efficiency of an acme versus a square thread.

18.4 TORQUE AND BALL SCREWS

Ball screws are a unique adaptation of the lead screw concept in which the principles of a ball bearing are adapted to a lead screw assembly, resulting in significantly lower friction than that which exists in a conventional lead screw. Remember that a lead screw uses sliding friction. Despite the introduction of lubricants and smooth surfaces, that friction is still significant. Figure 18.4 shows a cutaway view of a ball lead screw where it can be seen that the balls roll between the nut and the screw, resulting in a rolling action. Because these balls are typically hardened, as may be the surface of the screw, high efficiencies can be obtained, typically in excess of 90%. The actual expected efficiency ideally should be obtained from manufacturers' literature, but in the absence of a published value, we would typically use 90%. We can, therefore, solve for the efficiency of a ball lead screw assembly using this 90% or the published value as follows:

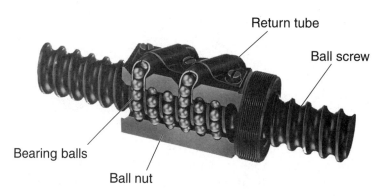

FIGURE 18.4 Ball screw assembly. (Courtesy of Linear Industries Ltd.)

Ball screw torque (18.11)

$$T_{up} = \frac{FL}{2\pi e} \quad (18.11)$$

If $e = .9$,

$$T_{up} = .177 FL \quad (18.12)$$

Because of the low friction in ball screw assemblies, they are almost never self-locking, and therefore they would not be suitable where self-locking is a desired property. However, because of this low friction, they lend themselves to the ability to create what is commonly called **back driving** where, by placing a load on the nut, you can cause the screw assembly to spin. This can be determined from the following relationship.

$$T_{bd} = \frac{FLe}{2\pi} \quad (18.13)$$

back driving

A unique property of ball screws due to their low friction whereby rotary motion can be created from linear motion. When the nut is pushed longitudinally along the ball screw, the ball-screw assembly rotates.

If $e = .9$,

$$T_{bd} = .143\ FL \quad (18.14)$$

Ball screw back driving (18.13)

$$T_{bd} = \frac{FLe}{2\pi}$$

A ball lead screw assembly features some of the principles of bearings, and in some applications there may be some preload between the nut and the rollers where tight clearance is required. There are often concerns about wear over time. Manufacturers' literature should be consulted for applications where extreme long life is required; the ball lead screw assembly may need to be de-rated in order to extend the life of the system. The following example problem shows the selection of a ball lead screw in order to determine the appropriate torque and the power necessary if a ball lead screw were substituted for the square thread used in Example Problem 18.1.

Example Problem 18.4

In Example Problem 18.1, substitute a ball screw and determine the torque and power (a 1½-inch ball screw has 2 threads per inch). Assume $e = .9$.

$$T = .177FL \qquad (18.12)$$

$$T = .177 \frac{5,800 \text{ lb}}{2}(.5) \text{ in.}$$

$$T = 257 \text{ in.-lb}$$

Find the power. For obtaining the same rate, however, the drive speed would need to be changed again by the ratio of the leads.

$$n = 175\frac{2}{3} = 117 \text{ rpm}$$

$$P = \frac{Tn}{63,000} \qquad (2.6)$$

$$P = \frac{257(117)}{63,000}$$

$$P = .48 \text{ hp}$$

Note that this system would likely need a brake system because this ball screw would not be self-locking. Also, note that collars or bearings that allow the rotation of a lead screw system can reduce efficiency or increase the required power.

18.5 SUMMARY

After completing this chapter, you should understand the principles of power screws and how the different types of power screws and/or ball screws have performance characteristics that can be useful for different design situations. You should understand the principles of efficiency for different types of power screws, and you should know when a power screw would be self-locking. In addition, you should now understand the relationship between torque and power for different power screws and be able to size appropriate drive systems for specific power, torque, and rotational speed requirements.

18.6 PROBLEMS

1. A jack with a 2-inch square-thread screw is supporting a load of 12 tons. It is expected that the coefficient of friction could vary from .10 to .15.
   a. What is the torque required to raise this load?
   b. What is the torque required to lower this load?
   c. Will this always be self-locking?
2. If an acme thread is loaded in tension and needs to lift a load of 25,000 pounds, select a screw size that would have a tensile stress below 15,000 psi.
3. For the screw size selected in Problem 2, find the torque to raise this load if $f = .15$.
4. In Problem 3, determine the power required if a desired rate of travel is 1.5 feet per second.
5. A square-thread screw, with a 1¾-inch diameter, exerts a force of 22,000 pounds.
   a. What is the axial stress in this screw?
   b. What torque would be required to lift this load if $f = .2$?
   c. What is the efficiency of this system?
6. For Problem 5, solve part b using Equation (18.4).
7. For the bottle jack shown in Figure 18.5, using a 1½-inch square thread, $f = .15$, and $F = 7,500$ pounds, determine the torque required to raise this load.

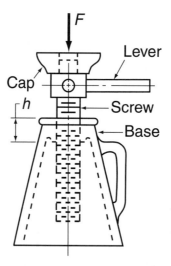

FIGURE 18.5 Bottle jack for Problems 7 through 11.

8. In Problem 7, if the maximum force on the lever is expected to be 100 pounds, how long should this lever be?
9. Solve Problems 7 and 8 using a 1½-inch acme thread instead of a square thread.
10. In Problem 7, if the contact stress in the thread should be below 5,000 psi, how long should the thread engagement be?
11. In Problem 7, using the principles learned about clutches and brakes, if the surface where the cap meets the lever has a $D_i = 1$ inch, $D_o = 2$ inches, and $f = .1$, how much torque is added to the torque to turn the screw if the cap does not rotate? What is the total torque?
12. A ½-inch ball screw moves a 50-pound box, which slides across a flat surface. If the coefficient of friction between the box and the surface is .4, what torque is needed to turn this screw?
13. In Problem 12, if the box oscillates back and forth a combined distance of 24 inches in a time of 2 seconds, find the necessary power and rotational speed of the drive.

18.7 DESIGN PROBLEM

14. Design an elevator system using square-thread screws for the following parameters:

    - Total weight 3 tons.
    - Total travel 20 feet.
    - Time to travel 20 feet is 10 seconds.
    - Screw and nut material is AISI 1040 CR steel with a SF = 2.

    Select a screw size and the number of screws, assuming the screws are supported from the top (no column action):
    a. Determine the torque.
    b. Determine the power.
    c. Determine the input speed.
    d. Determine the length of the nut.
    e. Make a sketch of your design.

18.8 CUMULATIVE PROBLEM

15. In the truck design problem from previous chapters, the trailer legs (which are used to support the trailer when it is disconnected from the tractor) are raised and lowered by two 1½-inch square-thread screws. They are driven by bevel gears and a crank handle that is 15 inches long. If the force on the handle should not exceed 80 pounds, decide on a ratio for the bevel gears. Assume, because of the placement of this jack, that 50% of the trailer weight and load would need to be raised. Assume a coefficient of friction of .15.

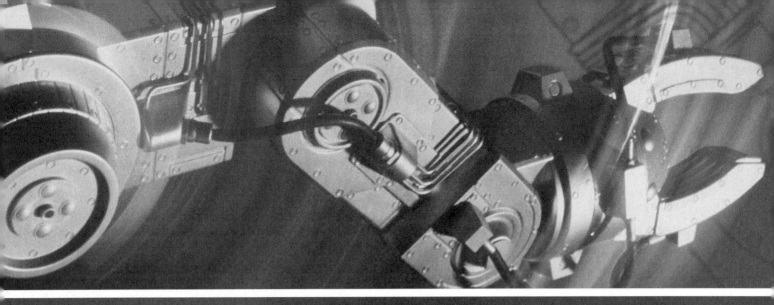

CHAPTER 19

Plain Surface Bearings

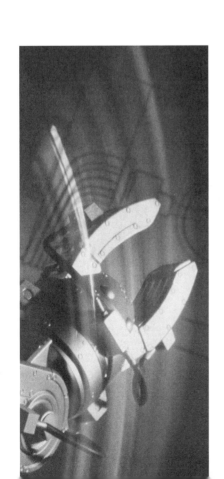

plain surface bearing
A term that refers to a smooth-surface bearing that relies on sliding between the shaft and the bearing, typically with some form of lubrication.

sleeve bearing
Another term for a plain-surface bearing in which a thin cylindrical material is used to allow for shaft rotation.

journal bearing
Another term for a plain-surface or sleeve bearing.

viscosity
A measure of the flow characteristics of a fluid. With bearings, it is often used as an indication of the fluid's lubrication properties.

minimum film thickness
The terminology used for determining the thickness that many lubricants try to maintain due to their internal capillary forces. It can also be thought of as the natural thickness of the fluid on a perfectly flat and smooth surface.

Plain surface bearings are often referred to as **sleeve bearings** or **journal bearings**, having been in existence in one form or another ever since the invention of the wheel. Early horse-drawn carriages used a bearing system constructed of leather (lubricated with animal fat) or wood. Later, whale oil was used as a lubricant for similar types of applications. Certainly modern technology has changed lubrications significantly, but many of the principles of operation of a plain surface bearing that applied then apply now. In almost all cases, plain surface bearings rely on a thin film of lubrication. This lubrication creates either sliding between the surfaces (whether metal, plastic, or, in the early days, wood or leather) in the case of a thin film bearing, or a **hydroplaning** effect, where the surfaces do not touch and the shaft rotates on a slightly thicker film of oil. The latter effect is similar to what can happen to an automobile on a rainy day when a film of water can exist between the wheels of the car and the road so that no physical contact is made between them. If you have ever experienced this phenomenon, you know that the friction decreases very dramatically and rapidly.

There are two basic properties of significance in the design of these types of bearings; the more significant is the **viscosity** of the lubricating fluid. If we envision two surfaces sliding against each other, whether they are round like a conventional bearing or simply two flat surfaces sliding, and if a film of lubrication exists between them, shearing forces are created within the fluid. This shearing force is what is referred to as the viscosity of the fluid. A fluid that can be sheared more easily—that is, one with lower viscosity—could be considered slipperier. The other factor that is important to lubricating surfaces for shafts is a property called the **minimum film thickness.** This is caused by the attraction between molecules in the fluid, which in turn causes a thin layer to be maintained between the surfaces. As we will learn later in this chapter, if the forces or associated pressures get too high, the lubrication can be forced out from between the surfaces, creating direct surface contact, which would negate the principles of bearings of this type. In general, the viscosity of a fluid is higher as the Society of Automotive Engineers Index (SAE) number increases. For instance, an SAE 40-weight oil would have a higher viscosity than a 10-weight oil. However, it is also possible that the minimum film thickness could be greater for the heavier oil; hence, the heavier oil may have some additional benefits as a lubricant in some applications, despite likely having slightly higher friction. As lubricants get heavier (for instance, in 80-weight oil or grease), the viscosity typically increases, and the minimum film thickness generally increases with viscosity. Many newer automobiles are using lighter-weight oils, which require greater accuracy in the manufacture of the sleeve- or plain surface–type bearings.

Temperature can also have a significant effect on the ability of materials to lubricate. The viscosity decreases with temperature in most fluids. This needs to be considered in order to ascertain the appropriate loading for bearings of this type.

Lastly, one of the principles of these types of bearings relates to the motion of the shaft or the surface that tends to "pull" the oil into position between the shafts because of the friction with the shaft surfaces. Most damage related to this occurs during the period in which the shaft initially starts rotating. This damage occurs because, when the shaft is not rotating, or when it may be oscillating and not making complete revolutions at a reasonable speed, the oil may dissipate from between the shafts, resulting in a condition where metal-on-metal contact could exist at the next startup. An interesting example of this is with regard to the purchase of a used car. We often hear of the used car that was owned by a little old lady who only drove it short distances to church on Sunday versus a car that may have greater mileage but was used extensively in highway driving. When we consider the effects of these two types of operation on the shaft-bearing assemblies (for example, in the engine), the little old lady's car that was only driven for short distances and left sitting for a week at a time probably had greater wear on these components. The oil probably leaked out from between the shaft and sleeve bearings during the week. Every Sunday when the car was started, damage then occurred between the shafts and bearings. This was likely compounded by the fact that the car was driven only short distances, so contaminants in the oil such as moisture may not have been as effectively cleaned.

In this chapter, we will focus mainly on using standard bearings and on the basic principles for selecting the correct size of the bearing. We will not go deeply into the principles of design for these types of bearings. That is a specialized field. However, mechanical designers regularly specify and use standard plain surface-style bearings.

19.1 OBJECTIVES

This chapter is intended to give you a basic understanding of the principles and operation of plain surface-style bearings and experience in selecting and using them. After completing this chapter, you should be able to:

- Understand the three modes of operation of a plain surface bearing: boundary, mixed-film, and full-film, or hydrodynamic operation.

- Recognize when these three forms of operation may be present.
- Gain an understanding of which types of materials are appropriate for different bearing applications.
- Understand how to use the pressure-velocity relationship in the selection and sizing of bearings.
- Use the principles of operation and the selection criteria to specify the appropriate bearings for individual applications.
- Understand the basic principles of lubrication and how they apply to the design process.
- Recognize how the hydrostatic bearing system operates and the associated pressures that may be necessary for bearings of this type.
- Understand how the principles of this chapter apply to modifications of standard bearings such as thrust bearings.

19.2 TYPES OF PLAIN SURFACE BEARINGS

Bearing design is complicated by the fact that, not only is the size of the parts pertinent and the type of lubrication, but the surface finish can also greatly affect the performance of the selected lubricant. Depending on these parameters, a journal or plain surface bearing can operate in any of the following forms.

Boundary Lubrication

In this type of lubrication system, actual contact exists between the surfaces. Although there is a film of lubricant that could be externally supplied, or lubricant exists from the manufacturing of the bearing, or the materials have natural lubricity, there is direct contact between the shaft and the bearing. Therefore, the surface coefficient of friction for **boundary lubrication** is usually the highest of the four types.

> **boundary lubrication**
> When there is actual contact between the surfaces but the sliding forces are minimized by the presence of lubrication.

Mixed-Film Lubrication

In this situation, there may be boundary lubrication or full-film lubrication. In general, because it is a transition condition, the actual type of lubrication may vary at different times, different temperatures, and under different conditions. Because the actual type of lubrication is often not known in this area, usually this type of system is avoided.

Full-Film Lubrication

This is often referred to as **hydrodynamic lubrication** where, once the machine is moving, a full film of lubricant supports the load. There is no actual contact, other than potentially at startup, between the two surfaces. In this case, the friction between the lubrication and the rotating shaft or bearing, because of **capillary forces,** tends to pull the oil between these surfaces, and the internal molecular attraction of the oil film creates a full intermediate barrier based on the principles of minimum film thickness discussed previously. This type of bearing lubrication is usually the preferred choice for long-term operation; other than during startup conditions, the only wear of the mating parts is by erosion from the lubricant flowing past the surfaces. Hence, this type of system can last for an extensive period of time. It is often used with some form of continuous oil supply system such as in an automobile engine.

Hydrostatic Bearing Lubrication

In a **hydrostatic** bearing lubrication system, an external pressurization source pumps oil under pressure between the shaft and the bearing so that the shaft is lifted from the bearing and rides on this pressurized fluid. This type of system is often used where very large loads are present, such as in a turbine generator system in a power plant or some very large stationary diesel systems where an external pump is activated before the shaft or journal rotates. Motion begins after this pressurization lifts the shaft to ride on a film of oil. Because the oil is forced in under pressure, this type of system can have a far greater contact load than a similar hydrodynamic system.

There are also bearings that incorporate more than one of these principles. For instance, some large engines have a pumped-in source of oil that creates somewhat of a combination of a hydrostatic and hydrodynamic situation. However, the basic principle of operation is typically hydrodynamic because, while the oil may be constantly supplied under pressure, it is not technically providing journal lift. A journal bearing in an automobile may fit this general description.

There are many factors to consider in creating an acceptable design for a bearing. In this chapter we will assume that in most cases the shaft is rotating, the bearing is stationary, and the type of bearing likely will be predetermined. The machine designer's main task will be to undertake the sizing calculations.

Figure 19.1 shows some different methods for lubricating sleeve-type bearing systems.

> **hydrodynamic lubrication**
> The phenomenon that occurs when the shaft or other smooth object is supported by the lubrication properties of the fluids and no actual contact is made between the metallic surfaces.
>
> **capillary forces**
> The forces that tend to hold a fluid to itself. These are the internal molecular attraction forces in a fluid that aid in lubrication.
>
> **hydrostatic**
> The phenomenon of creating a pressure in a fluid in order to physically lift the object; similar to, for example, the puck in an air hockey game.

LUBRICATION—BOST-BRONZ

All standard BOST-BRONZ bearings, bars and plates are impregnated with a high grade, oxidation-resistant mineral oil of SAE30 (ISO 100) viscosity. If properly stored, BOST-BRONZ parts retain their oil supply indefinitely. To prevent loss of lubricant, BOST-BRONZ should be stored in non-absorbent materials (metal, plastic, or suitably lined containers, etc.) The bearings should be covered to keep out dirt and dust.

REMOVING LUBRICANT: If it becomes necessary to remove the oil from BOST-BRONZ, for example to replace with another type or viscosity of lubricant, the following procedure may be used.:

Immerse parts in a good grade of oil solvent, such as lead-free gasoline, naptha, carbon tetrachloride or alcohol. Change solvent often, until solvent appears clear. Agitation will hasten the process.

RE-OILING: BOST-BRONZ parts may be re-impregnated by submerging in oil (pre-heated to about 150°F) for approximately 30 minutes. More time should be allowed for larger parts.

SUPPLEMENTARY LUBRICATION

The following designs illustrate simple, effective arrangements for providing supplementary lubrication.

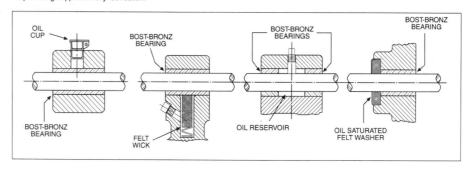

LUBRICATION—BEAR-N-BRONZ

The maintenance of an oil film between the shaft and bearing surfaces is extremely important, serving to reduce friction, dissipate heat, and retard wear by minimizing any metal to metal contact.

Lubricant is usually supplied into the bearing from an oil cup or fitting through an oil hole.

The drawings below illustrate two typical methods.

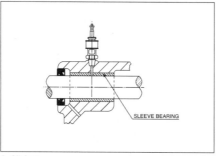

A. Oil Cup

Oil is fed from the oil cup to the bearing by gravity.

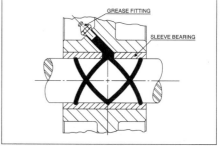

B. Oil or Grease Fitting

Lubricant is fed through the fitting under pressure and distributed through grooves by the rotation of the shaft.

FIGURE 19.1 Types of lubrication systems. (Courtesy of Boston Gear)

19.3 SELECTION CRITERIA AND USE OF THE PRESSURE-VELOCITY FACTOR

The suitability of a bearing is influenced by the rotational velocity, the type of material used for both the shaft and the bearing, clearances between the journal and the bearing, the temperature of operation, the type of lubrication, surface finish, and other factors. However, of greatest importance is the presence of the oil film between the bearing and shaft surfaces. This film reduces friction and, in some cases, dissipates heat, therefore reducing the potential for wear by minimizing or eliminating metal-to-metal contact. As discussed previously, the most important periods of operation are often during starting and stopping, when the loading conditions may cause the bearing surfaces to come into direct contact with each other. These starting and stopping conditions probably cause more wear than do long periods of continuous operation. The selection of an appropriate size and material for bearing applications can usually best be determined by calculating the product of the **contact pressure** and the velocity *(PV)* and comparing it to published values of suitability. Using this product of the pressure and the velocity, typically with limits on both of them individually, is the most straightforward way to select bearings of this type, and that is what will be presented in this chapter.

Typically the length of a bearing should be at least equal to the diameter, and up to two times the diameter is often preferable. However, many applications have space constraints that do not allow this to occur. The accuracy of manufacture of both the shaft and the bearing is critical and, as the bearing gets longer, factors such as straightness and surface finish can become even more important. In the past, very soft lead materials were often used for bearings, and they still are in some cases because of their ability to form to the mating surface. As discussed previously, in a hydrostatic design there is no contact between the journal and the shaft, and this soft lead material minimizes the accuracy required in manufacturing as it basically conforms to the shape of the shaft.

It is not only during starting and stopping that lubrication might be deficient. Oscillatory motion or cyclical conditions can also prevent a full film of oil from being maintained. When this happens, the metal-to-metal contact that likely occurs can cause significant wear. Because of this potential interruption of the oil film, significantly lower values of *PV* should be used.

In addition to factors such as speed, materials, clearances, and temperatures, the relationship between speed and oil viscosity should be considered. Generally, the following guidelines should be used to determine the

contact pressure
The ratio of force to the surface area. In the case of a round shaft, the surface area is typically calculated as the diameter times the length required to compensate for the angle of loading.

appropriate viscosity as it relates to speed. It should be noted that these are general guidelines and, for example, some recent engine designs use far lower-viscosity oils than would have been previously recommended. For high-speed applications, a fairly light oil such as an SAE 10 or 20 can minimize the internal friction and help assure proper oil distribution to the bearing/shaft surfaces. For moderate speeds, a heavier oil such as an SAE 20, 30, or 40 would typically be recommended unless very high temperatures are present. Under very low-speed conditions, a very heavy oil, possibly with pressure additives or grease, may be considered in order to prevent film failure and give sufficient lubrication for proper operation.

19.4 SHAFT CONSIDERATIONS

The surface of the shaft can be important in applications where frequent starting, stopping, or oscillatory motion exists. Generally, all bearing applications tend to result in some of these phenomena. For most applications, the shaft should be hardened and have a surface finish in the range of 4 to 12 rms. Roundness should be carefully controlled, as should the appropriate clearance. Medium-carbon steels generally offer the best results. Materials with a potential for **galling,** such as 300-series stainless or in some cases aluminum, may not give good service as galling will tend to create surface roughness. Care should be taken to minimize nicks, gouges, or anything that will tend to wipe the surface during rotation because this will also prevent proper oil distribution. The pressure times velocity ratio discussed previously can be found as follows:

$$PV = \frac{F}{Ld} \frac{\pi dn}{12} = \frac{\pi F n}{12L} \qquad (19.1)$$

where P = pressure (psi)

V = surface velocity (ft/min)

L = bearing length (inches)

d = shaft diameter (inches)

F = load on shaft (pounds)

n = speed (rpm)

Table 19.1 from the Boston Gear Company shows some recommended values for their bearing products. Note that, in general, the pressure, velocity, and pressure velocity factors all should be met.

> **galling**
> A phenomenon whereby a surface gets smeared or dragged, resulting in rough surface finishes and results in poor frictional properties.

TABLE 19.1
RECOMMENDED VALUES FOR BEARING PRODUCTS

Material	Max. PV	Max. P	Max. V
Bear-N-Bronz	75,000	3,000	750
Bost-Bronz	50,000	2,000	1,200
Bost-Bronz (Thrust washers)	10,000	2,000	1,200
F-1—Glass-filled Teflon	20,000	1,000	400
TN—Teflon-filled nylon	10,000	800	300
AF—Teflon-filled acetal	8,000	750	300
GS—Nylatron	4,000	500	300
D—Delrin or Celcon	3,000	480	300
N—Nylon	3,000	480	300
UHMW-PE	2,300	1,400	100
Nyloil	16,000	2,000	400
UHMW-PE with internal wear strip	4,000	1,400	100
Nyloil with internal wear strip	16,000	2,000	400

Courtesy of Boston Gear Company

Example Problem 19.1

Determine the length required for a nylon sleeve bearing under the following conditions: ¾-inch shaft diameter, $n = 500$ rpm, and a load on the bearing of 20 pounds.

If

$$(PV) = \frac{\pi F n}{12L} \quad (19.1)$$

Then

$$L = \frac{\pi F n}{12(PV)}$$

$$PV = 3,000 \quad \text{(from Table 19.1)}$$

$$L = \frac{\pi 20(500)}{12(3,000)}$$

$$L = .872 \text{ inch}$$

Use a ⅞-inch or longer bearing.
Check the maximum pressure and maximum velocity.

Maximum pressure:

$$P = \frac{F}{A} = \frac{20 \text{ lb}}{\frac{3}{4} \times \frac{7}{8}}$$

$$P = 30.5 \quad \text{(OK)}$$

Maximum velocity:

$$V = \frac{\pi D n}{12}$$

$$V = \frac{\pi (\frac{3}{4}) 500}{12}$$

$$V = 98 \quad \text{(OK)}$$

Therefore, a ⅞-inch-long nylon bearing will suffice.

19.5 SHAFT CLEARANCES

Shaft clearances can vary depending on size and lubrication but, for many materials, the graphs in Figure 19.2 can aid in determining what an appropriate clearance should be.

19.6 WEAR

Wear of sleeve-style bearings can vary significantly depending on the surface finish, loading, and other factors discussed in this chapter. The calculations and methods offered in this section should be considered general guidance only because variations in many of these factors can significantly affect the wear properties. Generally, wear is defined as the volumetric loss of the material over time, but temperature, deformation, and other factors can affect this as well. Usually, the process of estimating wear involves finding a wear-rate factor based on the type of material and relating it to the pressure-velocity factor mentioned previously. For metallic bearings with full-film or hydrostatic lubrication where the shaft rides on a film of oil, under ideal conditions, its life can typically be assumed to be infinite. Wear can be calculated as follows:

$$K = \frac{W}{FVT} \tag{19.2}$$

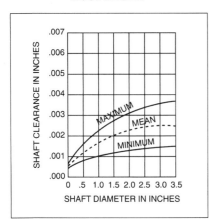
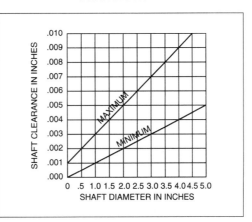
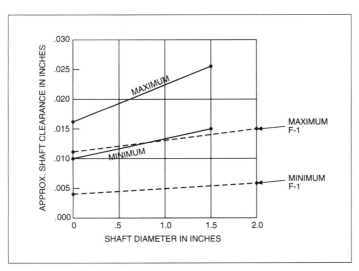

FIGURE 19.2 Recommended shaft clearances. (Courtesy of Boston Gear)

where W = volume of material lost (in.^{3})

F = load (pounds)

V = linear or surface speed (ft/min)

T = time (hours)

Units for the wear factor (K) would be

$$K = \frac{\text{in.}^3 \, \text{min}}{\text{lb ft hr}}$$

These would obviously be very small values as the rate of wear would be very small.

Equation 19.2 can be rewritten in terms of expected lifetime if

$$W = DLt$$

where t = thickness of wear

and

$$P = \frac{F}{LD}$$

then

$$T = \frac{t}{KPV} \tag{19.3}$$

where T = hours of running time

K = wear factor from Table 19.2

Table 19.2 shows wear factors for typical materials, and Example Problem 19.2 calculates the expected wear for the bearing in Example Problem 19.1.

TABLE 19.2
WEAR RATE FACTORS

Typical Materials	K
Delrin or Celcon (D)	50×10^{-10}
Nylatron GS (GS)	35×10^{-10}
Teflon-filled acetal (AF)	17×10^{-10}
Teflon-filled nylon (TN)	13×10^{-10}
Glass-filled Teflon (F-1)	12×10^{-10}
Nylon	12×10^{-10}

Example Problem 19.2

If, in the bearing from Example Problem 19.1, a wear of .010 inch is acceptable, what is the estimated life in hours of operation?

K is found from Table 19.2.

$$T = \frac{t}{KPV} \quad (19.3)$$

$$T = \frac{.010}{(12 \times 10^{-10})(30.5)(98)}$$

$$T = 2,800 \text{ hours}$$

19.7 BEARING MATERIALS

Many materials can be used for the construction of bearings. The following describes some of the more common materials, along with their advantages and disadvantages. For full-film or hydrostatic bearings, the materials from which the bearings are made are somewhat irrelevant, except for starting considerations. In actuality, however, because bearings are always started and stopped and because boundary or mixed-film conditions often exist, specific materials are often called for. The most common types of materials for bearings are described in the following paragraphs.

Steel

Steel is frequently used because of its machinability and its ability to obtain a good surface finish. In the case of automobile bearings, steel can be used in a relatively thin form. It offers good durability in situations where regular lubrication is present. It has reasonable heat transfer capabilities and good dimensional stability, and it can be manufactured with a very smooth finish.

Bronze

Bronze typically would be cast and could be alloyed with several different materials such as copper, tin, lead, zinc, or aluminum, either singly or in combination. When combined with lead, bronze has good natural lubricity and is often used in such applications as pumps, machinery, and home appliances. The tin and aluminum bronzes tend to have higher strength and hardness, which gives them the ability to carry higher loads, but they do not have as high a level of natural lubricity.

Babbitt

Babbitts are typically lead-based, and lead is usually a very high percentage of the overall composition. Babbitts can be alloyed with copper or antimony. Because of their softness, they have a good ability to form themselves to the shape of the shaft when minor irregularities occur. They have good resistance to seizure and good natural lubricity. They are, however, relatively low strength. Babbitts may be formed in place as in some cases they are poured from liquid lead materials around the in-place shaft.

Aluminum

Aluminum is frequently used because it can have a relatively hard finish and, in some cases, may be plated to enhance the surface. It can be made with a relatively smooth finish.

Porous Materials

Some of the more common sleeve-type bearings are made of porous materials, often of powdered metal construction, in which a deliberate attempt is made to create voids where lubrication can be forced into place to later "ooze" out to provide lubrication. Many sleeve-type bearings that are not intended to have regular lubrication are made in this manner because this once-applied lubrication can slowly migrate to the surface and provide long-term lubricating properties. Some of the porous materials can even have the oils resupplied by soaking them in hot oil solutions. Such bearings are also good for low-speed reciprocating or oscillating applications.

Plastics

Many different types of plastics are used in bearing applications. Some of the plastics have fairly good natural lubricities such as nylon, Teflon, Delrin, and other similar compositions. Some can be made with some porosity so that impregnated lubricants can be added. In general, however, for low-force applications, plastics offer good natural lubricity, and plastic bearings are typically made of phenolics, polycarbonates, acetals, nylons, Teflons, and delrin-type materials.

19.8 HYDROSTATIC BEARINGS

A hydrostatic bearing uses pressurized lubricant to lift the shaft away from the bearing before motion is initiated in order to provide a film of oil on which the shaft rides. In some cases, this is done only during starting to prevent the startup-related wear to which hydrodynamic bearings are prone. In other cases, the pressure-released oil flow is continued during operation. In general, the forces that are created can be estimated in the same way we determine surface pressures or contact stresses. However, a factor needs to be included for the reduced pressure as the oil approaches the ends of the bearings, due to leakage at the ends. Thus, it is necessary to de-rate the force available. This factor can be significant, depending on the clearance in the sealing of the ends of the hydrostatic bearing. The analysis of this is beyond the scope of this text, but it is introduced here so that you may understand that this type of bearing is commonly used and that the loads that can be created to lift the shaft from the bearing can be significant. Equipment such as turbine shafts, large coolant pumps in nuclear plants, and other large, critical applications may use this type of bearing system.

19.9 THRUST BEARINGS

Thrust bearings can be constructed to use any of the types of lubrications discussed earlier in the chapter, including boundary, mixed-film, full-film, or hydrostatic. Thrust bearing construction materials can be the same as for typical sleeve-bearing construction. The only difference is that they are usually made of two smooth surfaces to counteract thrust. Small thrust bearings carry light loads and can be plastic-type materials, whereas very large thrust bearings (for instance, in vertically mounted shafts and large pumps) may even be of the hydrostatic type. In light-load applications, sometimes sleeve bearings with shoulders can resist small amounts of thrust loading, but typically separate thrust bearings are specified.

19.10 SUMMARY

Bearings are used in many applications, and this chapter was intended to provide a summary of the techniques and methods involved in selecting bearings for rotating, sliding, or thrust-bearing applications. You should now be able to recognize when a sleeve bearing may be suitable and, in concert with the following chapter on rolling bearings, be able to determine which type of bearing is more appropriate for your design problem.

19.11 PROBLEMS

1. Find P, V, and PV values for a shaft loaded as follows: F = 100 lb, D = 1 inch, L = 1 inch, and n = 800 rpm. Select a bearing material.
2. For the shaft in Problem 1, estimate the life for a wear of .010 inch.
3. What initial clearance should be used in light of the allowed wear from Problem 2? Explain your answer.
4. What would the maximum allowable load be for a ¾-inch-diameter Bost-Bronz bearing, 1 inch long, rotating at 1,750 rpm?
5. In Figure 19.3, if the shaft rotates at 600 rpm, find the minimum length for the bearings if they are made from Delrin.
6. In Problem 4, estimate maximum life and recommend a shaft clearance.
7. A design requires a life of 1,000 hours, and plastic bearings are desired. If the shaft is ⅝ inch in diameter, L = 1 inch, F = 25 pounds, and n = 600 rpm, select a material.
8. Design a sleeve bearing for a rear-wheel-drive automobile differential that is splash lubricated by the differential fluid. The shaft diameter is 1 inch, the automobile weight is 3,600 pounds, the tire diameter is 24 inches, the design should use Bear-N-Bronze material, and the design condition is for 70 mph. Solve for length and specify the range of acceptable clearances.

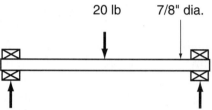

FIGURE 19.3 Shaft and bearings for Problems 4 and 5.

19.12 CUMULATIVE PROBLEMS

9. Determine the necessary length for both bearings in Figure 17.6 for the shaft size found in Chapter 17, Problem 16, assuming a Bost-Bronz material is used.
10. Determine the necessary length if Bear-N-Bronz is substituted in Problem 9.
11. Determine the expected wear life for both bearings and both materials found in Problems 9 and 10, based on the maximum range of clearances from Figure 19.2.
12. Which material would you recommend for maximum life from Problems 9 through 11?

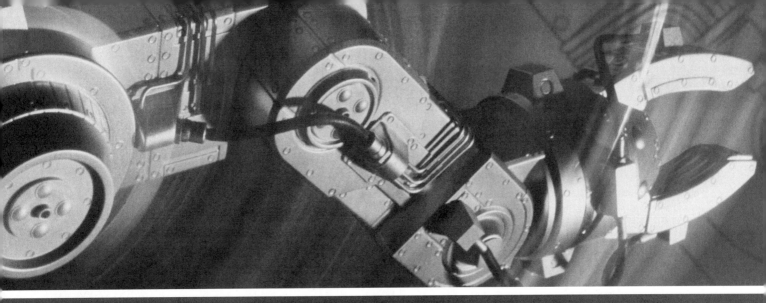

CHAPTER 20

Ball and Roller Bearings

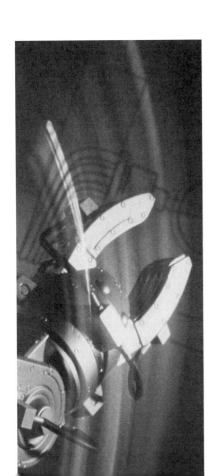

rolling contact bearings
A bearing that, instead of having sliding friction, contains balls or rollers that roll like wheels.

Conrad deep-groove bearing
A type of bearing that has a radial groove cut into both the inner and outer races to guide the ball assemblies and also to allow for minor thrust loading.

denting
The phenomenon in which, due to high static or low-speed loads, the balls in a ball bearing form a recess in the race over time.

The last chapter discussed sleeve or journal bearings in which either friction between the surfaces was present or motion occurred on a film of oil. This chapter will discuss ball and roller bearings, often referred to as **rolling contact bearings,** which, instead of having sliding motion, roll in the same manner as a wheel or a series of miniature wheels. We can illustrate the difference between these bearing types by comparing a sled to a wagon. When a sled is pulled on a flat surface, a greater force is needed to start it in motion than to keep it in motion. In many ways, a rolling contact bearing solves that problem. Rolling contact bearings have significantly lower values of friction, especially during starting, when the friction is approximately the same as it is during motion. A rolling contact bearing has an equivalent coefficient of friction (f) ranging from .001 to .005, which is significantly lower than the values that were found in Chapter 19 for plain surface bearings. The most common type of ball bearing is the **Conrad deep-groove bearing** which, in addition to having some capacity for carrying a radial load, can also resist limited amounts of thrust loading.

Further complicating the selection of bearings is the fact that different manufacturers use different systems for specifying bearings grades. For instance, some manufacturers refer to their bearings as "precision" or "commercial," while others use the terms "ground" or "unground." Still others use a set of ABEC criteria that typically ranges from 1 to 7. However, it is important to recognize that the most important criterion is the fact that each bearing manufacturer supplies the load and life ratings for its own bearings. As we will discuss later in the section on expected life, manufacturers specify rated lifetimes using different systems for the life of the bearings. Some manufacturers specify bearings by hours of expected life at different speeds, some by when 10% of the bearings will have failed, and some by when 50% of the bearings will have failed. When reviewing manufacturers' literature, you must be very careful to understand the basis that they are using.

Bearings are generally manufactured in both inch series and metric series. However, even when designing in inches, metric bearings are often used because they lend themselves to fitting on standard-sized shafts where the shaft may be turned down slightly to provide a mounting shoulder for a bearing. To a greater extent than with most engineering products, English-unit and metric-unit bearings are commonly intermixed in designs.

In this chapter, we will learn how to determine the suitability of ball bearings in different applications and the effects of a static load on a bearing of this type. We will also examine phenomena such as **denting** that can damage bearing races, especially during very slow motion or when the bearing is stopped. We will learn how these types of bearings are constructed, including the fact that many of them are made with an **interference fit** between the inner and outer races so that there is a preload that reduces radial motion. This interference is similar to the way a close or tight fit would act in a sleeve- or journal-style bearing. Because of this interference fit and the fact that there are significant stresses in rolling contact bearings, this type of bearing system has a finite life. This is typically factored into the design so that we can determine the acceptability of a certain size bearing, the effect of the load as compared to the maximum allowable on the life of the bearing, and how a variation in load can affect the overall lifetime of a bearing of this type. Lastly, this chapter will review some other types of rolling contact bearings such as roller bearings, tapered roller bearings, ball or roller thrust bearings, and needle bearings.

> **interference fit**
>
> As it applies to bearings, this occurs when the total of the dimensions for the inner race, the balls, and the outer race are greater than the space available. This results in a residual load in these parts, but it makes for a tighter bearing assembly.

20.1 OBJECTIVES

This chapter is intended to teach the principles of rolling contact bearings, including how to select the appropriate type and how to do the necessary sizing analysis. After completing this chapter, you should be able to:

- Recognize the different types of rolling contact bearings and the applications where they may be suitable.
- Evaluate different application considerations, including temperature conditions, bearing installation tolerances, classes of bearings, and other factors involved in using rolling contact bearings.
- Calculate the equivalent load on a Conrad-type ball bearing and the relationship of this load to the expected lifetime of the bearing.
- Factor in the effect of thrust loading on the equivalent radial load and the life of a bearing.
- Understand the principles of rolling contact bearings versus sleeve or journal bearings.
- Understand the principles of roller bearings, tapered roller bearings, thrust bearings, and needle bearings.
- Select the appropriate type of bearings and their sizes for the design being undertaken.
- Understand how to use the manufacturer's data for their product's capacity and lifetime.
- Understand what types of lubrication will be appropriate for different bearing systems.

20.2 LIFE EXPECTANCY OF BALL BEARINGS

The life expectancy of ball bearings is not uniform and is based on many factors, including manufacturing defects, surface finish, and material problems. Manufacturers typically rate their bearings based on the point at which 10% of the bearings would have failed. This is called the B_{10} or L_{10} life of the bearing. At this point, it would be expected that 90% of the bearings would still be providing suitable service. Figure 20.1 shows a typical graph of bearing failures. Note that some of the bearings fail very early in their lifetime, followed by a period of relatively few failures, followed by a period during which a greater number of bearings are then starting to fail. The 50% failure point is also noted; in some cases manufacturers specify this point as well or provide estimates of the relationship between these points. Many manufacturers specify load ratings based on a million revolutions of life with either a B_{10} or L_{10} life specified. Other manufacturers specify hours of life at different speeds and associated load ratings. The considerable range in failure points is due to many factors. Contamination can be a significant one because relatively minute particles of dirt or other impurities can significantly affect the life of a bearing. Minor differences in surface finish of either race can have a significant effect. Because most bearings are rather highly stressed, any defect in the material can also affect the life. The expected life of a bearing in terms of revolutions can be determined as follows:

$$L = \frac{Tn\ 60\ \text{min}}{10^6\ \text{hr}} \qquad (20.1)$$

where T = hours

n = revolutions per minute

L = millions of revolutions

We can adjust the life of a bearing, based on the ratio of the desired life to the design life, as follows:

$$\frac{L_d}{L_c} = \left(\frac{C_d}{P_d}\right)^k \qquad (20.2)$$

where k = 3 for ball bearings

k = 3.3 for roller bearings

L_d = desired life

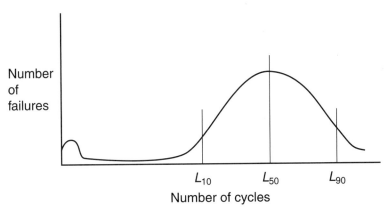

FIGURE 20.1 Typical bearing failure curve.

L_c = table life
C_d = dynamic rating (from manufacturer)
P_d = design load

Formula (20.2) is often easier to use in the following form:

$$C_d = P_d\left(\frac{L_d}{L_c}\right)^{1/k} \quad (20.3)$$

Bearing load correction (20.3)

Note L_c from Table 20.1 would be 10^6 revolutions.
If the expected life is desired, it can be found from the following form:

$$L_d = \left(\frac{C_d}{P_d}\right)^k L_c \quad (20.4)$$

Bearing life correction (20.4)

If the table value was for a million revolutions, this formula would become

$$L_d = \left(\frac{C_d}{P_d}\right)^k 10^6 \quad (20.5)$$

Table 20.1 shows the rated radial capacity for a million revolutions for different sizes of bearings, as well as a static load for the same bearings. The radial load would be the load at which the bearing could be expected to last

TABLE 20.1
BEARING SELECTION DATA FOR SINGLE-ROW, DEEP-GROOVE, CONRAD-TYPE BALL BEARINGS

Bearing Number	Nominal Bearing Dimensions*						Bearing Weight lb	Basic Static Load Rating C_s lb	Basic Dynamic Load Rating C_d lb
	d		D		B				
	mm	in.	mm	in.	mm	in.			
6200	10	.393 7	30	1.181	9	.354 3	.07	520	885
6201	12	.472 4	32	1.259 8	10	.393 7	.08	675	1,180
6202	15	.590 6	35	1.378 0	11	.433 1	.10	790	1,320
6203	17	.669 3	40	1.574 8	12	.472 4	.14	1,010	1,660
6204	20	.787 4	47	1.850 4	14	.551 2	.23	1,400	2,210
6205	25	.984 3	52	2.047 2	15	.590 6	.29	1,610	2,430
6206	30	1.181 1	62	2.440 9	16	.629 9	.44	2,320	3,350
6207	35	1.378 0	72	2.834 6	17	.669 3	.64	3,150	4,450
6208	40	1.574 8	80	3.149 6	18	.708 7	.82	3,650	5,050
6209	45	1.771 7	85	3.346 5	19	.748 0	.89	4,150	5,650
6210	50	1.968 5	90	3.543 3	20	.787 4	1.02	4,650	6,050
6211	55	2.165 4	100	3.937 0	21	.826 8	1.36	5,850	7,500
6212	60	2.362 2	110	4.330 7	22	.866 1	1.73	7,250	9,050
6213	65	2.559 1	120	4.724 4	23	.905 5	2.18	8,000	9,900
6214	70	2.755 9	125	4.921 3	24	.944 9	2.31	8,800	10,800
6215	75	2.952 8	130	5.118 1	25	.984 3	2.64	9,700	11,400
6216	80	3.149 6	140	5.511 8	26	1.023 6	3.09	10,500	12,600
6217	85	3.346 5	150	5.905 5	28	1.102 4	3.97	12,300	14,600
6218	90	3.543 3	160	6.299 2	30	1.181 1	4.74	14,200	16,600
6219	95	3.740 2	170	6.692 9	32	1.259 8	5.73	16,300	18,800
6220	100	3.937 0	180	7.086 6	34	1.338 6	6.94	18,600	21,100
6221	105	4.133 9	190	7.480 3	36	1.417 3	8.15	20,900	23,000
6222	110	4.330 7	200	7.874 0	38	1.496 1	9.59	23,400	24,900
6224	120	4.724 4	215	8.464 6	40	1.574 8	11.4	26,200	26,900

*d = bearing bore diameter; D = bearing outside diameter; B = bearing width

raveling
A similar process to galling; the difference is that in raveling the material may be pushed or displaced and in general can damage the bearing race by roughening the surfaces.

for a million revolutions when there are no other factors acting on it as determined from Formula (20.1). This value can be modified, if necessary, for a different lifetime using Formulas (20.2) or (20.3). The static rating would be of concern if the load is present when the bearing is not rotating or is rotating at extremely low speeds, due to the possible denting of the races. This effect is often referred to as **raveling**. Figure 20.2 shows a deep-groove, single-row ball bearing.

FIGURE 20.2 Cutaway view of a ball bearing. (Courtesy of NSK Corp.)

Example Problem 20.1

Table 12.1 lists the basic dynamic load for bearing number 6208 as 5,050 pounds. What would the expected life be for this bearing if it were subjected to a radial load of 2,400 pounds? If the shaft turns at 1,750 rpm, how many hours would this bearing last based on the L_{10} design life?

$$L_d = \left(\frac{C_d}{P_d}\right)^k 10^6 \qquad (20.5)$$

$$L_d = \frac{(5,050)^3}{2,400} 10^6$$

$$L_d = 9.3 \times 10^6 \text{ rev}$$

$$T = \frac{9.3 \times 10^6 \text{ rev}}{1,750 \text{ rev/min}} \frac{\text{hr}}{60 \text{ min}}$$

$$T = 88 \text{ hr}$$

Example Problem 20.2

For the conditions in Example Problem 20.1, if an L_{10} life of 200 hours is needed, select a bearing from Table 20.1 to meet this criterion.

$$L = \frac{Tn}{10^6} \frac{60 \text{ min}}{\text{hr}} \quad (20.1)$$

$$L = \frac{200 \text{ hr}}{10^6} \frac{1{,}750 \text{ rev}}{\text{min}} \frac{60 \text{ min}}{\text{hr}}$$

$$L = 21 \times 10^6 \text{ revolutions}$$

$$C_d = P_d \left(\frac{L_d}{L_c}\right)^k \quad (20.3)$$

$$C_d = 2{,}400 \text{ lb} \left(\frac{21}{1}\right)^{1/3}$$

$$C_d = 6{,}622 \text{ lb}$$

Bearing number 6211 has a dynamic load rating of 7,500 pounds, so this bearing would be acceptable. If the bearing bore is too large, a heavier series could be substituted.

In many cases, a bearing is subject to both a radial and a thrust load. As the manufacturers' capacity tables are based on radial loads only, we must combine the effect of the thrust load with the radial load to obtain an **equivalent combined radial load**. Note that this total radial load would never be lower than the radial load plus the factor, if necessary, for the outer race rotating, as follows:

$$P = VXR + YF_t \quad (20.6)$$

where P = combined equivalent radial load
V = 1.0 inner race rotating
    1.2 outer race rotating
X = radial factor (usually .56)
R = radial load
Y = thrust factor
F_t = thrust load

---

equivalent combined radial load
The method whereby the thrust loading is factored in to result in an equivalent radial load.

Bearing thrust equivalent (20.6)

$P = VXR + YF_t$

For a deep-groove, Conrad bearing with an approximate 5° contact angle, the radial factor used is typically .56. This factor is generally used for almost all typical Conrad deep-groove bearings. The rotation factor would be 1.0 if the inner race rotates (i.e., the shaft is the member turning) or 1.2 if used, for instance, to support a roller where the shaft is stationary and the outside race rotates. For a typical Conrad deep-groove bearing, if the shaft rotates, Formula (20.6) would become

$$P = (1.0)\ .56R + YF_t$$

In order to find the thrust factor, we would take the ratio of the thrust to the static load capacity of the bearing selected and interpolate, if necessary, from Table 20.2 to find this value.

$$\frac{F_t}{C_s} \quad (20.7)$$

where F_t = thrust on bearing
C_s = static rating (from manufacturer)

The difficulty in this analysis is that the static load factor is not known until you have selected a bearing. This forces some iteration; you must first select a bearing and verify its suitability and, if it is not suitable, you select the next larger size or series and repeat the analysis. Note that, in practice, the next larger series would be selected if the shaft is to remain the same size. (See Figure 20.3.)

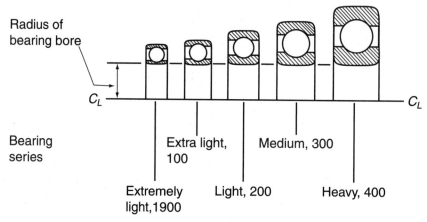

FIGURE 20.3 Standard bearing sizes.

Example Problem 20.3

Select a bearing from Table 20.1 to meet the following criteria:

$$R = 1{,}200 \text{ pounds}$$
$$F_t = 500 \text{ pounds}$$
$$n = 1{,}500 \text{ rpm—shaft rotates}$$
$$L_{10} = 5{,}000 \text{ hours}$$

First, in order to select a bearing, assume a thrust factor of $Y = 1.6$ (average of table).

$$P = .56R + YF_t \qquad (20.6)$$
$$P = .56 \cdot 1{,}200 \text{ lb} + 1.6 \bullet 500 \text{ lb}$$
$$P = 1{,}472 \text{ lb}$$

TABLE 20.2

THRUST FACTOR FOR DEEP-GROOVE BALL BEARINGS									
F_t/C_s	.014	.028	.056	.084	.11	.17	.28	.42	.56
Y	2.3	1.99	1.71	1.55	1.45	1.31	1.15	1.04	1.00

F_t = thrust load; C_s = basic static load rating

Find the life.

$$L = \frac{1,500 \text{ rpm } 5,000 \text{ hr } 60 \text{ min}}{10^6 \qquad \qquad \text{hr}}$$

$$L = 450 \quad (10^6 \text{ revolutions})$$

Adjust the rating for life.

$$C_d = P_d \left(\frac{L_d}{L_c}\right)^{1/k} \qquad (20.3)$$

$$C_d = 1,472 \left(\frac{450}{1}\right)^{1/3}$$

$$C_d = 11,280 \text{ lb}$$

From Table 20.1 (bearing number 6215),

$$C_d = 11,400 \text{ lb}$$
$$C_s = 9,700 \text{ lb}$$

Verify the assumption for Y.

$$\frac{F_t}{C_s} = \frac{500}{9,700} = .0515 \qquad (20.7)$$

Interpolate to find factor.

$$Y = \left(\frac{.056 - .0515}{.056 - .028}\right)(1.99 - 1.71) + 1.71$$

$$Y = 1.76$$

Verify:

$$P = .56R + YF_t$$
$$P = .56(1,200) + 1.76(500)$$
$$P = 1,550 \text{ lb}$$

$$C_d = P_d \left(\frac{L_d}{L_c}\right)^{1/k}$$

$$C_d = 1,550 \left(\frac{450}{1}\right)^{1/3}$$

$$C_d = 11,875 \text{ lb}$$

This is not acceptable. Try the next larger size bearing. Try bearing number 6216:

$$\frac{F_t}{C_s} = \frac{500}{10,500} = .048$$

Interpolate:

$$Y = \left(\frac{.056 - .048}{.056 - .028}\right)(1.99 - 1.71) + 1.71$$

$$Y = 1.79$$

$$P = .56(1,200 \text{ lb}) + 1.79(500 \text{ lb})$$

$$P = 1,567 \text{ lb}$$

$$C_d = 1,567\left(\frac{450}{1}\right)^{1/3}$$

$$C_d = 12,005 \text{ lb}$$

Bearing number 6216 meets the design criteria.

Example Problem 20.4 demonstrates the same process where the shaft is stationary and the outside rotates for an application using metric-size shafting.

Example Problem 20.4

Estimate the life for a 6200 series bearing that has an inside diameter of 30 mm on a stationary shaft where the outside race rotates. There is a radial load of 1,000 pounds and a thrust load of 300 pounds.

Find the thrust factor as the bearing is known.

$$\frac{F_t}{C_s} = \frac{300}{2,320} = .129 \qquad (20.7)$$

Interpolate from Table 20.2.

$$Y = \left(\frac{.17 - .129}{.17 - .11}\right)(1.45 - 1.31) + 1.31$$

$$Y = 1.41$$

$$P = VXR + YF_t \tag{20.6}$$

$$P = 1.2(.56)1,000 + 1.41(300)$$

$$P = 1,095 \text{ lb}$$

$$L_d = \left(\frac{C_d}{P_d}\right)^k 10^6 \tag{20.5}$$

$$L_d = \left(\frac{3,350}{1,095}\right)^3 10^6$$

$$L_d = 28.6 \times 10^6 \text{ revolutions}$$

The design process for many roller bearings can be similar, but with slightly different factors applied. However, as noted previously, different manufacturers often slightly modify this process, especially with regard to their tabular values for expected lifetime. Almost all manufacturers in their catalogs have an engineering section where the performance and life expectancy are demonstrated. However, the basic theory is similar.

20.3 TYPES OF ROLLING CONTACT BEARINGS

There are many types of rolling contact bearings and many very specialized applications where bearings are created for a specific application. This section will discuss some of the more common types of rolling contact bearings.

Roller Bearings

A **roller bearing** is somewhat similar in principle to a ball bearing, except that it is created with rollers so that the contact area is significantly enlarged, and hence the load capacity can be far greater. One modification of a roller bearing is a spherical roller, which has some additional specialized capability. Note that a straight roller bearing typically has no thrust capacity. Figure 20.4 depicts a spherical roller bearing.

> **roller bearing**
> Utilizes the same principles and is similar to a ball bearing, except by using a roller the contact area is increased.

Tapered Roller Bearing

A **tapered roller bearing** has the basic properties of a roller bearing with the exception that, because of the taper, some thrust capacity can be easily accommodated. A good example of a tapered roller bearing appears on the front wheel spindles of a typical automobile, where tapered roller bearings

> **tapered roller bearing**
> A roller bearing made in a conical shape, so that it can support thrust loads as well as radial loads.

FIGURE 20.4 Spherical roller bearing. (Courtesy of Emerson Power Transmission Corporation. All Rights Reserved.)

allow for the radial load from the rotating wheel as well as the thrust load when the car goes around a corner. Figure 20.5 depicts a tapered roller bearing.

Needle Bearing

A **needle bearing** is basically a roller bearing, except that the manufacturer does not usually supply the inner race; the shaft itself serves as the inner race of the bearing. A needle bearing can be smaller because of this feature. However, greater care needs to be taken with the surface finish and hardness of the shaft in order to accommodate the needle bearing's rollers. The rollers in a needle bearing can be very tiny when space is critical. For example, a universal joint in a rear-wheel-drive automobile typically uses needle bearings to create a relatively small rotating assembly. Figure 20.6 shows an example of a needle roller bearing.

> **needle bearings**
> Roller bearings that lack an inner race; the shaft is used as the inner race in order to save space.

Ball and Roller Bearings 439

FIGURE 20.5 Tapered roller bearing. (Courtesy of NSK Corp.)

FIGURE 20.6 Needle bearing. (Courtesy of Emerson Power Transmission Corporation. All Rights Reserved.)

Thrust Bearings

thrust bearings
Bearings that are intended primarily to resist thrust loads unlike most ball or roller bearings, whose primary purpose is resisting radial loading.

Thrust bearings can be ball thrust, roller thrust, or other variations such as tapered rollers. Basically, the same principles are involved as with other roller bearings with the exception that these bearings are intended primarily for resisting thrust or for situations where relatively little or no radial force is being applied. Thrust bearings could be used, for example, in a clutch assembly where there is no significant radial load but there are large thrust loads. Figure 20.7 shows a thrust-bearing assembly.

FIGURE 20.7 Thrust bearing. (Courtesy of NSK Corp.)

20.4 MOUNTED BEARINGS

Many styles of bearings come premounted in different assemblies or frames: for example, **pillow blocks, rod end bearings,** and **cam followers** are common types of mounted bearings. Some of the more common types are described in this section.

pillow blocks
A term designating a bearing mounted in a housing.

rod end bearing
Similar to a pillow block bearing but mounted in an assembly that would typically be attached to the end of a bar or shaft.

cam follower
A term for a bearing mounted on a bolt or shoulder screw that was originally intended for activating a switch or valve against a cam but may be used for other purposes as well.

Pillow Blocks

A pillow block is a common term for a flange-mounted bearing that can be mounted on a flat surface. The bearings that would be used in a pillow block are the same as the bearings that might be designed into any shaft housing assembly. They simply come premanufactured with a frame assembly so that they can be bolted into place. Many options are available

FIGURE 20.8 Common pillow block types. (Courtesy of Boston Gear)

for pillow block housings. Construction materials could be sheet metal, aluminum, cast iron, or steel. Built-in lubrication points such as grease fittings could also be included. Figure 20.8 shows a few different types of pillow block bearing assemblies.

Rod End

A rod end bearing is used, for example, in a connecting rod in an automobile steering assembly or in other similar applications. It is somewhat similar to a pillow block in the sense that the bearing is premounted for use where either a threaded connection may be available for mounting to the end of a tube or rod or where a pin assembly would be connected. Figure 20.9 depicts several typical rod end bearing assemblies.

Cam Followers

Cam followers are another form of mounted bearing. Most types of bearings could be mounted within the cam follower, whether a roller, needle, or ball bearing. A built-in wheel can be included, or the outer race of the bearing itself can be used as an activator. That is why the term "cam follower"

FIGURE 20.9 Rod end bearings. (Courtesy of Boston Gear)

applies in this case. Figure 20.10 depicts a typical cam follower bearing assembly.

Linear

linear bearing
A bearing assembly that allows motion axially along a shaft as opposed to radially; in some cases may be combined with a radial bearing.

Unlike a ball bearing, which rotates around a shaft, **linear bearings** are intended to move longitudinally along the shaft. The balls follow a path where they follow the rod, leave through a built-in chase in which they are transported back into position, and then once again follow along the length of the bearing assembly. Their applications could include working as support guides for many types of assemblies, such as those in elevators. They may be open- or closed-style, depending on how the rods are mounted. Figure 20.11 shows open and closed linear bearing assemblies.

Ball and Roller Bearings | **443**

FIGURE 20.10 Cam follower. (Courtesy of Emerson Power Transmission Corporation. All Rights Reserved.)

FIGURE 20.11 Open- and closed-style linear bearing. (Courtesy of Boston Gear)

20.5 LUBRICATION AND BEARING SEALING METHODS

An important aspect of the use of rolling contact bearings is how they are lubricated and sealed, both for keeping the lubrication in the bearings and for keeping contamination out.

Lubrication

Bearings can be lubricated with grease, oil, or other products, depending on the application. In some cases, bearings, since they are simply rolling, can be used dry. The author at one time designed underwater robotics equipment in which stainless steel bearings were used. The oil was washed off before use and the bearings were allowed to simply run in the water without additional lubrication. However, lubrication has multiple purposes. Certainly one is to aid in the rolling action, but lubrication may also provide some heat dissipation properties for the bearing. Generally, low-speed applications are more apt to be lubricated with grease, whereas higher-speed applications tend to use oil. There are many lubrication options and various methods for doing it.

Bearing Sealing

There are two main ways in which bearings are sealed: these are referred to as **shields** or **seals.** Many bearings come with a combination of both or a multiple series of either. Figure 20.12 depicts the types of seals or shields that may be available for bearing assemblies. These seals and/or shields have two purposes. The first is to keep the lubrication in the bearing, and the second is to keep contaminations out. Neither seals or shields are considered waterproof or full-sealing mechanisms because small contaminations will invariably migrate in. This is especially true during high-temperature conditions when the oil loses some of its lubricity. Oil may also tend to migrate out of the bearings.

---

shield
A type of sealing mechanism for a bearing that uses a small cover plate attached to one or the other of the races to prevent the entry of contaminants. In some cases it is free floating.

seals
Similar to a shield but typically made of elastomeric material to try to prevent, like a shield, the entrance of contaminants. May aid in keeping lubrication in the bearing assembly.

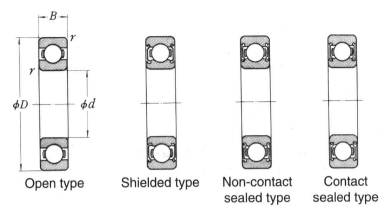

FIGURE 20.12 Bearing seals and shields. (Courtesy of NSK Corp.)

20.6 SUMMARY

Roller bearings are used in many applications along with sleeve bearings, which were discussed in Chapter 19. This chapter was intended to summarize the techniques and methods involved in selecting rolling surface bearings for rotating, sliding, and thrust applications. You should now be able to recognize where a rolling surface bearing may be suitable as opposed to a sleeve or journal bearing, and to determine which type of bearing would be appropriate. You should also be able to use manufacturers' literature to select the correct size of bearing for the application.

20.7 PROBLEMS

1. A radial bearing has a basic dynamic rating of 1,560 pounds for an L_{10} life of 10^6 revolutions. What is the expected L_{10} life for a radial load of 750 pounds?
2. For bearing number 6204 in Table 20.1, determine the expected life in hours if the bearing is subjected to a radial load of 1,200 pounds at 800 rpm.
3. For bearing number 6212 in Table 20.1, if a desired life of 1,500 hours at 1,200 rpm is needed, what radial load could be applied for that period?
4. Bearing number 6215 is subjected to a radial load of 800 pounds and a 270-pound thrust load. Determine the equivalent combined radial load if the shaft rotates.

5. For the conditions in Problem 4, estimate the expected L_{10} life in revolutions and in years of use if the bearing operates at 600 rpm for 12 hours per day.
6. For the conditions in Problems 4 and 5, if a 15-year L_{10} life is desired, select a bearing for this situation.
7. Select a bearing from Table 20.1 for the following conditions: a radial load of 350 pounds, a thrust load of 50 pounds, an L_{10} life of 500 hours, and an inner race rotating at 1,800 rpm.

20.8 CUMULATIVE PROBLEMS

8. An alternate design for the automobile in Problem 8 in Chapter 19 is to use bearing number 6206 from Table 20.1. Estimate the L_{10} life in miles for this bearing.
9. In Problem 8, if a life of 120,000 miles is desired, select a bearing.
10. In Chapter 19, Problem 9, substitute ball bearings from Figure 20.1 with a bore size equal to or greater than the shaft size found previously. Assume only one bearing carries the thrust load from the bevel gear and a life of 100,000 miles is desired.
11. Would you recommend the ball bearing design from Problem 10 or the sleeve bearing design from Chapter 19, Problem 9?

CHAPTER 21

The Design Process and Design Projects

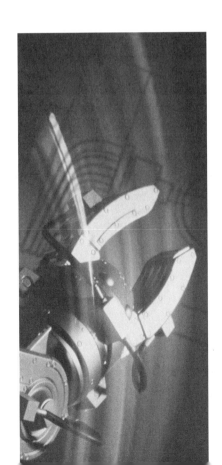

intellectual property
The term that is applied to an idea or a creation, based on the fact that by being its creator, the individual should have rights to that idea or creation.

patent
A form of protecting legal rights that is created by a government to insure that someone creating an invention can have the sole right to its use for some period of time.

copyright
The method by which ownership is established for a creative work of art, music, or text, similar to what a patent is for an invention.

technology-pushed
A term that is used when advances in engineering and technology encourage a company to upgrade its products.

market-pulled
When a perceived need in the marketplace encourages the creation of new products.

Throughout this text, you have been developing the skills you will need to select different machine elements to be used in the design of machines. This chapter will focus on the design process: how to undertake the design of different types of machines or other products, the process that is involved, and the different steps used by many firms in their development processes. One section of this chapter will discuss how to stimulate the creative process. We will examine some tools that engineers in the field use to try to stimulate the mind to find new and different ways of solving problems.

This chapter will also discuss the protection of the creative process and the concept of **intellectual property:** how **patents** and **copyrights** are awarded and the process for undertaking the protection of intellectual rights through patents. Lastly, this chapter will offer suggestions for some design projects that your instructor may assign so that you can use at least some of the knowledge you have gained in this course.

The need to develop new products is often considered to be either **technology-pushed** or **market-pulled.** For example, companies may need to develop or upgrade products to recapture market share, to maintain market share, or to enter into a new market. Technology-pushed development often occurs because the product that is being offered by a company either offers old technology or is losing market share. Market-pulled development occurs when a company identifies a need and attempts to fill that need. The life cycle of a product such as an automobile is a good example of this. New models of cars are often introduced with a great deal of fanfare, and sales increase during the first few years of the market cycle. Some products have long-term sustainability, whereas others, like automobiles, after some initial excitement, tend to decrease fairly rapidly. Soon a new model of automobile needs to be created to replace the older model, whose dwindling sales are due either to its image as old technology or to new features in competing products.

Albert Einstein once stated: "The formulation of a problem is often more essential than its solution which may be merely a matter of mathematical or experimental skill. To raise new questions, new possibilities, to regard old questions from a new angle requires creative imagination and marks real advances." When we think of the need to develop a new product, often, as stated by Einstein, the identification of the product or the need for the product is as difficult as the development of the product. In the section of this chapter on stimulating the creative process, we will try to define how this need comes to be developed and ways in which ideas can be formulated and brought to fruition. We can think of many unique products that have been developed over the years (for example, the electric light bulb). It is often stated that Thomas Edison accomplished two things with

the development of the light bulb. First and foremost, perhaps, he invented a light bulb, but he also discovered 1,800 ways how not to build one. When we think of the process of creativity, we need to recognize the role of innovation and implementation and of basic science.

It is interesting to consider the original concept of intellectual property rights in Article I of the U.S. Constitution, which empowered Congress to "promote the progress of science and useful arts by securing for limited times to authors and inventors the exclusive right to their respective writings and discoveries." This idea of owning intellectual property, either by persons or by corporations, is one of the driving forces that encourages innovation and development because the process often only returns the investment over long periods of time. You should be aware of how this process works and how to protect your intellectual property when and if it is created.

21.1 OBJECTIVES

This last chapter is intended to summarize the design process, to review how designs are typically formulated and some methods for stimulating the creative process, and to develop a basic understanding of patents and copyrights and how we protect the creative process. In addition, this chapter offers suggested design projects, if this course includes a design project, or you may be allowed to create your own design projects. After completing this chapter, you should be able to:

- Understand the basic process of design and some of the typical steps, including the conceptual design, preliminary design, prototyping, and redesign, and understand that many of these steps are often repeated.
- Understand how the creative process works and how we can stimulate that process.
- Understand how patents and copyrights are issued and what rights they confer on the inventor.

21.2 REVIEW OF THE MACHINE DESIGN PROCESS

The design process can best be understood as being twofold: identifying the problem and identifying the need for the design. The problem could be as simple as needing to create a fixture for a specialized manufacturing task, needing a new product for a specialized application, or needing a product in a specific area identified by the marketing department. For example, it could be a new model of automobile for a market niche where the company feels it does not have an appropriate product. Identifying the problem is often as difficult as solving it, but, typically, the solution to a design problem starts with a conceptual idea.

Once the idea is formulated, sketches are usually created. These may be created by an individual or by a group. At some point, these sketches are turned into design drawings from which a prototype is often created. This prototype could be for a part of a design if it is a complicated piece of machinery, a model (either to scale or a full-size prototype) suitable for testing. Throughout this process, there are often many iterations. For example, the structural analysis may reveal that certain aspects of the design need to be a different size or shape. The task of creating design drawings, along with the analytical part of the design, may have many stops and starts and changes along the way. The designer needs to recognize that no single concept in and of itself would ultimately be a final design.

Typically, once the design and the prototype are done, some form of testing is undertaken. This could be a complete test of an entire assembly or a test of individual pieces. It may include consumer testing or durability testing, depending on the type of product or the industry. During this prototype and testing stage, it is usually recognized that it is a feasible concept that should be pursued further, one that should be scrapped, or, more likely than not, one for which significant revisions are needed. At this stage, most designs undergo significant redesign. Some parts may be scrapped, different concepts substituted, and significant revisions made.

Typically in the design process, once new design drawings have been made or updated and many of the problems in the first-phase design and prototyping are recognized, a second prototype may be created. Often this is created in more detail than the earlier prototypes, and this step in the design may be much closer to the final product. Many different professions may be involved in the design process. For instance, electrical or computer engineers may create control systems, and manufacturing engineers may consider the manufacturability of the design. This newly created prototype likely will undergo testing similar to that used for the initial prototype; however, unlike the initial prototype testing, which is often limited to feasibility testing, this testing may now include complete functionality testing and perhaps testing of durability of the individual pieces. If it is a consumer product, it may also be subjected to some consumer testing to determine whether consumers like the product, find it convenient, and would be willing to buy it.

Depending on the industry, a third level of redesign and prototyping may be included. If it is, the process may be somewhat similar to the initial process, and it may involve creating additional prototypes or possibly

including longer-term consumer testing. In some cases, a prototype could be upgraded and become the final product.

It is important to note that various industries accomplish these types of activities quite differently. A mass-production industry may modify existing products rather than starting from scratch. For example, a new car model may use a frame and drive system from an existing model. Consumer products often require a greater level of testing for durability, safety, and other aspects than industrial products. In other fields, such as custom machinery design, many of these steps may be omitted, and prototyping may not be done at all. Some products may need to go through certification testing by government agencies or trade associations. However, the general process of creating a **conceptual design** and turning it into a product is somewhat similar in all industries, varying more than anything else by the number of iterations, types of iterations, and whether models or full-scale products can be prototyped. If we think of the differences between a product such as a small mechanism for which the complete product could be made and tested versus something like a highway bridge, we can begin to understand how the design process varies, whether or not prototypes or scale models are made. In some cases **computer modeling** may be appropriate. We can envision some of these steps by reviewing Figure 21.1.

conceptual design
The creative process in which a concept evolves to solve a problem or a need.

computer modeling
Creating a visual depiction of a proposed idea, often in three dimensions, in order to have a better feel for what the product would look like. Computer models often cost less than creating a physical prototype or a model.

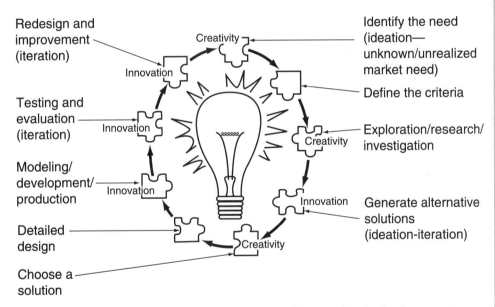

FIGURE 21.1 The design process. (Courtesy of Design Tools for Engineering Teams, Delmar Learning)

21.3 STIMULATING THE CREATIVE PROCESS

We often assume that creativity is an innate ability that some people are born with and others are not. In fact, the creative design process can be stimulated by different activities and can be enhanced by understanding the basic attributes of the process. Furthermore, understanding how the human mind processes information can help us to understand this process. Think of the scenario where you meet someone that you have not seen in years and cannot remember her name, although you certainly know her name. You are troubled by the fact that it has escaped your memory. Hours or even days later while you are relaxing, unexpectedly that person's name pops into your head. What happened during that period of time? One way to think of the subconscious part of our memory or our mind is to compare it to a computer system that is working on problems but only has access to the output screen during certain periods of time. For example, when we met the person discussed earlier walking down the street, the fact that it bothered us that we could not remember her name was placed in the subconscious part of our mind. The subconscious part of the mind worked on the problem, and at some period of relaxation when the subconscious and conscious parts of the mind were linked together, as occurs during dreaming, the name reappeared. The subconscious part of the mind is thinking continuously, usually without us knowing it except during daydreaming, and when dreaming, the subconscious may be constantly working on a problem. However, if we have not placed a problem in the subconscious part of the mind to be worked on and then had that period of future relaxation where the two consciousnesses seem to be linked together like the computer and the computer screen, the solution will not appear. Think about your own experiences; if you had been trying to recall a person's name, the name probably would not have occurred to you while you were busily crossing a street or actively engaged in another activity, but only during a period of relaxation. We can use this skill to help us solve problems by thinking about them before we go to sleep at night or during periods of relaxation. When those ideas appear, we must take time to write them down because, ironically, in many cases after they reappear in the conscious part of our mind, if we do not make a written note of them, they are likely to disappear again. Understanding this process leads us to some of the techniques that can be used to stimulate creativity, both in how we think of a problem and how we go about solving it.

Attributes of the Problem

As was discussed in Chapter 1, when we are presented with an engineering problem to solve, certain activities can encourage the process of creativity. First, we must think of the problem to be solved in terms of its basic attributes. We must take care not to be thinking of potential solutions at this point but of simply trying to define the problem in its most basic terms. For instance, to join two boards together, we could think either in terms of using a hammer and nails or we could think solely in terms of a process for joining the two boards together. Although this process might include using nails, it might include other options as well. If we define this problem only in terms of nails, we have eliminated possibilities such as gluing, screwing, or other potentially creative and new solutions. It also helps during the defining of the problem or listing of the attributes not to know how others have solved this or similar problems. This can be difficult because often a supervisor may present you with a problem to be solved and present it as an "oh, by the way" and "here is how company X solved this problem" or "here is how I think you might want to approach this." It is very difficult not to entertain those ideas, but by putting aside preconceived notions about what the potential solution is, we certainly allow our minds greater freedom to explore possible methods of solving the problem at hand.

Brainstorming

Brainstorming can be one of the most enlightening activities that engineers undertake in their day-to-day jobs. This can be a fanciful process as the basic concept of brainstorming involves simply trying to come up with as many ideas as possible while placing no judgment upon the ideas. Inherent in this process is simply trying to create a large quantity of ideas without regard to quality. It is very important not to evaluate their feasibility as we list the ideas. We may allow ideas to piggyback or build on each other and we can simply let our minds run free to put together the different ideas. In some cases, they may grow and multiply and become closer to our ideal solutions.

Sometimes matrices are used where we may list possible solutions along the x and y axes and try to join them together in a somewhat organized manner to see if, for example, column A combined with line B offers some additional ideas or leads our mind to thinking in a slightly different manner in an attempt to create additional options for the design. Many other techniques can be useful, such as looking at sketches upside down or through the back of the paper, thinking of the task at hand as inside out, or thinking about how different species or cultures may approach the idea at hand. For instance, many mobile robotic systems are based somewhat

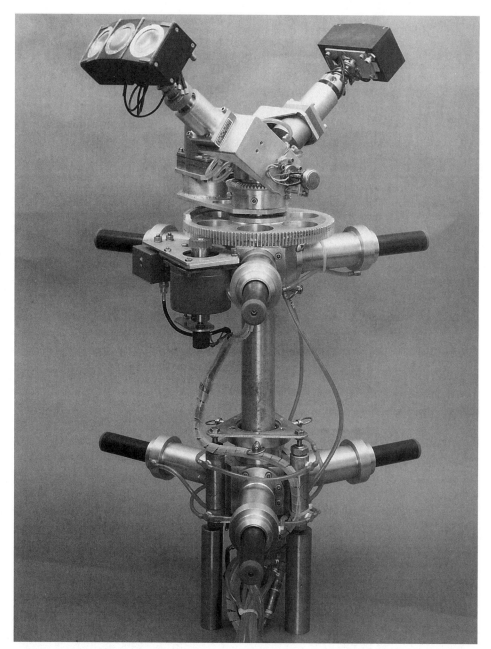

FIGURE 21.2 Six-legged climbing robot.

loosely on the principles of how different insects walk or climb. Figures 21.2 and 21.3 show a mobile robotic system designed by the author that uses the principle of shimmying like a six-legged insect might move. Other

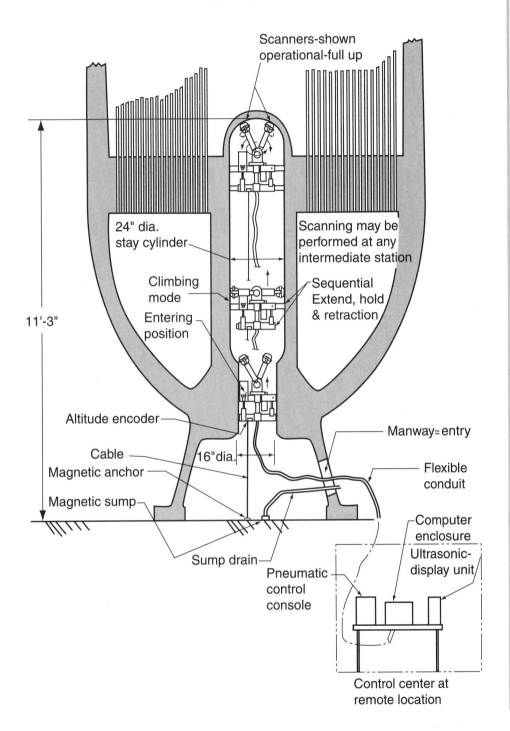

FIGURE 21.3
Six-legged climbing robot in nuclear steam generator.

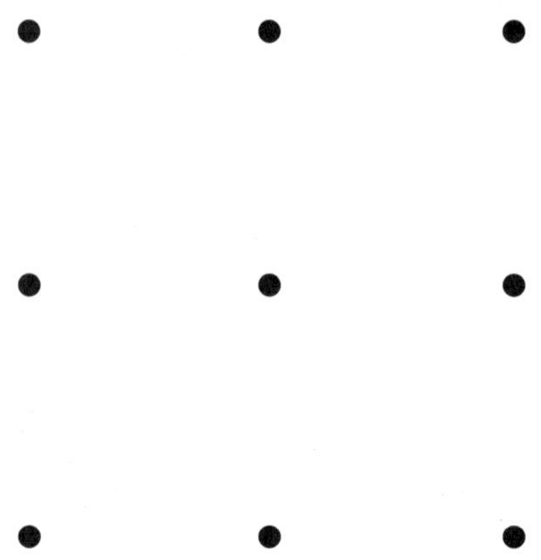

FIGURE 21.4 Try to connect the nine dots with four straight lines without lifting the pencil from the paper.

• • •

• • •

• • •

designs might try to duplicate the features of snakes or other animals in order to somewhat think "outside of the box" in trying to resolve the problem. If we attempt the task in Figure 21.4, which is to connect all of the dots with four straight lines without lifting the pencil from the paper, many of us will struggle with this task. However, if we consider that we can go outside of this box in order to accomplish this task, the solution becomes relatively straightforward. Figure 21.5 shows this potential solution. Many ideas that have been produced over the years have been conceived by thinking of problems in unique ways or thinking of them "outside of the box." This process is important to understand because it can help us to become more creative in the mechanical design process.

Creativity Sessions

creativity sessions
A more formal process of brainstorming that involves a group of people brainstorming about an idea or problem in order to come up with possible remedies or solutions.

The previous section discussed brainstorming. **Creativity sessions** are really brainstorming in a group setting. However, they have some unique attributes that individual brainstorming may not include. Like brainstorming, the idea is to create a quantity of ideas, not quality, and no judgment should be undertaken in the first part of the process. However, for a creativity session to be productive, the members involved (and they can number from two on upward) usually need to be of equal status in the corporate structure. The presence of a vice president among working engineers would usually inhibit the free flow of ideas. People would be concerned with impressing management in a setting such as this. In addition, the cre-

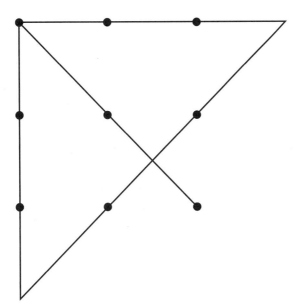

FIGURE 21.5 Solution to Figure 21.4.

ativity session needs to have people with open minds who can let the process work freely. One "Doubting Thomas" in the group finding fault with every idea will poison the whole process and make the creativity session less than productive. This is why the leader of a creativity session needs to clearly understand his or her role. The leader's task must be to keep the group from becoming judgmental and to create a climate of trust and confidence among the people in the group in order to encourage the free exchange of ideas. Members must feel free to piggyback on each other's ideas and to allow the ideas to grow and expand, sometimes in a very silly, irrelevant manner. The true process of innovation often starts with things that seem like silly ideas. The fear of criticism, in the form of ridicule, fear of failure, or punishment, can be very counterproductive in the creativity session process.

Evaluation of Ideas

At some point in the creative process, the ideas need to be evaluated. This can involve refining ideas, combining different ideas, doing some structural analysis or other engineering analysis, and then selecting two or three of the most feasible ideas for further review and study. Once the most promising ideas have been selected, then either detailed sketches or drawings are often created, and some preliminary sizing analysis might be done.

Also, at this point, it is now appropriate to learn about how others have solved either this or similar problems. It is important in this process not to "poison" the mind with how others have solved the problem as, once that is known, it is very difficult to think of really new or different ideas. Often when a solution is in front of you it can block almost all vision around that possible solution. However, at this point in the process, it is important to introduce how others have approached this problem and to consider combining your solution with theirs or to evaluate whether your solution really offers anything of significance. This is often a very difficult process as inherently we have a great deal of **pride of ownership** in our ideas. It is difficult to see someone else's ideas as being better than the ones that we have worked hard on and have refined into what we believe would be an excellent solution. However, we often have to set our egos aside and try to evaluate what the best idea is, whether it is ours or someone else's. By undertaking many of the steps just described, we can often be more creative in the design task at hand.

> **pride of ownership**
> Refers, in the engineering world, to the sense that one's own ideas are inherently better than those created by others.

21.4 PATENTS, COPYRIGHTS, AND PROTECTION OF THE CREATIVE PROCESS

The protection of intellectual property usually takes one of two forms, copyrights or patents. Design professionals use patents far more than copyrights because patents typically cover the design of machinery and similar kinds of products. However, it is important to understand how both of the processes work.

Copyrights

A copyright is a legal right given to an author or authors of an original work. Copyrights were established by the United States Constitution, and the rules have changed over time. However, in general, a copyright could be used to establish ownership of:

- A literary work that may include a computer program
- Musical works such as tunes, melodies, and the associated words
- Plays or other dramatic works
- Pictures, graphics, or sculptures
- Movies or other motion pictures
- Certain architectural works and other intellectual creations

A copyright offers protection from the time at which the work is created in some permanent form (for example, when it is written on paper or recorded), not from the time when the idea is created. A copyright is submitted to the Copyright Office along with the current fee, and a claim is then established by that filing. The date of the claim for the copyright becomes the date when the work was put in a permanent form.

Patents

A patent is often referred to as a letters patent, and the process has changed recently to bring it into general conformity with the way European patents are issued. A few years ago, patents were issued based on the date of creation, similar to a copyright. The date when the idea was put into hard form such as a sketch, drawing, or description, which was often signed and witnessed, became the date on which the inventor could claim originality if there was a dispute regarding ownership. The system was recently changed to make it similar to the European system; the date of ownership is now the date on which an application is filed for a patent with the United States Patent Office.

Along with this change, which took effect in 1995, the Patent Office now allows a provisional application for a patent that does not require all of the sketches, drawings, claims, and other details necessary for a permanent patent. This provisional patent allows the inventor to document the date of the invention, with as much description as possible at that point, but not with all of the formal requirements. This provisional application is filed with the Patent Office and is in effect for a period of 1 year. When this is done, the inventor can apply the term **patent pending** to his or her invention.

A patent, which is the grant of a property right to an invention, can then be filed during this 1-year period. This patent would have a 20-year term from the date on which the application for the patent was filed. This application would typically include a description of the invention, drawings done in an appropriate format (the requirements can be obtained from the Patent Office), and certain claims with regard to how the product or invention is new, unique, and different. These claims may be many or few, depending on the scope and complexity of the invention. The patent law states that the invention needs to be useful; by that, it means it needs to solve a problem and have a use. It cannot simply be an abstract idea. Unlike a copyright, a patent is reviewed by a patent examiner at the Patent Office who may reject or accept the claims made in the application. Part of the process for applying for a patent is research into what is referred to as **prior art.** Prior art is the term used to describe anything similar that has been

> **patent pending**
> A term that can be applied to a product in order to inform others that a patent has been applied for and is expected to be received.

> **prior art**
> Refers to an idea that may have been created previously. In the patent process, it would prevent a patent from being issued because the idea is not unique.

done before in any form or fashion, regardless of whether it was previously patented. Hence, if the patent examiner, or a patent search that the inventor may do before filing an application, finds prior art, the idea is not considered unique and is not eligible to receive a patent. In order to obtain a patent, the applicant must pay a fee, and there are also periodic maintenance fees to keep the patent in effect. A patent may be issued to any person who "invents or discovers any new or useful process, machine, manufacture, or composition of matter, or any new and useful improvement thereof," subject to the current laws and regulations. Figure 21.6 shows the front page of a patent with a listing of some of the claims that were approved for this application.

One of the difficulties in the patent process is that a patent can be challenged later by other parties who claim that prior art did exist and, in many cases, that they own the prior art. When this happens, the irony is that the patent holder has to defend the patent in a court of law. This can be cost prohibitive to individual inventors and, in many cases, they may not have the resources to defend their patent. However, if prior art was appropriately researched, this rarely occurs.

A patent is always issued in the name of the inventor. However, the ownership of that patent can be transferred and, generally, if you are working for a firm or corporation in a design capacity, the firm would own the rights to anything invented in your capacity as their employee. This is often part of an employment agreement that you would sign before starting work. It is generally expected that you bring your creative ability to bear on your employment and that, since they are paying you for that employment, they would then own any inventions that you create in the course of your employment. If you invent something independently and apply for and obtain a patent, you could market it yourself, sell the rights to it, or do some combination thereof.

Trademarks

trademarks
A name or symbol used frequently to identify a product. A trademark allows its unique use by the creator under certain conditions.

Trademarks are not as applicable to the design process as patents are. They are more closely related to marketing a name, a trade name, or a specific type of product. A registered trademark allows only the owner of that trademark to market the product. Typically, a product that has a trademark is marked with an ® and, although not required, it can be registered with the federal government. If registered with the federal government, a trademark has a term of 10 years and can be renewed for additional 10-year terms. The registration of that trademark is cancelled if this renewal is not filed in the appropriate time period.

United States Patent [19]
Wentzell et al.

[11] 4,432,271
[45] Feb. 21, 1984

[54] **LOCOMOTION UNIT FOR A TOOL SUPPORT ADAPTED FOR PROGRESSION THROUGH PASSAGEWAYS**

[75] Inventors: **Timothy H. Wentzell**, South Windsor; **Charles B. Innes, Jr.**, Granby, both of Conn.

[73] Assignee: **Combustion Engineering, Inc.**, Windsor, Conn.

[21] Appl. No.: 293,541

[22] Filed: **Aug. 17, 1981**

[51] Int. Cl.³ F15B 13/06
[52] U.S. Cl. 91/527; 73/633; 92/2; 254/134.6; 376/249
[58] Field of Search .... 73/623, 633, 640; 91/508, 527; 92/2, 147, 148; 165/11 A; 254/106, 107, 134.6; 376/249, 231

[56] **References Cited**
U.S. PATENT DOCUMENTS

2,992,812	7/1961	Rasmussen et al.	254/107
3,780,571	12/1973	Wiesener	73/67.8 S
3,809,607	5/1974	Murray et al.	376/249
3,862,578	1/1975	Schluter	376/249

Primary Examiner—Martin P. Schwadron
Assistant Examiner—Richard S. Meyer
Attorney, Agent, or Firm—Arthur L. Wade

[57] **ABSTRACT**

A locomotion unit is remotely controlled to advance stepwise through a passageway by frictional engagement with the surfaces of the passageway walls. A support section is mounted on the locomotion device to bring special tools to bear at predetermined positions within the passageway to which the support is brought by the locomotion unit.

7 Claims, 5 Drawing Figures

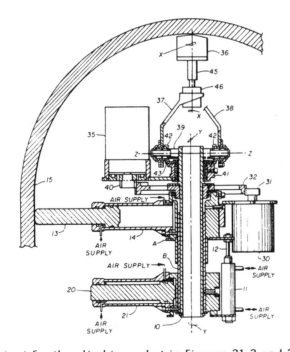

FIGURE 21.6 Patent for the climbing robot in Figures 21.2 and 21.3.

21.5 SUMMARY

After completion of this chapter, you should have a better understanding of the design process, including how the mind works and the role of innovation in design. In addition, we have studied some of the techniques for stimulating the creative process. Lastly, you should have a better understanding of patents, copyrights, trademarks, and the protection of the creative process, and especially how to apply for a patent in order to protect the inventor's rights.

21.6 PROBLEMS (DESIGN PROJECTS)

Clutch System

Design a clutch, including springs, pressure plates, friction disks, and spline to allow movement of the clutch for an automobile that displaces 250 hp at 2,800 rpm. Forces for disengagement, friction factors, and loads on the individual parts should be delineated as well. In addition, select a bearing to allow the clutch disk to be pressed into engagement as necessary, and include a sketch of your layout and design.

Hydraulic Elevator System

Design a hydraulic elevator for a 4-story building capable of transporting 10 people at a time. Determine appropriate cylinder sizes, pressures, storage reservoir sizes, and pump and motor sizes. In addition, if you have taken Fluid Mechanics, select a pump system to provide power for your elevator.

Electric Elevator

Design a lead-screw-powered elevator system for lifting freight in a 2-floor warehouse that is capable of lifting 10,000 pounds over an elevation change of 10 feet. Select lead screws, determine the appropriate power for the lead screws, and select a gearbox from available literature or design a gear system to provide appropriate input speeds. It should take approximately 30 to 60 seconds to move between these two elevations.

Garage Door Opener 1

Design a lead-screw-powered garage door opener to open a conventional 8-foot-wide garage door. Determine appropriate operating speeds and forces, and select a lead screw and motor assembly for this system.

Garage Door Opener 2

Design a garage door system for an application similar to the previous problem, except select a chain drive that would result in similar operating parameters. Select a gearbox, motor, and chain, and determine the appropriate forces.

Garage Door Opener 3

Design a garage door system that uses water power for use in an environment where no electric power is available. Assume that freezing is not a constraint and that the motive means for opening and closing your garage door is simply the supply or dumping of water. Include a sketch of your layout, and explain how this system operates. Also, determine the volume of water and the appropriate forces.

Car Jack

Design a floor jack to lift either the front or rear end of a car. Estimate what the required forces would be, and determine the sizes of the parts, the motive means, and the force needed by the user to operate the jack system.

Winch

Design an electrically operated winch with a load capacity of 2,000 pounds. Decide upon drum size, motive means, and the appropriate reduction, whether it be chains, belts, gears, or motor size. Include a sketch and layout of your design. Determine whether your winch design is self-locking and, if not, include a brake assembly.

Spur-Gear Reducer

Design a spur-gear reduction gearbox to reduce an input speed of 3,600 rpm to 180 rpm from a 5-hp motor. Select gear, shafts, and bearings to accomplish this. A complete stress analysis of all gears, shafting, and loads on the bearings should be included.

Bevel Gearbox Design

For the design considerations in the previous problem, include instead the ability to use a bevel gear in order to make the output shaft 90° from the input.

Pogo Stick

Design a pogo stick for a child who weighs up to 120 pounds, including the spring design, structural design, and the size of tubes for the leg. Estimate the performance characteristics of this design, including the amount of compression in the spring and the height to which the pogo stick would jump.

Clock Design

Design a clock that uses an 1,800 rpm input motor speed, including all of the gear ratios for each of the 3 hands. A structural analysis of the gears is not required for this design, but the layout needed to produce in this design should include the ratio for each of the hands on the clock.

APPENDIXES

Illustrations and Tables

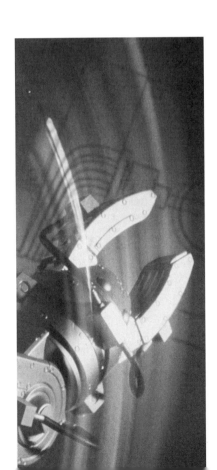

APPENDIX 1
Categorization of Stress Types (Basic Stress Theory)

			Basic stress theory		
Method of molecule separation	**Type**	**Example**	**Stress**	**Deflection**	**Stress allowable**
Push together or pull apart	Axial		$S = \dfrac{F}{A}$	$\delta = \dfrac{FL}{AE}$	S_y or S_u
	Bending		$S = \dfrac{MC}{I}$ or $\dfrac{M}{Z}$ I or Z from App. 3 Calculate M from App. 2	δ from App. 2	S_y or S_u
Slide by	Torsion		$S = \dfrac{TC}{J}$ or $\dfrac{T}{Z'}$ J or Z' from App. 3	$\theta = \dfrac{TL}{JG}$	S_s if known; otherwise use .5 or .6 S_y or S_u
	Shear		$S = \dfrac{F}{A}$ $S = \dfrac{VQ}{IB}$	Usually don't care!	S_s if known; otherwise use .5 or .6 S_y or S_u

APPENDIX 2 Beam Moment and Deflections

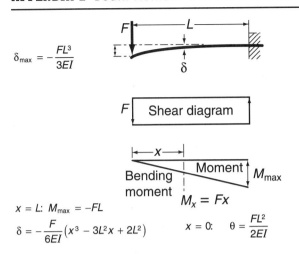

$\delta_{max} = -\dfrac{FL^3}{3EI}$

$x = L: M_{max} = -FL$

$\delta = -\dfrac{F}{6EI}(x^3 - 3L^2x + 2L^2)$ $x = 0:$ $\theta = \dfrac{FL^2}{2EI}$

$M_x = Fx$

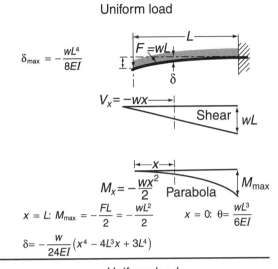

Uniform load

$\delta_{max} = -\dfrac{wL^4}{8EI}$

$V_x = -wx$

$M_x = -\dfrac{wx^2}{2}$ Parabola

$x = L: M_{max} = -\dfrac{FL}{2} = -\dfrac{wL^2}{2}$ $x = 0:$ $\theta = \dfrac{wL^3}{6EI}$

$\delta = -\dfrac{w}{24EI}(x^4 - 4L^3x + 3L^4)$

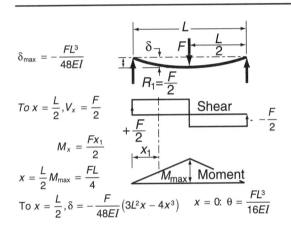

$\delta_{max} = -\dfrac{FL^3}{48EI}$

$R_1 = \dfrac{F}{2}$

To $x = \dfrac{L}{2}, V_x = \dfrac{F}{2}$

$M_x = \dfrac{Fx_1}{2}$

$x = \dfrac{L}{2}, M_{max} = \dfrac{FL}{4}$

To $x = \dfrac{L}{2}, \delta = -\dfrac{F}{48EI}(3L^2x - 4x^3)$ $x = 0:$ $\theta = \dfrac{FL^3}{16EI}$

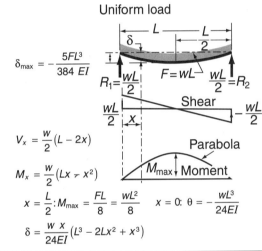

Uniform load

$\delta_{max} = -\dfrac{5FL^3}{384\,EI}$

$R_1 = \dfrac{wL}{2}$ $F = wL$ $\dfrac{wL}{2} = R_2$

$V_x = \dfrac{w}{2}(L - 2x)$

$M_x = \dfrac{w}{2}(Lx - x^2)$

$x = \dfrac{L}{2}: M_{max} = \dfrac{FL}{8} = \dfrac{wL^2}{8}$ $x = 0:$ $\theta = -\dfrac{wL^3}{24EI}$

$\delta = \dfrac{w\,x}{24EI}(L^3 - 2Lx^2 + x^3)$

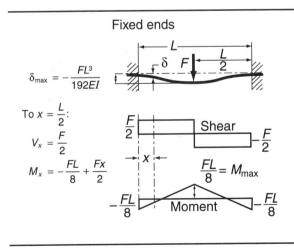

Fixed ends

$\delta_{max} = -\dfrac{FL^3}{192EI}$

To $x = \dfrac{L}{2}:$

$V_x = \dfrac{F}{2}$

$M_x = -\dfrac{FL}{8} + \dfrac{Fx}{2}$

$\dfrac{FL}{8} = M_{max}$

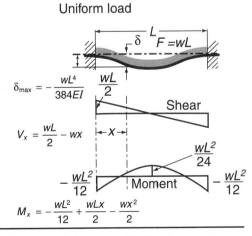

Uniform load

$\delta_{max} = -\dfrac{wL^4}{384EI}$

$V_x = \dfrac{wL}{2} - wx$

$M_x = -\dfrac{wL^2}{12} + \dfrac{wLx}{2} - \dfrac{wx^2}{2}$

F lb = applied force; w = pounds per inch of length; $F = wL$, where L in. = length; E psi = modulus of elasticity, tension; I in.^{4} = moment of inertia; δ in. = deflection; θ radians = slope.

APPENDIX 2
Beam Moment and Deflections (Concluded)

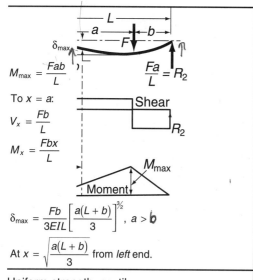

$M_{max} = \dfrac{Fab}{L}$

$\dfrac{Fa}{L} = R_2$

To $x = a$:

$V_x = \dfrac{Fb}{L}$

$M_x = \dfrac{Fbx}{L}$

$\delta_{max} = \dfrac{Fb}{3EIL}\left[\dfrac{a(L+b)}{3}\right]^{3/2}$, $a > b$

At $x = \sqrt{\dfrac{a(L+b)}{3}}$ from *left* end.

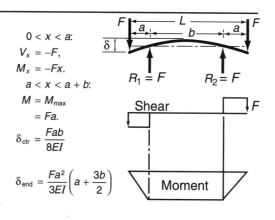

$0 < x < a$:
$V_x = -F$,
$M_x = -Fx$.

$a < x < a + b$:
$M = M_{max} = Fa$.

$\delta_{ctr} = \dfrac{Fab}{8EI}$

$\delta_{end} = \dfrac{Fa^2}{3EI}\left(a + \dfrac{3b}{2}\right)$

Uniform strength, cantilever

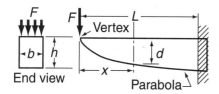

$M_x = Fx = \dfrac{sI}{c}$. For $s = C$,

$d^2 = \dfrac{6F}{bs}x = \dfrac{x}{L}h^2$

At $x = 0$: $\delta_{max} = -\dfrac{8FL^3}{bEh^3}$

Uniform strength, simple beam

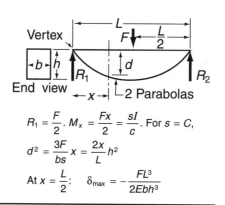

$R_1 = \dfrac{F}{2}$. $M_x = \dfrac{Fx}{2} = \dfrac{sI}{c}$. For $s = C$,

$d^2 = \dfrac{3F}{bs}x = \dfrac{2x}{L}h^2$

At $x = \dfrac{L}{2}$: $\delta_{max} = -\dfrac{FL^3}{2Ebh^3}$

APPENDIX 3
Properties of Common Shapes

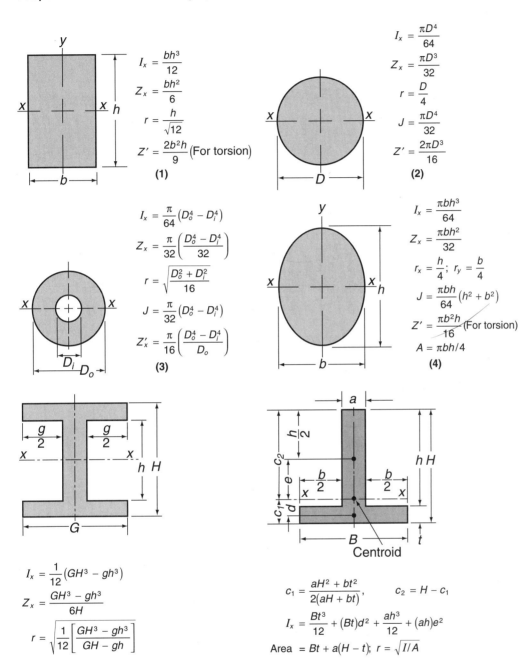

(1) Rectangle:
$I_x = \dfrac{bh^3}{12}$
$Z_x = \dfrac{bh^2}{6}$
$r = \dfrac{h}{\sqrt{12}}$
$Z' = \dfrac{2b^2h}{9}$ (For torsion)

(2) Circle:
$I_x = \dfrac{\pi D^4}{64}$
$Z_x = \dfrac{\pi D^3}{32}$
$r = \dfrac{D}{4}$
$J = \dfrac{\pi D^4}{32}$
$Z' = \dfrac{2\pi D^3}{16}$

(3) Hollow circle:
$I_x = \dfrac{\pi}{64}(D_o^4 - D_i^4)$
$Z_x = \dfrac{\pi}{32}\left(\dfrac{D_o^4 - D_i^4}{32}\right)$
$r = \sqrt{\dfrac{D_o^2 + D_i^2}{16}}$
$J = \dfrac{\pi}{32}(D_o^4 - D_i^4)$
$Z'_x = \dfrac{\pi}{16}\left(\dfrac{D_o^4 - D_i^4}{D_o}\right)$

(4) Ellipse:
$I_x = \dfrac{\pi b h^3}{64}$
$Z_x = \dfrac{\pi b h^2}{32}$
$r_x = \dfrac{h}{4}$; $r_y = \dfrac{b}{4}$
$J = \dfrac{\pi b h}{64}(h^2 + b^2)$
$Z' = \dfrac{\pi b^2 h}{16}$ (For torsion)
$A = \pi b h / 4$

(5) I-beam:
$I_x = \dfrac{1}{12}(GH^3 - gh^3)$
$Z_x = \dfrac{GH^3 - gh^3}{6H}$
$r = \sqrt{\dfrac{1}{12}\left[\dfrac{GH^3 - gh^3}{GH - gh}\right]}$

(6) T-section:
$c_1 = \dfrac{aH^2 + bt^2}{2(aH + bt)}$, $c_2 = H - c_1$
$I_x = \dfrac{Bt^3}{12} + (Bt)d^2 + \dfrac{ah^3}{12} + (ah)e^2$
Area $= Bt + a(H - t)$; $r = \sqrt{I/A}$

I_x = moment of inertia about the axis $x - x$, J = polar moment of inertia about the centroidal axis, $Z = I/c$ = rectangular section modulus, about $x - x$, $Z' = J/c$ = polar section modulus, $r = \sqrt{I/\text{area}}$ = radius of gyration.

APPENDIX 4
Properties of Common Steels

Material Designation (AISI Number)	Condition	Tensile Strength (ksi)	Tensile Strength (MPa)	Yield Strength (ksi)	Yield Strength (MPa)	Ductility (Percent Elongation in 2 in.)	Brinell Hardness (HB)
1020	Hot-rolled	55	379	30	207	25	111
1020	Cold-drawn	61	420	51	352	15	122
1020	Annealed	60	414	43	296	38	121
1040	Hot-rolled	72	496	42	290	18	144
1040	Cold-drawn	80	552	71	490	12	160
1040	OQT 1300	88	607	61	421	33	183
1040	OQT 400	113	779	87	600	19	262
1050	Hot-rolled	90	620	49	338	15	180
1050	Cold-drawn	100	690	84	579	10	200
1050	OQT 1300	96	662	61	421	30	192
1050	OQT 400	143	986	110	758	10	321
1117	Hot-rolled	62	427	34	234	33	124
1117	Cold-drawn	69	476	51	352	20	138
1117	WQT 350	89	614	50	345	22	178
1137	Hot-rolled	88	607	48	331	15	176
1137	Cold-drawn	98	676	82	565	10	196
1137	OQT 1300	87	600	60	414	28	174
1137	OQT 400	157	1,083	136	938	5	352
1144	Hot-rolled	94	648	51	352	15	188
1144	Cold-drawn	100	690	90	621	10	200
1144	OQT 1300	96	662	68	469	25	200
1144	OQT 400	127	876	91	627	16	277
1213	Hot-rolled	55	379	33	228	25	110
1213	Cold-drawn	75	517	58	340	10	150
12L13	Hot-rolled	57	393	34	234	22	114
12L13	Cold-drawn	70	483	60	414	10	140
1340	Annealed	102	703	63	434	26	207
1340	OQT 1300	100	690	75	517	25	235
1340	OQT 1000	144	993	132	910	17	363
1340	OQT 700	221	1,520	197	1,360	10	444
1340	OQT 400	285	1,960	234	1,610	8	578
3140	Annealed	95	655	67	462	25	187
3140	OQT 1300	115	792	94	648	23	233
3140	OQT 1000	152	1,050	133	920	17	311
3140	OQT 700	220	1,520	200	1,380	13	461
3140	OQT 400	280	1,930	248	1,710	11	555
4130	Annealed	81	558	52	359	28	156
4130	WQT 1300	98	676	89	614	28	202

APPENDIX 4
Properties of Common Steels (Continued)

Material Designation (AISI Number)	Condition	Tensile Strength (ksi)	Tensile Strength (MPa)	Yield Strength (ksi)	Yield Strength (MPa)	Ductility (Percent Elongation in 2 in.)	Brinell Hardness (HB)
4130	WQT 1000	143	986	132	910	16	302
4130	WQT 700	208	1,430	180	1,240	13	415
4130	WQT 400	234	1,610	197	1,360	12	461
4140	Annealed	95	655	60	414	26	197
4140	OQT 1300	117	807	100	690	23	235
4140	OQT 1000	168	1,160	152	1,050	17	341
4140	OQT 700	231	1,590	212	1,460	13	461
4140	OQT 400	290	2,000	251	1,730	11	578
4150	Annealed	106	731	55	379	20	197
4150	OQT 1300	127	880	116	800	20	262
4150	OQT 1000	197	1,360	181	1,250	11	401
4150	OQT 700	247	1,700	229	1,580	10	495
4150	OQT 400	300	2,070	248	1,710	10	578
4340	Annealed	108	745	68	469	22	217
4340	OQT 1300	140	965	120	827	23	280
4340	OQT 1000	171	1,180	158	1,090	16	363
4340	OQT 700	230	1,590	206	1,420	12	461
4340	OQT 400	283	1,950	228	1,570	11	555
5140	Annealed	83	572	42	290	29	167
5140	OQT 1300	104	717	83	572	27	207
5140	OQT 1000	145	1,000	130	896	18	302
5140	OQT 700	220	1,520	200	1,380	11	429
5140	OQT 400	276	1,900	226	1,560	7	534
5150	Annealed	98	676	52	359	22	197
5150	OQT 1300	116	800	102	700	22	241
5150	OQT 1000	160	1,100	149	1,030	15	321
5150	OQT 700	240	1,650	220	1,520	10	461
5150	OQT 400	312	2,150	250	1,720	8	601
5160	Annealed	105	724	40	276	17	197
5160	OQT 1300	115	793	100	690	23	229
5160	OQT 1000	170	1,170	151	1,040	14	341
5160	OQT 700	263	1,810	237	1,630	9	514
5160	OQT 400	322	2,220	260	1,790	4	627
6150	Annealed	96	662	59	407	23	197
6150	OQT 1300	118	814	107	738	21	241
6150	OQT 1000	183	1,260	173	1,190	12	375
6150	OQT 700	247	1,700	223	1,540	10	495
6150	OQT 400	315	2,170	270	1,860	7	601

APPENDIX 4
Properties of Common Steels (Concluded)

Material Designation (AISI Number)	Condition	Tensile Strength (ksi)	Tensile Strength (MPa)	Yield Strength (ksi)	Yield Strength (MPa)	Ductility (Percent Elongation in 2 in.)	Brinell Hardness (HB)
8650	Annealed	104	717	56	386	22	212
8650	OQT 1300	122	841	113	779	21	255
8650	OQT 1000	176	1,210	155	1,070	14	363
8650	OQT 700	240	1,650	222	1,530	12	495
8650	OQT 400	282	1,940	250	1,720	11	555
8740	Annealed	100	690	60	414	22	201
8740	OQT 1300	119	820	100	690	25	241
8740	OQT 1000	175	1,210	167	1,150	15	363
8740	OQT 700	228	1,570	212	1,460	12	461
8740	OQT 400	290	2,000	240	1,650	10	578
9255	Annealed	113	780	71	490	22	229
9255	Q&T 1300	130	896	102	703	21	262
9255	Q&T 1000	181	1,250	160	1,100	14	352
9255	Q&T 700	260	1,790	240	1,650	5	534
9255	Q&T 400	310	2,140	287	1,980	2	601

Note: Properties common to all carbon and alloy steels:
Poisson's ratio: .27
Shear modulus: 11.5×10^6 psi; 80 GPa
Coefficient of thermal expansion: 6.5×10^{-6} °F^{-1}
Density: .283 lb/in.^{3}; 7,680 kg/m^3
Modulus of elasticity: 30×10^6 psi; 207 GPa

APPENDIX 5.1
Properties of American Standard Steel Channels: C-Shapes*

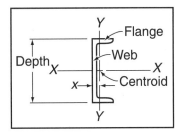

Designation	Area (in.²)	Depth (in.)	Web Thickness (in.)	Flange Width (in.)	Flange Average Thickness (in.)	Axis X–X I (in.⁴)	Axis X–X Z (in.³)	Axis Y–Y I (in.⁴)	Axis Y–Y Z (in.³)	x (in.)
C15 × 50	14.7	15.00	.716	3.716	.650	404	53.8	11.0	3.78	.798
C15 × 40	11.8	15.00	.520	3.520	.650	349	46.5	9.23	3.37	.777
C12 × 30	8.82	12.00	.510	3.170	.501	162	27.0	5.14	2.06	.674
C12 × 25	7.35	12.00	.387	3.047	.501	144	24.1	4.47	1.88	.674
C10 × 30	8.82	10.00	.673	3.033	.436	103	20.7	3.94	1.65	.649
C10 × 20	5.88	10.00	.379	2.739	.436	78.9	15.8	2.81	1.32	.606
C9 × 20	5.88	9.00	.448	2.648	.413	60.9	13.5	2.42	1.17	.583
C9 × 15	4.41	9.00	.285	2.485	.413	51.0	11.3	1.93	1.01	.586
C8 × 18.75	5.51	8.00	.487	2.527	.390	44.0	11.0	1.98	1.01	.565
C8 × 11.5	3.38	8.00	.220	2.260	.390	32.6	8.14	1.32	.781	.571
C6 × 13	3.83	6.00	.437	2.157	.343	17.4	5.80	1.05	.642	.514
C6 × 8.2	2.40	6.00	.200	1.920	.343	13.1	4.38	.693	.492	.511
C5 × 9	2.64	5.00	.325	1.885	.320	8.90	3.56	.632	.450	.478
C5 × 6.7	1.97	5.00	.190	1.750	.320	7.49	3.00	.479	.378	.484
C4 × 7.25	2.13	4.00	.321	1.721	.296	4.59	2.29	.433	.343	.459
C4 × 5.4	1.59	4.00	.184	1.584	.296	3.85	1.93	.319	.283	.457
C3 × 6	1.76	3.00	.356	1.596	.273	2.07	1.38	.305	.268	.455
C3 × 4.1	1.21	3.00	.170	1.410	.273	1.66	1.10	.197	.202	.436

*Data are taken from a variety of sources. Sizes listed represent a small sample of the sizes available.
I = moment of inertia; Z = section modulus.

APPENDIX 5.2

Properties of Steel Angles, Equal Legs, and Unequal Legs: L-Shapes*

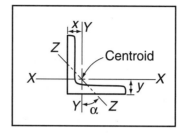 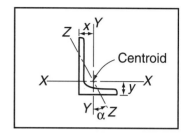

Designation	Area (in.²)	Weight per Foot (lb)	Axis X-X			Axis Y-Y			Axis Z-Z	
			I (in.⁴)	Z (in.³)	y (in.)	I (in.⁴)	Z (in.³)	x (in.)	r (in.)	α (deg)
L8 × 8 × 1	15.0	51.0	89.0	15.8	2.37	89.0	15.8	2.37	1.56	45.0
L8 × 8 × ½	7.75	26.4	48.6	8.36	2.19	48.6	8.36	2.19	1.59	45.0
L8 × 4 × 1	11.0	37.4	69.6	14.1	3.05	11.6	3.94	1.05	.846	13.9
L8 × 4 × ½	5.75	19.6	38.5	7.49	2.86	6.74	2.15	.859	.865	14.9
L6 × 6 × ¾	8.44	28.7	28.2	6.66	1.78	28.2	6.66	1.78	1.17	45.0
L6 × 6 × ⅜	4.36	14.9	15.4	3.53	1.64	15.4	3.53	1.64	1.19	45.0
L6 × 4 × ¾	6.94	23.6	24.5	6.25	2.08	8.68	2.97	1.08	.860	23.2
L6 × 4 × ⅜	3.61	12.3	13.5	3.32	1.94	4.90	1.60	.941	.877	24.0
L4 × 4 × ½	3.75	12.8	5.56	1.97	1.18	5.56	1.97	1.18	.782	45.0
L4 × 4 × ¼	1.94	6.6	3.04	1.05	1.09	3.04	1.05	1.09	.795	45.0
L4 × 3 × ½	3.25	11.1	5.05	1.89	1.33	2.42	1.12	.827	.639	28.5
L4 × 3 × ¼	1.69	5.8	2.77	1.00	1.24	1.36	.599	.896	.651	29.2
L3 × 3 × ½	2.75	9.4	2.22	1.07	.932	2.22	1.07	.932	.584	45.0
L3 × 3 × ¼	1.44	4.9	1.24	.577	.842	1.24	.577	.842	.592	45.0
L2 × 2 × ⅜	1.36	4.7	.479	.351	.636	.479	.351	.636	.389	45.0
L2 × 2 × ¼	.938	3.19	.348	.247	.592	.348	.247	.592	.391	45.0
L2 × 2 × ⅛	.484	1.65	.190	.131	.546	.190	.131	.546	.398	45.0

*Data are taken from a variety of sources. Sizes listed represent a small sample of the sizes available.
Z–Z is axis of minimum moment of inertia (I) and radius of gyration (r).
I = moment of inertia; Z = section modulus; r = radius of gyration.

APPENDIX 5.3

Properties of Steel Wide-Flange Shapes: W-Shapes*

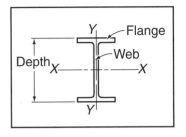

Designation	Area (in.²)	Depth (in.)	Web Thickness (in.)	Flange Width (in.)	Flange Thickness (in.)	Axis X–X I (in.⁴)	Axis X–X Z (in.³)	Axis Y–Y I (in.⁴)	Axis Y–Y Z (in.³)
W24 × 76	22.4	23.92	.440	8.990	.680	2,100	176	82.5	18.4
W24 × 68	20.1	23.73	.415	8.965	.585	1,830	154	70.4	15.7
W21 × 73	21.5	21.24	.455	8.295	.740	1,600	151	70.6	17.0
W21 × 57	16.7	21.06	.405	6.555	.650	1,170	111	30.6	9.35
W18 × 55	16.2	18.11	.390	7.530	.630	890	98.3	44.9	11.9
W18 × 40	11.8	17.90	.315	6.015	.525	612	68.4	19.1	6.35
W14 × 43	12.6	13.66	.305	7.995	.530	428	62.7	45.2	11.3
W14 × 26	7.69	13.91	.255	5.025	.420	245	35.3	8.91	3.54
W12 × 30	8.79	12.34	.260	6.520	.440	238	38.6	20.3	6.24
W12 × 16	4.71	11.99	.220	3.990	.265	103	17.1	2.82	1.41
W10 × 15	4.41	9.99	.230	4.000	.270	69.8	13.8	2.89	1.45
W10 × 12	3.54	9.87	.190	3.960	.210	53.8	10.9	2.18	1.10
W8 × 15	4.44	8.11	.245	4.015	.315	48.0	11.8	3.41	1.70
W8 × 10	2.96	7.89	.170	3.940	.205	30.8	7.81	2.09	1.06
W6 × 15	4.43	5.99	.230	5.990	.260	29.1	9.72	9.32	3.11
W6 × 12	3.55	6.03	.230	4.000	.280	22.1	7.31	2.99	1.50
W5 × 19	5.54	5.15	.270	5.030	.430	26.2	10.2	9.13	3.63
W5 × 16	4.68	5.01	.240	5.000	.360	21.3	8.51	7.51	3.00
W4 × 13	3.83	4.16	.280	4.060	.345	11.3	5.46	3.86	1.90

* Data are taken from a variety of sources. Sizes listed represent a small sample of the sizes available.
I = moment of inertia; *Z* = section modulus.

APPENDIX 5.4

Properties of American Standard Steel Beams: S-Shapes*

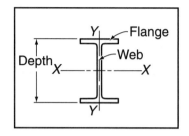

Designation	Area (in.²)	Depth (in.)	Web Thickness (in.)	Flange Width (in.)	Flange Average Thickness (in.)	Axis X–X I (in.⁴)	Axis X–X Z (in.³)	Axis Y–Y I (in.⁴)	Axis Y–Y Z (in.³)
S24 × 90	26.5	24.00	.625	7.125	.870	2,250	187	44.9	12.6
S20 × 96	28.2	20.30	.800	7.200	.920	1,670	165	50.2	13.9
S20 × 75	22.0	20.00	.635	6.385	.795	1,280	128	29.8	9.32
S20 × 66	19.4	20.00	.505	6.255	.795	1,190	119	27.7	8.85
S18 × 70	20.6	18.00	.711	6.251	.691	926	103	24.1	7.72
S15 × 50	14.7	15.00	.550	5.640	.622	486	64.8	15.7	5.57
S12 × 50	14.7	12.00	.687	5.477	.659	305	50.8	15.7	5.74
S12 × 35	10.3	12.00	.428	5.078	.544	229	38.2	9.87	3.89
S10 × 35	10.3	10.00	.594	4.944	.491	147	29.4	8.36	3.38
S10 × 25.4	7.46	10.00	.311	4.661	.491	124	24.7	6.79	2.91
S8 × 23	6.77	8.00	.441	4.171	.426	64.9	16.2	4.31	2.07
S8 × 18.4	5.41	8.00	.271	4.001	.426	57.6	14.4	3.73	1.86
S7 × 20	5.88	7.00	.450	3.860	.392	42.4	12.1	3.17	1.64
S6 × 12.5	3.67	6.00	.232	3.332	.359	22.1	7.37	1.82	1.09
S5 × 10	2.94	5.00	.214	3.004	.326	12.3	4.92	1.22	.809
S4 × 7.7	2.26	4.00	.193	2.663	.293	6.08	3.04	.764	.574
S3 × 5.7	1.67	3.00	.170	2.330	.260	2.52	1.68	.455	.390

*Data are taken from a variety of sources. Sizes listed represent a small sample of the sizes available.
I = moment of inertia; Z = section modulus.

APPENDIX 5.5
Properties of Steel Structural Tubing: Square and Rectangular*

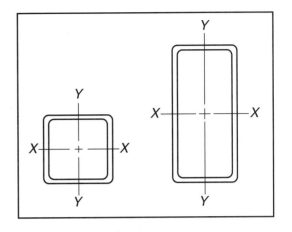

Size	Area (in.²)	Weight per Foot (lb)	Axis X–X I (in.⁴)	Axis X–X Z (in.³)	Axis X–X r (in.)	Axis Y–Y I (in.⁴)	Axis Y–Y Z (in.³)	Axis Y–Y r (in.)
8 × 8 × ½	14.4	48.9	131	32.9	3.03	131	32.9	3.03
8 × 8 × ¼	7.59	25.8	75.1	18.8	3.15	75.1	18.8	3.15
8 × 4 × ½	10.4	35.2	75.1	18.8	2.69	24.6	12.3	1.54
8 × 4 × ¼	5.59	19.0	45.1	11.3	2.84	15.3	7.63	1.65
8 × 2 × ¼	4.59	15.6	30.1	7.52	2.56	3.08	3.08	.819
6 × 6 × ½	10.4	35.2	50.5	16.8	2.21	50.5	16.8	2.21
6 × 6 × ¼	5.59	19.0	30.3	10.1	2.33	30.3	10.1	2.33
6 × 4 × ¼	4.59	15.6	22.1	7.36	2.19	11.7	5.87	1.60
6 × 2 × ¼	3.59	12.2	13.8	4.60	1.96	2.31	2.31	.802
4 × 4 × ½	6.36	21.6	12.3	6.13	1.39	12.3	6.13	1.39
4 × 4 × ¼	3.59	12.2	8.22	4.11	1.51	8.22	4.11	1.51
4 × 2 × ¼	2.59	8.81	4.69	2.35	1.35	1.54	1.54	.770
3 × 3 × ¼	2.59	8.81	3.16	2.10	1.10	3.16	2.10	1.10
3 × 2 × ¼	2.09	7.11	2.21	1.47	1.03	1.15	1.15	.742
2 × 2 × ¼	1.59	5.41	.766	.766	.694	.766	.766	.694

*Data are taken from a variety of sources. Sizes listed represent a small sample of the sizes available.
I = moment of inertia; Z = section modulus; r = radius of gyration.

APPENDIX 5.6

Properties of American National Standard Schedule 40 Welded and Seamless Wrought Steel Pipe

	Diameter (in.)		Wall Thickness (in.)	Cross-Sectional Area of Metal (in.^{2})	Properties of Sections			
Nominal	Actual Inside	Actual Outside			Moment of Inertia I (in.^{4})	Radius of Gyration (in.)	Section Modulus Z (in.^{3})	Polar Section Modulus Z' (in.^{3})
⅛	.269	.405	.068	.072	.001 06	.122	.005 25	.010 50
¼	.364	.540	.088	.125	.003 31	.163	.012 27	.024 54
⅜	.493	.675	.091	.167	.007 29	.209	.021 60	.043 20
½	.622	.840	.109	.250	.017 09	.261	.040 70	.081 40
¾	.824	1.050	.113	.333	.037 04	.334	.070 55	.141 1
1	1.049	1.315	.133	.494	.087 34	.421	.132 8	.265 6
1¼	1.380	1.660	.140	.669	.194 7	.539	.234 6	.469 2
1½	1.610	1.900	.145	.799	.309 9	.623	.326 2	.652 4
2	2.067	2.375	.154	1.075	.665 8	.787	.560 7	1.121
2½	2.469	2.875	.203	1.704	1.530	.947	1.064	2.128
3	3.068	3.500	.216	2.228	3.017	1.163	1.724	3.448
3½	3.548	4.000	.226	2.680	4.788	1.337	2.394	4.788
4	4.026	4.500	.237	3.174	7.233	1.510	3.215	6.430
5	5.047	5.563	.258	4.300	15.16	1.878	5.451	10.90
6	6.065	6.625	.280	5.581	28.14	2.245	8.496	16.99
8	7.981	8.625	.322	8.399	72.49	2.938	16.81	33.62
10	10.020	10.750	.365	11.91	160.7	3.674	29.91	59.82
12	11.938	12.750	.406	15.74	300.2	4.364	47.09	94.18
16	15.000	16.000	.500	24.35	732.0	5.484	91.50	183.0
18	16.876	18.000	.562	30.79	1,172	6.168	130.2	260.4

APPENDIX 5.7
Properties of Aluminum Association Standard Channels

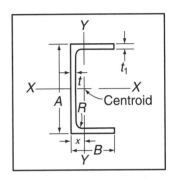

Size				Flange Thickness	Web Thickness	Fillet Radius	Section Properties‡						
Depth	Width						Axis X–X			Axis Y–Y			
A	B	Area*	Weight†	t_1	t	R	I	Z	r	I	Z	r	x
(in.)	(in.)	(in.²)	(lb/ft)	(in.)	(in.)	(in.)	(in.⁴)	(in.³)	(in.)	(in.⁴)	(in.³)	(in.)	(in.)
2.00	1.00	.491	.577	.13	.13	.10	.288	.288	.766	.045	.064	.303	.298
2.00	1.25	.911	1.071	.26	.17	.15	.546	.546	.774	.139	.178	.391	.471
3.00	1.50	.965	1.135	.20	.13	.25	1.41	.94	1.21	.22	.22	.47	.49
3.00	1.75	1.358	1.597	.26	.17	.25	1.97	1.31	1.20	.42	.37	.55	.62
4.00	2.00	1.478	1.738	.23	.15	.25	3.91	1.95	1.63	.60	.45	.64	.65
4.00	2.25	1.982	2.331	.29	.19	.25	5.21	2.60	1.62	1.02	.69	.72	.78
5.00	2.25	1.881	2.212	.26	.15	.30	7.88	3.15	2.05	.98	.64	.72	.73
5.00	2.75	2.627	3.089	.32	.19	.30	11.14	4.45	2.06	2.05	1.14	.88	.95
6.00	2.50	2.410	2.834	.29	.17	.30	14.35	4.78	2.44	1.53	.90	.80	.79
6.00	3.25	3.427	4.030	.35	.21	.30	21.04	7.01	2.48	3.76	1.76	1.05	1.12
7.00	2.75	2.725	3.205	.29	.17	.30	22.09	6.31	2.85	2.10	1.10	.88	.84
7.00	3.50	4.009	4.715	.38	.21	.30	33.79	9.65	2.90	5.13	2.23	1.13	1.20
8.00	3.00	3.526	4.147	.35	.19	.30	37.40	9.35	3.26	3.25	1.57	.96	.93
8.00	3.75	4.923	5.789	.41	.25	.35	52.69	13.17	3.27	7.13	2.82	1.20	1.22
9.00	3.25	4.237	4.983	.35	.23	.35	54.41	12.09	3.58	4.40	1.89	1.02	.93
9.00	4.00	5.927	6.970	.44	.29	.35	78.31	17.40	3.63	9.61	3.49	1.27	1.25
10.00	3.50	5.218	6.136	.41	.25	.35	83.22	16.64	3.99	6.33	2.56	1.10	1.02
10.00	4.25	7.109	8.360	.50	.31	.40	116.15	23.23	4.04	13.02	4.47	1.35	1.34
12.00	4.00	7.036	8.274	.47	.29	.40	159.76	26.63	4.77	11.03	3.86	1.25	1.14
12.00	5.00	10.053	11.822	.62	.35	.45	239.69	39.95	4.88	25.74	7.60	1.60	1.61

*Areas listed are based on nominal dimensions.
†Weights per foot are based on nominal dimensions and a density of .098 lb/in.³ which is the density of alloy 6061.
‡I = moment of inertia; Z = section modulus; r = radius of gyration.
Source: Aluminum Association, *Aluminum Standards and Data,* 11th ed., Washington, DC © 1993, p. 187. Courtesy of Aluminum Association.

APPENDIX 5.8

Properties of Aluminum Association Standard I-Beams

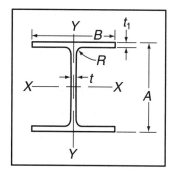

Size				Flange Thickness	Web Thickness	Fillet Radius	Section Properties[‡]					
Depth	Width						Axis X–X			Axis Y–Y		
A	B	Area*	Weight[†]	t_1	t	R	I	Z	r	I	Z	r
(in.)	(in.)	(in.^{2})	(lb/ft)	(in.)	(in.)	(in.)	(in.^{4})	(in.^{3})	(in.)	(in.^{4})	(in.^{3})	(in.)
3.00	2.50	1.392	1.637	.20	.13	.25	2.24	1.49	1.27	.52	.42	.61
3.00	2.50	1.726	2.030	.26	.15	.25	2.71	1.81	1.25	.68	.54	.63
4.00	3.00	1.965	2.311	.23	.15	.25	5.62	2.81	1.69	1.04	.69	.73
4.00	3.00	2.375	2.793	.29	.17	.25	6.71	3.36	1.68	1.31	.87	.74
5.00	3.50	3.146	3.700	.32	.19	.30	13.94	5.58	2.11	2.29	1.31	.85
6.00	4.00	3.427	4.030	.29	.19	.30	21.99	7.33	2.53	3.10	1.55	.95
6.00	4.00	3.990	4.692	.35	.21	.30	25.50	8.50	2.53	3.74	1.87	.97
7.00	4.50	4.932	5.800	.38	.23	.30	42.89	12.25	2.95	5.78	2.57	1.08
8.00	5.00	5.256	6.181	.35	.23	.30	59.69	14.92	3.37	7.30	2.92	1.18
8.00	5.00	5.972	7.023	.41	.25	.30	67.78	16.94	3.37	8.55	3.42	1.20
9.00	5.50	7.110	8.361	.44	.27	.30	102.02	22.67	3.79	12.22	4.44	1.31
10.00	6.00	7.352	8.646	.41	.25	.40	132.09	26.42	4.24	14.78	4.93	1.42
10.00	6.00	8.747	10.286	.50	.29	.40	155.79	31.16	4.22	18.03	6.01	1.44
12.00	7.00	9.925	11.672	.47	.29	.40	255.57	42.60	5.07	26.90	7.69	1.65
12.00	7.00	12.153	14.292	.62	.31	.40	317.33	52.89	5.11	35.48	10.14	1.71

*Areas listed are based on nominal dimensions.
[†]Weights per foot are based on nominal dimensions and a density of .098 lb/in.^{3}, which is the density of alloy 6061.
[‡]I = moment of inertia; Z = section modulus; r = radius of gyration.
Source: Aluminum Association, *Aluminum Standards and Data*, 11th ed., Washington, DC © 1993, p. 187. Courtesy of Aluminum Association.

APPENDIX 6

Stress Concentration Factors: **(a)** Stepped Shaft in Tension **(b)** Stepped Shaft in Bending **(c)** Stepped Shaft in Torsion (Continued)

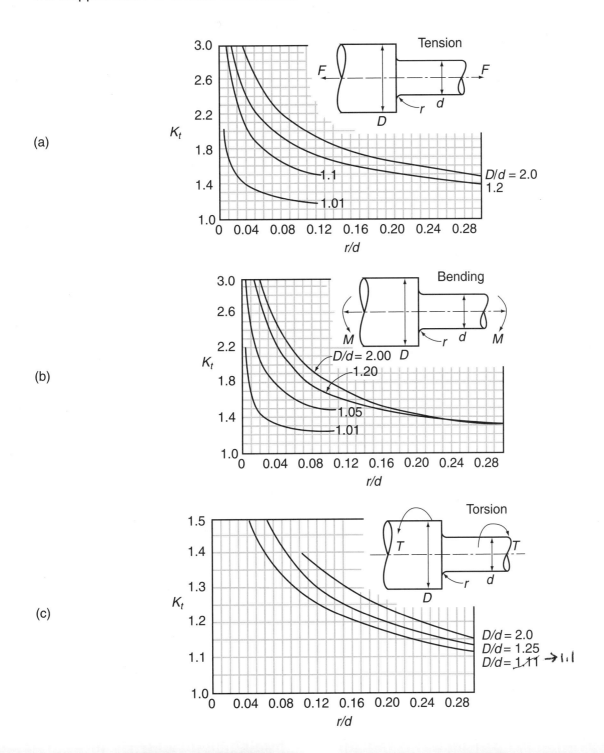

APPENDIX 6

(d) Round Shaft with Cross Hole (Continued)

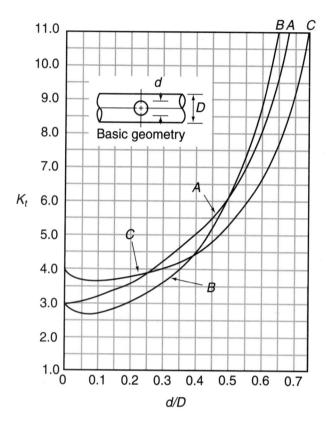

Note: K_t is based on the nominal stress in a round bar without a hole (gross section).

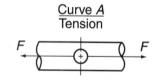

Curve A
Tension

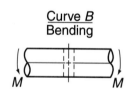

Curve B
Bending

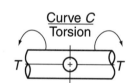

Curve C
Torsion

APPENDIX 6
(g) Flat Bar with Central Hole (Concluded)

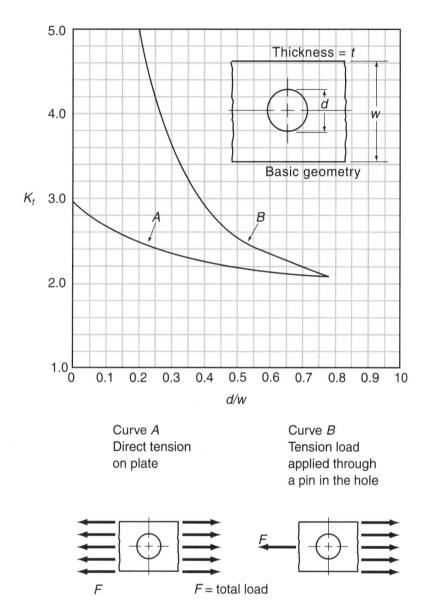

Curve A
Direct tension on plate

Curve B
Tension load applied through a pin in the hole

APPENDIX 6
(e) & (f) Stepped Bar with Radius (Continued)

(e)

(f)

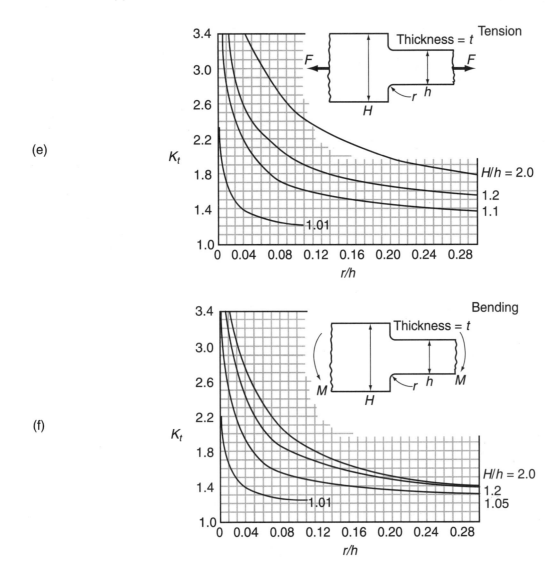

APPENDIX 7
Typical Properties of Some Non-Ferrous Metals

Material (ASTM No.)	Ultimate Strength (ksi) S_u	S_{us}	Tensile Yield (ksi) S_y	Endurance Strength S_n (ksi at no. cycles)	Modulus of Elasticity $E \times 10^6$	Density lb/in.^{3}	Thermal Expansion in./in. °F $\alpha \times 10^{-6}$
Copper Alloys							
Admiralty brass (B111)	53		22	18 at 10^7	15	.308	11.2
Aluminum bronze (B150-2)	120		70	32 at 3×10^8	16	.274	9
Aluminum bronze (B150-1)	100			28 at 8×10^7	15	.274	9.2
Commercial bronze (B134-2)	72	42	62	21 at 15×10^6	17	.318	10.2
Free-cutting brass (B16)	55	32	44	20 at 10^8	14	.307	11.4
Naval brass (B124-3)	70	43	48	15 at 3×10^8	15	.304	11.8
Phosphor bronze (B139C)	80		65	31 at 10^8	16	.318	10.1
Yellow brass (B36-8)	55	36	40	11 at 10^8	15	.306	11.3
Yellow brass (B36-8)	61	40	50		15	.306	11.3
Aluminum Alloys							
3003-H14	22	14	21	9 at 5×10^8	10	.099	12.9
2014-T6	70	42	60	18 at 5×10^8	10.6	.101	12.8
2024-T4	68	41	47	20 at 5×10^8	10.6	.098	13.0
6061-T6	45	30	40	14 at 5×10^8	10	.100	12.7
Magnesium Alloys							
AZ61A-F	45	20	33	17 at 5×10^8	6.5	.065	14.4
AZ80A-T5	50	23	36	16 at 5×10^8	6.5	.065	14.4
Lead and Tin Alloys							
Babbitt (B23-46T-8)	10	$S_{yc} = 3.4$		3.9 at 2×10^7	4.2	.36	13.3
Babbitt (B23-46T-8)	5.4	$S_{yc} = 1.7$				.36	13.3
Tin babbitt (B23-49-1)	9.3	$S_{yc} = 4.4$		3.8 at 2×10^7	7.3	.265	
Miscellaneous							
Titanium (B265, gr. 5) annealed	135		130	42	15	.160	5.8
Titanium (B265, gr. 5) hardened	170		158	61	15	.160	5.8
Zinc (AC41A)	47.6	38		8 at 10^8	2	.24	15.2

APPENDIX 8

Typical Properties of Some Stainless Steels

Material (AISI No.)	Ultimate Strength S_u (ksi)	Tensile Yield S_y (ksi)	Endurance Limit S_n (ksi)	Modulus of Elasticity $E \times 10^6$	Density lb/in.^{3}	Thermal Expansion in./in. °F $\alpha \times 10^{-6}$
301, 1/4 hard	125	75	30	28	.286	9.4
302, annealed	90	37	34	28	.286	9.6
302, 1/4 hard	125	75	70	28	.286	9.6
303, annealed	90	35	35	28	.286	9.6
304, annealed	85	35		28	.286	9.6
316, cold worked	90	60	40	28	.286	8.9
321, annealed	87	35	38	28	.290	9.3
347, annealed	90	40	39	28	.286	9.3
403, 410, heat treated	110	85	58	29	.279	5.7
410, cold worked	100	85	53	29	.279	5.7
416, annealed	75	40	40	29	.278	5.7
430, annealed	75	45	40	29	.277	5.8
431, OQT 1000	150	130		29	.28	6.5
17-7 PH rod	175	155	41	29	.276	5.6

APPENDIX 9
Properties of Cast Iron

Material Designation (ASTM No.)	Grade	Tensile Strength (ksi)	Tensile Strength (MPa)	Yield Strength (ksi)	Yield Strength (MPa)	Ductility % Elongation (in 2 in.)	Modulus of Elasticity (10^6 psi)	Modulus of Elasticity (GPa)
Gray iron A48-83	20	20	138			<1	12	83
	25	25	172			<1	13	90
	30	30	207			<1	15	103
	40	40	276			<1	17	117
	50	50	345			<1	19	131
	60	60	414			<1	20	138
Ductile iron A536-84	60-40-18	60	414	40	276	18	22	152
	80-55-06	80	552	55	379	6	22	152
	100-70-03	100	689	70	483	3	22	152
	120-90-02	120	827	90	621	2	22	152
Malleable iron A47-84(R1989)	32510	50	345	32	221	10	25	172
	35018	53	365	35	241	18	25	172
A220-88	40010	60	414	40	276	10	26	179
	45006	65	448	45	310	6	26	179
	50005	70	483	50	345	5	26	179
	70003	85	586	70	483	3	26	179
	90001	105	724	90	621	1	26	179

APPENDIX 10

Wire Gages and Diameters for Springs

Gage No.	U.S. Steel Wire Gage (in.)	Music Wire Gage (in.)	Brown & Sharp Gage (in.)	Preferred Metric Diameters (mm)
7/0	.4900			13.0
6/0	.4615	.004	.5800	12.0
5/0	.4305	.005	.5165	11.0
4/0	.3938	.006	.4600	10.0
3/0	.3625	.007	.4096	9.0
2/0	.3310	.008	.3648	8.5
0	.3065	.009	.3249	8.0
1	.2830	.010	.2893	7.0
2	.2625	.011	.2576	6.5
3	.2437	.012	.2294	6.0
4	.2253	.013	.2043	5.5
5	.2070	.014	.1819	5.0
6	.1920	.016	.1620	4.8
7	.1770	.018	.1443	4.5
8	.1620	.020	.1285	4.0
9	.1483	.022	.1144	3.8
10	.1350	.024	.1019	3.5
11	.1205	.026	.0907	3.0
12	.1055	.029	.0808	2.8
13	.0915	.031	.0720	2.5
14	.0800	.033	.0641	2.0
15	.0720	.035	.0571	1.8
16	.0625	.037	.0508	1.6
17	.0540	.039	.0453	1.4
18	.0475	.041	.0403	1.2
19	.0410	.043	.0359	1.0
20	.0348	.045	.0320	.90
21	.0327	.047	.0285	.80
22	.0286	.049	.0253	.70
23	.0258	.051	.0226	.65
24	.0230	.055	.0201	.60 or .55
25	.0204	.059	.0179	.50 or .55
26	.0181	.063	.0159	.45
27	.0173	.067	.0142	.45
28	.0162	.071	.0126	.40
29	.0150	.075	.0113	.40
30	.0140	.080	.0100	.35
31	.0132	.085	.00893	.35
32	.0128	.090	.00795	.30 or .35
33	.0118	.095	.00708	.30
34	.0104	.100	.00630	.28
35	.0095	.106	.00501	.25
36	.0090	.112	.00500	.22
37	.0085	.118	.00445	.22
38	.0080	.124	.00396	.20
39	.0075	.130	.00353	.20
40	.0070	.138	.00314	.18

APPENDIX 11

Mechanical Properties of Wire for Coil Springs

Material	Design Stress S_{sd} (ksi) (Light Load)	Min. Tensile S_u (ksi)	Max. "Solid" S_s (Approx. S_{ys})
Oil tempered ASTM A229		$\dfrac{146}{D_w^{.19}}$	$.6 S_u$
Hard drawn ASTM A227	Use .85 times service factors	$\dfrac{140}{D_w^{.19}}$	$.5 S_u$
Music wire ASTM A228		$\dfrac{190}{D_w^{.154}}$	$.5 S_u$ [190 ksi max.]
Cr-Si steel ASTM A401		$\dfrac{202}{D_w^{.107}}$	$.6 S_u$
Stainless steel (Cr-Ni) ASTM A313		$\dfrac{170}{D_w^{.14}}$ $\left[.01 < D_w < .13\right]$ $\dfrac{97}{D_w^{.41}}$ $\left[.13 < D_w < .375\right]$	$.47 S_u$
Beryllium copper	Use .8 times values for A229	160–200	$.5 S_u$
Spring brass	Use .35 times values for A229	$\dfrac{88}{D_w^{.1}}$ $\left[125 \text{ ksi max.}\right]$	$\dfrac{42}{D_w^{.26}}$ $\left[68 \text{ ksi max.}\right]$
Phosphor bronze	Use .5 times values for A229	$\dfrac{106}{D_w^{.08}}$ $\left[145 \text{ ksi max.}\right]$	$.45 S_u$

Compiled from a variety of sources.

APPENDIX 12

Stock Compression Springs: Music Wire and Stainless Steel (Courtesy of Associated Spring Raymond/Barnes Group Inc.)

STOCK COMPRESSION SPRINGS
Music Wire and Stainless Steel

SPEC — Associated Spring Raymond / BARNES Group Inc.

CATALOG NUMBER*	Outside Diameter		Wire Diameter		Free Length L, Approx.		Load, P at L_1		Length, L_1		Solid Height, Approx.		Spring Rate, R	
	in	mm	in	mm	in	mm	lb†	N†	in	mm	in	mm	lb/in†	N/mm†
C0480-081-2000					2.00	50.80			1.460	37.08	1.204	30.58	82.2	14.39
C0480-081-2250					2.25	57.15			1.637	41.58	1.334	33.88	72.3	12.66
C0480-081-2500					2.50	63.50			1.813	46.05	1.465	37.21	64.6	11.31
C0480-081-2750	0.480	12.19	0.081	2.06	2.75	69.85	44.34	197.22	1.990	50.55	1.595	40.51	58.3	10.21
C0480-081-3000					3.00	76.20			2.167	55.04	1.726	43.84	53.2	9.32
C0480-081-3250					3.25	82.55			2.343	59.51	1.857	47.17	48.9	8.56
C0480-081-3500					3.50	88.90			2.520	64.01	1.987	50.47	45.2	7.92
C0600-045-0500					0.50	12.70			0.257	6.53	0.153	3.89	24.7	4.33
C0600-045-0625					0.62	15.75			0.292	7.42	0.179	4.55	18.0	3.15
C0600-045-0750					0.75	19.05			0.330	8.38	0.205	5.21	14.0	2.45
C0600-045-0880					0.88	22.35			0.375	9.53	0.224	5.69	12.0	2.10
C0600-045-1000					1.00	25.40			0.400	10.16	0.251	6.38	10.0	1.751
C0600-045-1250					1.25	31.75			0.500	12.70	0.292	7.42	8.0	1.401
C0600-045-1380					1.38	35.05			0.540	13.72	0.306	7.77	7.1	1.243
C0600-045-1500			0.045	1.14	1.50	38.10	6.00	26.69	0.575	14.61	0.338	8.59	6.5	1.138
C0600-045-1750					1.75	44.45			0.764	19.41	0.366	9.30	6.7	1.173
C0600-045-2000					2.00	50.80			0.866	22.00	0.401	10.19	5.8	1.016
C0600-045-2250					2.25	57.15			0.967	24.56	0.436	11.07	5.2	0.911
C0600-045-2500					2.50	63.50			1.069	27.15	0.471	11.96	4.6	0.805
C0600-045-2750					2.75	69.85			1.170	29.72	0.540	13.72	3.8	0.665
C0600-045-3000					3.00	76.20			1.272	32.31	0.585	14.86	3.5	0.613
C0600-045-3500					3.50	88.90			1.475	37.47	0.659	16.74	3.0	0.525
C0600-049-0620					0.62	15.75			0.320	8.13	0.234	5.94	28.0	4.90
C0600-049-0750					0.75	19.05			0.375	9.53	0.255	6.48	22.4	3.92
C0600-049-1000					1.00	25.40			0.481	12.22	0.297	7.54	16.2	2.84
C0600-049-1250					1.25	31.75			0.587	14.91	0.338	8.59	12.7	2.22
C0600-049-1500					1.50	38.10			0.693	17.60	0.380	9.65	10.4	1.821
C0600-049-1750			0.049	1.24	1.75	44.45	8.40	37.37	0.800	20.32	0.422	10.72	8.8	1.541
C0600-049-2000					2.00	50.80			0.906	23.01	0.463	11.76	7.7	1.348
C0600-049-2250					2.25	57.15			1.012	25.70	0.505	12.83	6.8	1.191
C0600-049-2500					2.50	63.50			1.118	28.40	0.546	13.87	6.1	1.068
C0600-049-2750					2.75	69.85			1.224	31.09	0.588	14.94	5.5	0.963
C0600-049-3000					3.00	76.20			1.330	33.78	0.629	15.98	5.0	0.876
C0600-055-0625					0.62	15.75			0.325	8.26	0.226	5.74	40.0	7.00
C0600-055-0750					0.75	19.05			0.385	9.78	0.251	6.38	33.0	5.78
C0600-055-0880					0.88	22.35			0.450	11.43	0.276	7.01	28.0	4.90
C0600-055-1000					1.00	25.40			0.500	12.70	0.304	7.72	24.0	4.20
C0600-055-1250					1.25	31.75			0.580	14.73	0.369	9.37	18.0	3.15
C0600-055-1500	0.600	15.24	0.055	1.40	1.50	38.10	12.00	53.38	0.700	17.78	0.420	10.67	15.0	2.63
C0600-055-1750					1.75	44.45			0.856	21.74	0.509	12.93	13.0	2.28
C0600-055-2000					2.00	50.80			0.970	24.64	0.562	14.27	11.3	1.979
C0600-055-2250					2.25	57.15			1.083	27.51	0.614	15.60	10.0	1.751
C0600-055-2500					2.50	63.50			1.197	30.40	0.667	16.94	8.9	1.558
C0600-055-2750					2.75	69.85			1.311	33.30	0.701	17.81	8.3	1.453
C0600-055-3000					3.00	76.20			1.424	36.17	0.752	19.10	7.6	1.331
C0600-059-0620					0.62	15.75			0.357	9.07	0.298	7.57	53.7	9.40
C0600-059-0750					0.75	19.05			0.419	10.64	0.329	8.36	42.7	7.48
C0600-059-0880					0.88	22.35			0.481	12.22	0.360	9.14	35.4	6.20
C0600-059-1000					1.00	25.40			0.538	13.67	0.389	9.88	30.6	5.36
C0600-059-1250					1.25	31.75			0.657	16.69	0.449	11.40	23.8	4.17
C0600-059-1500			0.059	1.50	1.50	38.10	14.13	62.85	0.776	19.71	0.510	12.95	19.5	3.41
C0600-059-1750					1.75	44.45			0.895	22.73	0.570	14.48	16.5	2.89
C0600-059-2000					2.00	50.80			1.014	25.76	0.630	16.00	14.3	2.50
C0600-059-2250					2.25	57.15			1.133	28.78	0.690	17.53	12.6	2.21
C0600-059-2500					2.50	63.50			1.252	31.80	0.750	19.05	11.3	1.979
C0600-059-2750					2.75	69.85			1.371	34.82	0.811	20.60	10.2	1.786
C0600-059-3000					3.00	76.20			1.490	37.85	0.871	22.12	9.4	1.646
C0600-063-0625					0.62	15.75			0.335	8.51	0.280	7.11	62.0	10.86
C0600-063-0750					0.75	19.05			0.400	10.16	0.314	7.98	51.0	8.93
C0600-063-0880					0.88	22.35			0.465	11.81	0.343	8.71	44.0	7.70
C0600-063-1000			0.063	1.60	1.00	25.40	18.00	80.07	0.525	13.34	0.378	9.60	38.0	6.65
C0600-063-1250					1.25	31.75			0.670	17.02	0.435	11.05	31.0	5.43
C0600-063-1500					1.50	38.10			0.750	19.05	0.526	13.36	24.0	4.20
C0600-063-1750					1.75	44.45			0.935	23.75	0.632	16.05	20.8	3.64

†For stainless steel, multiply values by 0.833.
*When inquiring or ordering, use letter "M" or letter "S" as suffix on catalog numbers to designate music wire or stainless-steel wire, respectively.

APPENDIX 12
(Continued)

STOCK COMPRESSION SPRINGS
Music Wire and Stainless Steel

SPEC — Associated Spring Raymond / BARNES GROUP INC.

CATALOG NUMBER*	Outside Diameter		Wire Diameter		Free Length L, Approx.		Load, P at L_1		Length, L_1		Solid Height, Approx.		Spring Rate, R	
	in	mm	in	mm	in	mm	lb†	N†	in	mm	in	mm	lb/in†	N/mm†
C0720-065-1750					1.75	44.45			0.865	21.97	0.532	13.51	17.6	3.08
C0720-065-2000					2.00	50.80			0.978	24.84	0.584	14.83	15.2	2.66
C0720-065-2250			0.065	1.65	2.25	57.15	15.57	69.26	1.092	27.74	0.636	16.15	13.4	2.35
C0720-065-2500					2.50	63.50			1.205	30.61	0.689	17.50	12.0	2.10
C0720-065-2750					2.75	69.85			1.319	33.50	0.714	18.14	10.9	1.909
C0720-065-3000					3.00	76.20			1.432	36.37	0.793	20.14	9.9	1.733
C0720-067-0750					0.75	19.05			0.390	9.91	0.279	7.09	50.0	8.76
C0720-067-0880					0.88	22.35			0.425	10.80	0.315	8.00	40.0	7.00
C0720-067-1000					1.00	25.40			0.450	11.43	0.354	8.99	33.0	5.78
C0720-067-1250					1.25	31.75			0.580	14.73	0.403	10.24	27.0	4.73
C0720-067-1500					1.50	38.10			0.680	17.27	0.464	11.79	22.0	3.85
C0720-067-1750			0.067	1.70	1.75	44.45	18.00	80.07	0.750	19.05	0.538	13.67	18.0	3.15
C0720-067-2000					2.00	50.80			0.850	21.59	0.603	15.32	15.5	2.71
C0720-067-2250					2.25	57.15			1.113	28.27	0.669	16.99	14.9	2.61
C0720-067-2500					2.50	63.50			1.228	31.19	0.724	18.39	13.3	2.33
C0720-067-3000					3.00	76.20			1.460	37.08	0.835	21.21	11.0	1.926
C0720-067-3250					3.25	82.55			1.575	40.01	0.850	21.59	10.7	1.874
C0720-067-3500					3.50	88.90			1.691	42.95	0.902	22.91	10.0	1.751
C0720-072-0750					0.75	19.05			0.438	11.13	0.368	9.35	67.3	11.78
C0720-072-0880					0.88	22.35			0.435	11.05	0.365	9.27	48.0	8.41
C0720-072-1000					1.00	25.40			0.500	12.70	0.397	10.08	42.0	7.35
C0720-072-1250					1.25	31.75			0.650	16.51	0.448	11.38	35.0	6.13
C0720-072-1500					1.50	38.10			0.700	17.78	0.554	14.07	26.0	4.55
C0720-072-1750					1.75	44.45			0.810	20.57	0.628	15.95	22.0	3.85
C0720-072-2000			0.072	1.83	2.00	50.80	21.00	93.41	0.950	24.13	0.677	17.20	20.0	3.50
C0720-072-2250					2.25	57.15			1.166	29.62	0.751	19.08	19.1	3.34
C0720-072-2500					2.50	63.50			1.287	32.69	0.815	20.70	17.1	2.99
C0720-072-2750					2.75	69.85			1.409	35.79	0.878	22.30	15.4	2.70
C0720-072-3000					3.00	76.20			1.530	38.86	0.942	23.93	14.1	2.47
C0720-072-3250	0.720	18.29			3.25	82.55			1.629	41.38	1.005	25.53	13.0	2.28
C0720-072-3500					3.50	88.90			1.749	44.42	1.069	27.15	12.0	2.10
C0720-072-4000					4.00	101.60			2.015	51.18	1.182	30.02	10.6	1.856
C0720-081-0750					0.75	19.05			0.472	11.99	0.429	10.90	103.5	18.12
C0720-081-0880					0.88	22.35			0.541	13.74	0.470	11.94	84.7	14.83
C0720-081-1000					1.00	25.40			0.604	15.34	0.508	12.90	72.6	12.71
C0720-081-1250					1.25	31.75			0.736	18.69	0.587	14.91	55.9	9.79
C0720-081-1500					1.50	38.10			0.868	22.05	0.665	16.89	45.5	7.97
C0720-081-1750					1.75	44.45			1.000	25.40	0.744	18.90	38.3	6.71
C0720-081-2000			0.081	2.06	2.00	50.80	28.72	127.75	1.132	28.75	0.823	20.90	33.1	5.80
C0720-081-2250					2.25	57.15			1.264	32.11	0.902	22.91	29.1	5.10
C0720-081-2500					2.50	63.50			1.396	35.46	0.981	24.92	26.0	4.55
C0720-081-2750					2.75	69.85			1.528	38.81	1.060	26.92	23.5	4.12
C0720-081-3000					3.00	76.20			1.660	42.16	1.139	28.93	21.4	3.75
C0720-081-3500					3.50	88.90			1.924	48.87	1.297	32.94	18.2	3.19
C0720-081-4000					4.00	101.60			2.186	55.52	1.455	36.96	15.9	2.78
C0720-085-0750					0.75	19.05			0.508	12.90	0.434	11.02	139.0	24.34
C0720-085-0880					0.88	22.35			0.584	14.83	0.474	12.04	113.5	19.87
C0720-085-1000					1.00	25.40			0.654	16.61	0.512	13.00	97.1	17.00
C0720-085-1250					1.25	31.75			0.800	20.32	0.589	14.96	74.6	13.06
C0720-085-1500					1.50	38.10			0.946	24.03	0.666	16.92	60.6	10.61
C0720-085-1750					1.75	44.45			1.092	27.74	0.743	18.87	51.0	8.93
C0720-085-2000			0.085	2.16	2.00	50.80	33.59	149.41	1.237	31.42	0.821	20.85	44.0	7.70
C0720-085-2250					2.25	57.15			1.383	35.13	0.898	22.81	38.7	6.78
C0720-085-2500					2.50	63.50			1.529	38.84	0.975	24.77	34.6	6.06
C0720-085-2750					2.75	69.85			1.675	42.55	1.052	26.72	31.2	5.46
C0720-085-3000					3.00	76.20			1.821	46.25	1.130	28.70	28.5	4.99
C0720-085-3500					3.50	88.90			2.112	53.64	1.284	32.61	24.2	4.24
C0720-085-4000					4.00	101.60			2.404	61.06	1.439	36.55	21.0	3.68
C0720-096-0750					0.75	19.05			0.560	14.22	0.490	12.45	238.6	41.78
C0720-096-0880					0.88	22.35			0.646	16.41	0.537	13.64	193.6	33.90
C0720-096-1000			0.096	2.44	1.00	25.40	45.33	201.63	0.725	18.42	0.581	14.76	164.8	28.86
C0720-096-1250					1.25	31.75			0.890	22.61	0.671	17.04	125.9	22.05
C0720-096-1500					1.50	38.10			1.055	26.80	0.762	19.35	101.8	17.83
C0720-096-1750					1.75	44.45			1.220	30.99	0.852	21.64	85.5	14.97
C0720-096-2000					2.00	50.80			1.385	35.18	0.943	23.95	73.7	12.91

†For stainless steel, multiply values by 0.833.
*When inquiring or ordering, use letter "M" or letter "S" as suffix on catalog numbers to designate music wire or stainless-steel wire, respectively.

APPENDIX 12
(Concluded)

STOCK COMPRESSION SPRINGS
Music Wire and Stainless Steel

SPEC Associated Spring Raymond / BARNES GROUP INC

CATALOG NUMBER*	Outside Diameter in	Outside Diameter mm	Wire Diameter in	Wire Diameter mm	Free Length L, Approx. in	Free Length L, Approx. mm	Load, P at L_1 lb†	Load, P at L_1 N†	Length, L_1 in	Length, L_1 mm	Solid Height, Approx. in	Solid Height, Approx. mm	Spring Rate, R lb/in†	Spring Rate, R N/mm†
C0720-096-2250					2.25	57.15			1.549	39.34	1.034	26.26	64.7	11.33
C0720-096-2500					2.50	63.50			1.714	43.54	1.124	28.55	57.7	10.10
C0720-096-2750			0.096	2.44	2.75	69.85	45.33	201.63	1.879	47.73	1.215	30.86	52.1	9.12
C0720-096-3000					3.00	76.20			2.044	51.92	1.305	33.15	47.4	8.30
C0720-096-3500					3.50	88.90			2.374	60.30	1.486	37.74	40.3	7.06
C0720-096-4000					4.00	101.60			2.704	68.68	1.668	42.37	35.0	6.13
C0720-105-0750					0.75	19.05			0.585	14.86	0.537	13.64	354.7	62.11
C0720-105-0880					0.88	22.35			0.675	17.15	0.591	15.01	285.9	50.06
C0720-105-1000					1.00	25.40			0.758	19.25	0.640	16.26	242.5	42.46
C0720-105-1250					1.25	31.75			0.932	23.67	0.743	18.87	184.2	32.25
C0720-105-1500					1.50	38.10			1.105	28.07	0.846	21.49	148.5	26.00
C0720-105-1750					1.75	44.45			1.279	32.49	0.949	24.10	124.4	21.78
C0720-105-2000	0.720	18.29	0.105	2.67	2.00	50.80	58.60	260.65	1.452	36.88	1.052	26.72	107.0	18.74
C0720-105-2250					2.25	57.15			1.626	41.30	1.155	29.34	93.9	16.44
C0720-105-2500					2.50	63.50			1.799	45.69	1.258	31.95	83.6	14.64
C0720-105-2750					2.75	69.85			1.973	50.11	1.361	34.57	75.4	13.20
C0720-105-3000					3.00	76.20			2.146	54.51	1.464	37.19	68.7	12.03
C0720-105-3500					3.50	88.90			2.493	63.32	1.670	42.42	58.2	10.19
C0720-105-4000					4.00	101.60			2.840	72.14	1.876	47.65	50.5	8.84
C0720-112-0750					0.75	19.05			0.602	15.29	0.576	14.63	468.8	82.09
C0720-112-0880					0.88	22.35			0.695	17.65	0.636	16.15	375.9	65.82
C0720-112-1000					1.00	25.40			0.781	19.84	0.691	17.55	317.7	55.63
C0720-112-1250					1.25	31.75			0.960	24.38	0.805	20.45	240.3	42.08
C0720-112-1500					1.50	38.10			1.140	28.96	0.919	23.34	193.2	33.83
C0720-112-1750					1.75	44.45			1.319	33.50	1.034	26.26	161.6	28.30
C0720-112-2000			0.112	2.84	2.00	50.80	69.59	309.54	1.499	38.07	1.148	29.16	138.8	24.30
C0720-112-2250					2.25	57.15			1.678	42.62	1.262	32.05	121.7	21.31
C0720-112-2500					2.50	63.50			1.858	47.19	1.376	34.95	108.3	18.96
C0720-112-2750					2.75	69.85			2.037	51.74	1.491	37.87	97.6	17.09
C0720-112-3000					3.00	76.20			2.216	56.29	1.605	40.77	88.8	15.55
C0720-112-3500					3.50	88.90			2.575	65.41	1.834	46.58	75.3	13.19
C0720-112-4000					4.00	101.60			2.934	74.52	2.062	52.37	65.3	11.43
C0850-068-0750					0.75	19.05			0.445	11.30	0.292	7.42	35.9	6.29
C0850-068-0875					0.88	22.35			0.508	12.90	0.317	8.05	29.8	5.22
C0850-068-1000					1.00	25.40			0.571	14.50	0.342	8.69	25.5	4.47
C0850-068-1250					1.25	31.75			0.696	17.68	0.391	9.93	19.8	3.47
C0850-068-1500					1.50	38.10			0.822	20.88	0.441	11.20	16.1	2.82
C0850-068-1750			0.068	1.73	1.75	44.45	10.94	48.66	0.948	24.08	0.490	12.45	13.6	2.38
C0850-068-2000					2.00	50.80			1.074	27.28	0.540	13.72	11.8	2.07
C0850-068-2250					2.25	57.15			1.200	30.48	0.590	14.99	10.4	1.821
C0850-068-2500					2.50	63.50			1.325	33.66	0.639	16.23	9.3	1.628
C0850-068-3000					3.00	76.20			1.577	40.06	0.739	18.77	7.7	1.348
C0850-068-3500					3.50	88.90			1.828	46.43	0.838	21.29	6.5	1.138
C0850-068-4000					4.00	101.60			2.080	52.83	0.937	23.80	5.7	0.998
C0850-074-0875					0.88	22.35			0.530	13.46	0.355	9.02	40.1	7.02
C0850-074-1000					1.00	25.40			0.595	15.11	0.384	9.75	34.2	5.99
C0850-074-1250					1.25	31.75			0.727	18.47	0.443	11.25	26.5	4.64
C0850-074-1500	0.850	21.59	0.074	1.88	1.50	38.10	13.86	61.65	0.858	21.79	0.501	12.73	21.6	3.78
C0850-074-2000					2.00	50.80			1.120	28.45	0.618	15.70	15.8	2.77
C0850-074-2500					2.50	63.50			1.383	35.13	0.735	18.67	12.4	2.17
C0850-074-3000					3.00	76.20			1.645	41.78	0.852	21.64	10.2	1.786
C0850-074-3500					3.50	88.90			1.908	48.46	0.969	24.61	8.7	1.523
C0850-081-0750					0.75	19.05			0.486	12.34	0.365	9.27	67.7	11.85
C0850-081-0875					0.88	22.35			0.555	14.10	0.400	10.16	55.8	9.77
C0850-081-1000					1.00	25.40			0.624	15.85	0.435	11.05	47.5	8.32
C0850-081-1250					1.25	31.75			0.762	19.35	0.504	12.80	36.6	6.41
C0850-081-1500					1.50	38.10			0.900	22.86	0.573	14.55	29.7	5.20
C0850-081-1750			0.081	2.06	1.75	44.45	17.83	79.31	1.038	26.37	0.642	16.31	25.1	4.40
C0850-081-2000					2.00	50.80			1.176	29.87	0.712	18.08	21.6	3.78
C0850-081-2250					2.25	57.15			1.314	33.38	0.781	19.84	19.1	3.34
C0850-081-2500					2.50	63.50			1.452	36.88	0.850	21.59	17.0	2.98
C0850-081-2750					2.75	69.85			1.590	40.39	0.920	23.37	15.4	2.70
C0850-081-3000					3.00	76.20			1.728	43.89	0.989	25.12	14.0	2.45
C0850-081-3500					3.50	88.90			2.004	50.90	1.127	28.63	11.9	2.08

†For stainless steel, multiply values by 0.833.

APPENDIX 13

Values of Limiting Wear-Load Factor K_g

Combinations of Materials (BHN) and Life	S_{nsurf} (ksi)	K_g 14½°	K_g 20°
Both gears steel:			
Sum of BHN = 300, 10^6 cycles		63	86
Sum of BHN = 300, 10^7 cycles		40	54
Sum of BHN = 300, 4×10^7 cycles or more	50	30	41
Sum of BHN = 350	60	43	58
Sum of BHN = 400, 10^6 cycles		119	162
Sum of BHN = 400, 10^7 cycles		75	102
Sum of BHN = 400, 4×10^7 cycles or more	70	58	79
Sum of BHN = 450	80	76	103
Sum of BHN = 500	90	96	131
Sum of BHN = 550	100	119	162
Sum of BHN = 600, 10^6 cycles		292	400
Sum of BHN = 600, 10^7 cycles		185	252
Sum of BHN = 600, 4×10^7 cycles or more	110	144	196
Sum of BHN = 650	120	171	233
Sum of BHN = 700	130	196	270
Sum of BHN = 750	140	233	318
Sum of BHN = 800	150	268	366
Steel (450) and same	170	344	470
Steel (500), induction hardened, and same, 10^7 cycles		880	1190
Steel (500), induction hardened, and same, 10^8 cycles		670	920
Steel (500), induction hardened, and same, 10^{10} cycles		405	555
Steel (600), carburized case hardened, and same, 10^7 cycles		1230	1680
Steel (600), carburized case hardened, and same, 10^8 cycles		940	1280
Steel (600), carburized case hardened, and same, 10^{10} cycles		550	750
Steel (150) and cast iron	50	44	60
Steel (250 and over) and chilled phosphor bronze	83	128	175
Cast iron and same, 10^6 cycles		376	515
Cast iron and same, 10^7 cycles		212	290
Cast iron and phosphor bronze	83	170	234

Note: Specified BHNs are minimums. Values are for indefinite life unless otherwise indicated. Straight-line interpolations on the sum of BHNs are permissible when the difference in BHNs is less than 100 points.
Source: Compiled from a variety of sources.

APPENDIX 14
Conversion Factors (as a Product of Unity)

Length	$1 = \dfrac{25.4 \text{ mm}}{\text{inch}}$	$= \dfrac{\text{inch}}{25.4 \text{ mm}}$
	$1 = \dfrac{.3048 \text{ m}}{\text{ft}}$	$= \dfrac{3.281 \text{ ft}}{\text{m}}$
Area	$1 = \dfrac{645.16 \text{ mm}^2}{\text{in.}^2}$	$= \dfrac{\text{in.}^2}{645.16 \text{ mm}^2}$
	$1 = \dfrac{\text{m}^2}{10.76 \text{ ft}^2}$	$= \dfrac{10.76 \text{ ft}^2}{\text{m}^2}$
	$1 = \dfrac{144 \text{ in.}^2}{\text{ft}^2}$	$= \dfrac{\text{ft}^2}{144 \text{ in.}^2}$
Angle	$1 = \dfrac{57.3°}{\text{radian}}$	$= \dfrac{\text{radian}}{57.3°}$
Volume	$1 = \dfrac{16,387 \text{ mm}^3}{\text{in.}^3}$	$= \dfrac{\text{in.}^3}{16,387 \text{ mm}^3}$
	$1 = \dfrac{1,728 \text{ in.}^3}{\text{ft}^3}$	$= \dfrac{\text{ft}^3}{1,728 \text{ in.}^3}$
	$1 = \dfrac{1 \times 10^9 \text{ mm}^3}{\text{m}^3}$	$= \dfrac{\text{m}^3}{1 \times 10^9 \text{ mm}^3}$
	$1 = \dfrac{1 \text{ gal}}{231 \text{ in.}^3}$	$= \dfrac{231 \text{ in.}^3}{\text{gal}}$
Velocity	$1 = \dfrac{39.4 \text{ in./sec}}{\text{m/sec}}$	$= \dfrac{\text{m/sec}}{39.4 \text{ in./sec}}$
	$1 = \dfrac{3.28 \text{ ft/sec}}{\text{m/sec}}$	$= \dfrac{\text{m/sec}}{3.28 \text{ ft/sec}}$
	$1 = \dfrac{9.55 \text{ rpm}}{\text{rad/sec}}$	$= \dfrac{\text{rad/sec}}{9.55 \text{ rpm}}$
Stress or pressure	$1 = \dfrac{6.895 \text{ kPa}}{\text{psi}}$	$= \dfrac{\text{psi}}{6.895 \text{ kPa}}$
	$1 = \dfrac{6,895 \text{ Pa}}{\text{psi}}$	$= \dfrac{\text{psi}}{6,895 \text{ Pa}}$
	$1 = \dfrac{145 \text{ psi}}{\text{MPa}}$	$= \dfrac{\text{MPa}}{145 \text{ psi}}$

APPENDIX 14
Conversion Factors (as a Product of Unity) (Continued)

Bending moment; torque	$1 = \dfrac{Nm}{8.851 \text{ in.-lb}}$	$= \dfrac{8.851 \text{ in.-lb}}{Nm}$
	$1 = \dfrac{1.356 \text{ Nm}}{\text{ft-lb}}$	$= \dfrac{\text{ft-lb}}{1.356 \text{ Nm}}$
Force	$1 = \dfrac{4.448 \text{ N}}{\text{lb}}$	$= \dfrac{\text{lb}}{4.448 \text{ N}}$
Mass	$1 = \dfrac{\text{kg}}{2.205 \text{ lbm}}$	$= \dfrac{2.205 \text{ lbm}}{\text{kg}}$
Mass to force	$1 = \dfrac{32.2 \text{ lbm}}{\text{lb}}$	$= \dfrac{\text{lb}}{32.2 \text{ lbm}}$
	$1 = \dfrac{9.8 \text{ kg}}{N}$	$= \dfrac{N}{9.8 \text{ kg}}$
Density	$1 = \dfrac{25.68 \text{ gm/cc}}{\text{lb/in.}^3}$	$= \dfrac{\text{lb/in.}^3}{25.68 \text{ gm/cc}}$
	$1 = \dfrac{1{,}728 \text{ lb/in.}^3}{\text{lb/ft}^3}$	$= \dfrac{\text{lb/ft}^3}{1{,}728 \text{ lb/in.}^3}$
Moment of inertia	$1 = \dfrac{4.162 \times 10^5 \text{ mm}^4}{\text{in.}^4}$	$= \dfrac{\text{in.}^4}{4.162 \times 10^5 \text{ mm}^4}$
	$1 = \dfrac{2.403 \times 10^6 \text{ in.}^4}{\text{m}^4}$	$= \dfrac{\text{m}^4}{2.403 \times 10^6 \text{ in.}^4}$
Section modulus	$1 = \dfrac{1.639 \times 10^4 \text{ mm}^3}{\text{in.}^3}$	$= \dfrac{\text{in.}^3}{1.639 \times 10^4 \text{ mm}^3}$
Acceleration	$1 = \dfrac{39.4 \text{ in./sec}}{\text{m/sec}}$	$= \dfrac{\text{m/sec}}{39.4 \text{ in./sec}}$
	$1 = \dfrac{3.28 \text{ ft/sec}}{\text{m/sec}}$	$= \dfrac{\text{m/sec}}{3.28 \text{ ft/sec}}$
Energy	$1 = \dfrac{1.356 \text{ J}}{\text{ft-lb}}$	$= \dfrac{\text{ft-lb}}{1.356 \text{ J}}$

APPENDIX 14
Conversion Factors (as a Product of Unity) (Concluded)

Power $\quad\quad 1 = \dfrac{1.356 \text{ W}}{\text{ft-lb/sec}} \quad\quad = \dfrac{\text{ft-lb/sec}}{1.356 \text{ W}}$

$\quad\quad\quad\quad\quad 1 = \dfrac{745.7 \text{ W}}{\text{hp}} \quad\quad = \dfrac{\text{hp}}{745.7 \text{ W}}$

$\quad\quad\quad\quad\quad 1 = \dfrac{550 \text{ ft-lb/sec}}{\text{hp}} \quad\quad = \dfrac{\text{hp}}{550 \text{ ft-lb/sec}}$

$\quad\quad\quad\quad\quad 1 = \dfrac{33,000 \text{ ft-lb/min}}{\text{hp}} \quad\quad = \dfrac{\text{hp}}{33,000 \text{ ft-lb/min}}$

Spring rate $\quad 1 = \dfrac{175 \text{ N/m}}{\text{lb/in.}} \quad\quad = \dfrac{\text{lb/in.}}{175 \text{ N/m}}$

Selected Formulas

(1.1)	$\delta = \alpha L \Delta T$		Thermal expansion
(2.6)	$\text{hp} = \dfrac{Tn}{63,000}$		Power (T, in.-lb; n, rpm)
(3.1)	$S = \dfrac{F}{A}$		Stress/Pressure
(3.2)	$\delta = \dfrac{FL}{AE}$		Axial deflection
(3.3)	$S = \dfrac{Mc}{I} = \dfrac{M}{Z}$		Bending stress
(3.5)	$S_s = \dfrac{Tc}{J}$		Torsional stress
(3.6)	$S_s = \dfrac{T}{Z'}$		Torsional stress
(3.7)	$\theta = \dfrac{TL}{JG}$		Angular deflection
(3.13)	$F_{allowable} = \dfrac{P_c}{N} = \dfrac{\pi^2 EI}{N(L_e^2)}$		Euler column
(3.15)	$F_{allowable} = \dfrac{P_c}{N} = \dfrac{AS_y}{N}\left[1 - \dfrac{S_y\left(\dfrac{L_e}{r}\right)^2}{4\pi^2 E}\right]$		Johnson column
(4.2)	$S = \pm \dfrac{F}{A} \pm \dfrac{M}{Z}$		Combined normal stress
(4.4)	$\sigma = \dfrac{S}{2} \pm \left[S_s^2 + \left(\dfrac{S}{2}\right)^2\right]^{1/2}$		Principal stress theory

(4.5) $\tau = \left[S_s^2 + \left(\dfrac{S}{2}\right)^2\right]^{1/2}$ — Shear stress theory

(5.2) $S_{mean} = \dfrac{S_{max} + S_{min}}{2}$ — Mean stress

(5.3) $S_{alt} = \dfrac{S_{max} - S_{min}}{2}$ — Alternating stress

(5.5) $\dfrac{1}{N} = \dfrac{S_m}{S_y} + \dfrac{S_a}{S_n}$ — Soderberg equation

(5.7) $\dfrac{1}{N} = \left(\dfrac{S_m}{S_u}\right)^2 + \dfrac{S_a}{S_n}$ — Gerber theory

(5.8) $\dfrac{1}{N} = \dfrac{S_m}{S_u} + \dfrac{S_a}{S_n}$ — Modified Goodman

(5.9) $\dfrac{1}{N} = \dfrac{S_m}{S_y} + \dfrac{K_t S_a}{S_n}$ — Soderburg with concentration factor

(5.10) $S_n = S_n' \left(\dfrac{10^6}{N_c}\right)^{.09}$ — Endurance limit life adjustment

(6.1) $T \approx CDF_i$ — Fastener torquing

(6.2) $F = \dfrac{\delta E A}{L}$ — Fastener force

(6.5) $\Delta T = \dfrac{\delta}{\alpha L}$ — Thermal expansion

(6.6) $k_c = \dfrac{A_c E_c}{L_c}$ or $k_b = \dfrac{A_b E_b}{L_b}$ — Stiffness

(6.9) $F_t = F_i + \left(\dfrac{k_b}{k_b + k_c}\right) F_e$ — Force in bolt

(6.10) $F_c = F_i - \left(\dfrac{k_c}{k_b + k_c}\right) F_e$ — Force in flange

Selected Formulas

(7.3)	$U = KE = \tfrac{1}{2}mv^2$	Kinetic energy
(7.4)	$U = PE = mgh$	Potential energy
(7.5)	$\delta = \dfrac{W}{k} + \dfrac{W}{k}\left(1 + \dfrac{2hk}{W}\right)^{1/2}$	Impact-deflection
(7.6)	$k = \dfrac{F}{\delta} = \dfrac{AE}{L}$	Stiffness
(7.10)	$S = \dfrac{W}{A} + \dfrac{W}{A}\left(1 + \dfrac{2hEA}{LW}\right)^{1/2}$	Impact-stress
(7.12)	$W\left(\dfrac{v^2}{2g} + \delta\right) = \dfrac{F\delta}{2} = \dfrac{k\delta^2}{2}$	Impact
(8.10)	$S_s = K\dfrac{8FD_m}{\pi D_w^3}$	Spring stress
(8.11)	$(LF)\dfrac{Q}{D_w^x} \geq K\dfrac{8FD_m}{\pi D_w^3}$	
(8.12)	$\delta = \dfrac{8FD_m^3 N_a}{GD_w^4} = \dfrac{8FC^3 N_a}{GD_w}$	Spring deflection
(8.13)	$k = \dfrac{GD_w^4}{8D_m^3 N_a}$	Spring rate
(11.3)	$P_c = \dfrac{\pi D_p}{N}$	Circular pitch
(11.4)	$P_d = \dfrac{N_g}{D_g} = \dfrac{N_p}{D_p}$	Diametral pitch
(12.3)	$F_t = \dfrac{2T}{D_p}$	Transmitted force
(12.5)	$V_m = \dfrac{\pi D_p n}{12}$	Surface speed
(12.6)	$F_t = \dfrac{33{,}000 \text{ hp}}{V_m}$	Transmitted force

(12.9)	$F_s = \dfrac{S_n b Y}{P_d}$	Force capability
(12.10)	$F_d = \dfrac{600 + V_m}{600} F_t$	Commercial gear dynamic force
(12.11)	$F_d = \dfrac{1,200 + V_m}{1,200} F_t$	Carefully cut gear dynamic force
(12.12)	$F_d = \dfrac{78 + V_m^{1/2}}{78} F_t$	Precision gear dynamic force
(12.13)	$F_d = \dfrac{50 + V_m^{1/2}}{50} F_t$	Hobbed or shaved gear dynamic force
(12.16)	$F_w = D_p\, b\, Q\, K_g$	Gear wear
(12.17)	$Q = \dfrac{2 D_g}{D_g + D_p} = \dfrac{2 N_g}{N_g + N_p} = \dfrac{2 m_g}{m_g + 1}$	
(13.16)	$e = \dfrac{\tan \lambda (1 - f \tan \lambda)}{f + \tan \lambda}$	Worm efficiency
(13.18)	$Q = (1 - e)\, \text{hp}_{in}\, (2,544)$ Btu/hr	Heat generated
(14.6)	$L = 2C + 1.57(D_1 + D_2) + \dfrac{(D_2 - D_1)^2}{4C}$	Belt length
(14.7)	$C = \dfrac{B + \sqrt{B^2 - 32(D_2 - D_1)^2}}{16}$	Pulley spacing
(14.8)	$B = 4L - 6.28(D_2 + D_1)$	
(14.9)	$\theta = 180° \pm 2 \sin^{-1}\left(\dfrac{D_2 - D_1}{2C}\right)$	Wrap angle
(15.2)	$L = \dfrac{2T}{S_s b D}$	Key shear
(15.4)	$L = \dfrac{4T}{S_c t D}$	Key compression

Selected Formulas **501**

(16.2) $\quad T_f = fN\left(\dfrac{r_o + r_i}{2}\right)$ — Clutch/Brake torque

(16.3) $\quad P_f = \dfrac{T_f n}{63,000}$ — Frictional power (T, in.-lb; n, rpm)

(16.7) $\quad P_f = \dfrac{T_f n}{63,000} = \dfrac{U_f}{550 t}$ — Frictional energy (U, ft-lb; t, sec)

(17.1) $\quad T = \dfrac{63,000\ \text{hp}}{n}$ — Torque (T, in.-lb; n, rpm)

(17.3) $\quad D = \sqrt[3]{\dfrac{16 T}{\pi S_s}}$ — Shaft sizing

(17.9) $\quad \tau = \dfrac{5.1}{D^3}\left(T^2 + M^2\right)^{1/2}$ — Shaft shear stress

(17.10) $\quad \sigma = \dfrac{5.1}{D^3}\left[M + \left(T^2 + M^2\right)^{1/2}\right]$ — Shaft normal stress

(17.13) $\quad N_c = \dfrac{30}{\pi}\left(\dfrac{g_o \Sigma F \delta}{\Sigma F \delta^2}\right)^{1/2}$ — Natural frequency

(18.1) $\quad \tan \lambda = \dfrac{L}{\pi D_p}$ — Lead angle

(18.2) $\quad T_{up} = \dfrac{F D_p}{2}\left(\dfrac{L + \pi f D_p}{\pi D_p - f L}\right)$ — Power screw torque up

(18.3) $\quad T_{down} = \dfrac{F D_p}{2}\left(\dfrac{\pi f D_p - L}{\pi D_p + f L}\right)$ — Power screw torque down

(18.8) $\quad e = \dfrac{T'}{T} = \dfrac{\tan \lambda (1 - f \tan \lambda)}{\tan \lambda + f}$ — Power screw efficiency

(18.11) $\quad T_{up} = \dfrac{FL}{2\pi e}$ — Ball screw torque

(18.13) $\quad T_{bd} = \dfrac{FLe}{2\pi}$ — Ball screw back driving

(20.3) $$C_d = P_d\left(\frac{L_d}{L_c}\right)^{1/k}$$ Bearing load correction

(20.4) $$L_d = \left(\frac{C_d}{P_d}\right)^k L_c$$ Bearing life correction

(20.6) $$P = VXR + YF_t$$ Bearing thrust equivalent

Answers to Selected Problems

CHAPTER 2
Force, Work, and Power

1. a. $F = 1{,}000$ lb
   b. $W = 150{,}000$ ft-lb
   c. $P = 20.3$ kw
   d. $T = 9{,}000$ in.-lb
   e. $n = 191$ rpm
   f. $P = 27.3$ hp
3. a. $F = 1{,}497$ lb
   b. $d = 32$ ft
   c. $P = 43.5$ hp
   d. 60% more
5. a. $P = 26.7$ hp
   b. $P = 67$ hp
   c. $P = 93.7$ hp
7. a. $P = 90$ psi
   b. $W = 90{,}000$ ft-lb
   c. $P = 16.4$ hp
   d. $P = 8.7$ hp
   e. $P = 95$ psi
   f. Same as part b
9. a. $I = 36$ in.^{4} $Z = 12$ in.^{3}
   b. $I = 4$ in.^{4} $Z = 2.65$ in.^{3}
   c. $I = 12.5$ in.^{4} $Z = 6.26$ in.^{3}
11. a. $F = 33{,}930$ lb
    $T = 271{,}400$ in.-lb
    At wheel: $F = 22{,}600$ lb
    Friction possible: $F = 24{,}000$ lb
    No, wheel will not slip.
    b. $P = 3{,}000$ hp
13. $L = 4.2$ in.
15. $T = 30{,}140$ in.-lb

CHAPTER 3
Stress and Deformation

1. $D = .30$ in. Use ⅜ if not threaded.
   $\delta = .009$ in.
3. $x = 20$ in.
5. a. $S = 48$ lb/in.^{2} Insignificant
      $b = .0009$ in.   Also insignificant
   b. $S = 9{,}040$ lb/in.^{2}
      $\delta = .136$ in.
   c. $S_t = 9{,}090$ lb/in.^{2}
      $\delta_t = .137$ in.
   d. No, both affect results less than ½%.
7. $h = 10.08$
   Note that $\dfrac{10.08}{8} = 1.26$
   which is $\sqrt[3]{2}$
9. $T = 137$ in.-lb
   $\theta = 1.2$ radians or 68°
10. $P = 26$ hp
    $\theta = .067$ radians or 3.8°
11. $D = 1.2$ in.
    Savings is 2 lb-ft
17. $\delta = 3$ in.

503

CHAPTER 4
Combined Stress and Failure Theories

1. $S = 960$ lb/in.^{2}
3. $S = 1{,}270$ lb/in.^{2}
4. $\sigma = 7{,}860$ lb/in.^{2}
   $\tau = 4{,}030$ lb/in.^{2}
5. $S = 8{,}610$ lb/in.^{2}
7. $\sigma = 8{,}800$ lb/in.^{2}
9. $\tau = 3{,}030$ lb/in.^{2} Using maximum shear stress method
11. $\sigma = 24.2$ MPa   Yes, this is acceptable.
13. $S_{actual} = 48{,}300$ lb/in.^{2};
    $S_{allowable\ needed} \geq 193{,}200$ lb/in.^{2}

CHAPTER 5
Repeated Loading

1. $N = 3.7$
3. $N = 7.4$
5. $N = 2.2$ Yes, it meets the criteria.
7. $D = 9.8$ mm
9. $N = 1.36$ No, it does not meet criteria.
11. $D = 1.03$ in.
13. $N = 1.6$
14. $N = 1.19$ Very marginal
15. Appendix 6(f)   Use $h = 1.5$
17. $D = .57$ in.   Use ⅝ inch shaft
19. $D = .522$ in.   Use 9/16 inch shaft
22. $D = 2.7$ in.   Center OK
24. $F = 2{,}560$ N

CHAPTER 6
Fasteners and Fastening Methods

1. Yes $S_{actual} = 15{,}000$ lb/in.^{2}
         $S_{allowable} = 28{,}300$ lb/in.^{2}

3. a. $S = 410$ MPa
   b. $S = 530$ MPa
   c. Grade 10.9 or 12.9
   d. $T = 200$ Nm
   e. $\alpha = 90°$
5. a. $F = 14{,}200$ lb per stud or $170{,}400$ lb total
   b. $F_t = 171{,}100$ lb total
   c. No, almost no change
7. a. $\delta = .0064$ in.
   b. $S = 96{,}000$ lb/in.^{2}
   c. $F = 6{,}600$ lb
9. a. $F = 2{,}500$ lb
   b. $S = 12{,}800$ lb/in.^{2}
   c. $S = 13{,}600$ lb/in.^{2}
   d. $F_t = 14{,}400$ lb or $3{,}600$ lb/bolt
   e. $F_t = 8{,}400$ lb
   f. $\Delta F_b = 1{,}100$ lb/bolt
      $\Delta F_c = 1{,}600$ lb
10. a. $\delta_{bolt} = .034$ in.
       $\delta_{tube} = .029$ in.
       $\delta_{total} = .0625$ in.
    b. $S_{bolt} = 42{,}100$ lb/in.^{2}
       $S_{tube} = 12{,}000$ lb/in.^{2}
11. $S = 22{,}600$ lb/in.^{2}
    $N = 2.4$
13. Factor of safety for head shear is approximately double.
15. $\delta = .0026$ in.
    $\alpha = 17°$

CHAPTER 7
Impact and Energy Analysis

1. a. $\delta = .018$ in.
   b. $S = 18{,}000$ lb/in.^{2}
   c. No
3. $\delta = .331$ in.
   $S = 33{,}100$ lb/in.^{2}

5. a. $S = 21{,}700$ lb/in.^{2}
   b. $\delta = .05$ in.
   c. $S = 3{,}180$ lb/in.^{2}
   d. $\delta = .008$ in.
   e. $S = 14{,}200$ lb/in.^{2}
      $\delta = .10$ in.
      $S_s = 3{,}180$ lb/in.^{2} (same)
      $\delta = .023$ in.
7. a. $S = 17{,}350$ lb/in.^{2}
   b. $\delta = .4$ in.
8. a. $\delta = 7.2$ in.
   b. $S = 72{,}000$ lb/in.^{2}
   c. $F = 14{,}400$ lb
11. $h = .030$ in.

CHAPTER 8
Spring Design

1. $k = 10.4$ lb/in.
3. a. $N_a = 21.8$ coils
   b. $D_m = .826$ in.
   c. $k = .7$ lb/in.
   d. $L_s = 1.17$ in.
   e. $\lambda = 5°$
   f. $\delta = 3.83$ in.
   g. $F = 2.7$ lb
   h. $K = 1.08$
   i. $S_s = 52{,}100$ lb/in.^{2}
5. a. $k = 16$ lb/in.
   b. $L_f = 4$ in.
   c. C0720-081-4000
9. a. C0850-068-3000
   b. $F = 62$ N
   c. $F_s = 77.4$ N
11. $\theta = 2.5°$
12. $\delta = .18$ in.
13. $\delta = 1.23$ in.
    $k = 81$ lb/in.

CHAPTER 9
Electric Motors

1. U.S. frequency   60 cps
   European frequency   50 cps
3. a. When the motor runs at the input speed, that is, no induction slip. For example, a single winding motor would turn at 3,600 rpm.
   b. If it is a typical single winding induction motor, full-load speed would be about 3,450 rpm. Synchronous speed minus induction slip.
   c. See b ≈ 2,875 rpm.
   d. Startup torque.
   e. Maximum torque, but the point at which the speed drops rapidly.
   f. Torque at which motor "falls" out of synchronization; if load is not removed, motor likely stops.
5. a. 440 V, three-phase
   b. 440 V, three-phase
7. Totally enclosed (TENV) or totally enclosed fan-cooled (TEFC)
9. C–C is 4.5 inches above base.
   Shaft: $D = 1.125$
   Key: ¼ × ⅛
11. $T = 19.4$ Nm
    $T = 172$ in.-lb

CHAPTER 10
Pneumatic and Hydraulic Drives

1. a. $D = 2$ in.
   b. Needle valve
3. $F_8 = 30{,}200$ lb
   $F_7 = 23{,}100$ lb
   $F_6 = 17{,}000$ lb

5. a. 3 in.
   b. 141.4 in.^{3}
6. a. L_l = .54 in.
   b. P = 133 lb/in.^{2}
7. a. P = 70 lb/in.^{2}
   b. P = .5 MPa

CHAPTER 11
Gear Design

1. a. D_p = 3 in.
   b. P_c = .196 in.
   c. m = 1.59
   d. m = 1.5
   e. C–C = 2.0625
3. V_r = 17.25/1
   n_{out} = 100 rpm
   CCW
   T_o = 1,898 in.-lb
5. V_r = 9.6/1
   n_{out} = 375 rpm
   CCW
   T_o = 840 in.-lb
   C–C = 4.292 in.
7. V_r = 6,000/1
   n_{out} = .6 rpm
   CW
   T_o = 525,000 in.-lb

CHAPTER 12 Spur Gear Design and Selection

1. a. V_r = 2.5/1
   b. D_p = 2 in.
      D_g = 5 in.
   c. V_m = 916 ft/min
   d. C–C = 3.5 in.
   e. T_{in} = 180 in.-lb
      T_{out} = 450 in.-lb
   f. F_t = 180 lb
   g. F_n = 66 lb
   h. F_r = 192 lb
3. a. Y_p = .337
      Y_q = .421
   b. F_s = 1,334 lb (pinion)
      F_s = 1,666 lb (gear)
   c. F_s = 1,334 lb (i.e., lower value)
5. a. F_d = 455 lb
   b. Yes
   c. Yes
7. b = .433 in.
9. 12-pitch steel:
   21T = 2.06 hp at 600 rpm
   with 42T = 2.42 hp at 300 rpm
   or cast iron:
   48T = 2.41 hp at 600 rpm
   with 96T = 2.56 hp at 300 rpm
11. P = 6 hp versus about 3 in catalog
    Possibly hidden safety factor
    N_{sf} around 2 or wear considerations
13. Possible design:
    P_d = 12
    D_g = 7 in.
    D_p = 6 in.
    AISI 4140 annealed
    θ = 20°
    N_{sf} = 1.5
    b = ¾ in.
    Sum of BHN ≈ 600
15. a. F_d = 875 lb
    b. No, not acceptable
    c. 1⅞ in. wide
    d. No, not in recommended range
17. F_w = 1,047 lb
    No, they should be OK for wear.

CHAPTER 13
Helical, Bevel, and Worm Gears

1. a. $P_{dn} = 5.77$
   b. $V_m = 5{,}655$ ft/min
   c. $F_t = 400$ lb
3. a. $F_s = 6{,}680$ lb
   b. $F_d = 4{,}170$ lb
   c. $N_{sf} = 1.6$
5. 10-pitch hardened steel .875-inch face
   Input 10 tooth
   Output 30 tooth
   There are many other solutions.
7. a. $\lambda = 4.77°$
   b. $n_o = 96$ rpm
   c. C–C = 5.25 in.
9. a. $e = .67$ or 67%
   b. $T_o = 441$ in.-lb
11. $1,293/yr
12. 2 yrs
13. a. $n_{in} = 450$ rpm
    b. $n_{in} = 225$ rpm
15. Yes
17. $T_{in} = 83$ in.-lb
    $P_{in} = .6$ hp

CHAPTER 14
Belt and Chain Drives

1. $L = 43.6$ in.
2. $F_d = 117$ lb
   $F_f = 137$ lb
   $F_t = 157$ lb
   $V_m = 1{,}414$ ft/min
   $n_{out} = 600$ rpm
3. $\theta = 151°$ input
   $\theta = 209°$ output
5. $F_f = 198$ lb
   $F_t = 216$ lb
7. 2 belts
9. $P = 11.1$ hp
   $F = 243$ lb
   Ratio = 15/1
11. $n = 583⅓$ rpm intermediate shaft
    $n = 155.5$ rpm output shaft
    $T_{in} = 36$ in.-lb
    $T_{int} = 108$ in.-lb
    $T_{out} = 405$ in.-lb
    $V_m = 292$ ft/min
13. $F_f = 21.6$ lb Belt
    $F_d = 114$ lb Chain
15. $\sigma = 25{,}030$ lb/in.^{2}

CHAPTER 15
Keys and Couplings

1. $L_s = .81$ in.
   $L_c = 1.23$ in.
   Use 1¼-inch-long key
3. $S_s = 25{,}600$ lb/in.^{2}
   $S_c = 51{,}200$ lb/in.^{2}
5. $S_s = 257$ MPa
   $S_c = 513$ MPa
7. $L = .44$ inch for CD 1137
   No safety factor
   Use 1-inch long key
9. $T = 3440$ in.-lb
   $P = 98$ hp
10. $T = 62{,}800$ in.-lb
    $P = 900$ hp
11. $T = 7{,}700$ in.-lb

CHAPTER 16
Clutches and Brakes

1. a. $T_f = 360$ in.-lb
   b. $P_f = 11.4$ hp
3. $N = 1{,}200$ lb
5. $T_f = 47$ Nm
   $P_f = 2.95$ kw

7. T_f = 656 in.-lb
   P_f = 25 hp
9. a = 25 ft/s^2
11. A possible design
    D_o = 36 in. D_i = 24 in.
    f = .4
    N = 12,780 lb
    Piston diameter = 6 in.
    Pressure = 450 lb/in.^{2}
    U = 7.2 × 10^6 ft-lb
    ΔT = 387° F
    f_{hp} = 292 hp per brake

2. 1¾-inch diameter
3. T_{up} = 4,170 in.-lb
5. a. S = 14,300 lb/in.^{2}
   b. T_{up} = 4,950 in.-lb
   c. e = 28%
7. T_{up} = 1,170 in.-lb
9. T_{up} = 1,100 in.-lb
10. L = 1.1 inches long
11. T_f = 562 in.-lb
    T_t = 1,732 in.-lb
13. n = 3,600 rpm
    P = .04 hp

CHAPTER 17
Shaft Design

1. n_a = 1,150 rpm
   n_b = 776.7 rpm
   n_c = 383.3 rpm
3. S_s = 3,300 lb/in.^{2}
5. S_b = 9,440 lb/in.^{2}
7. σ = 10,400 lb/in.^{2}
   Results are very similar.
9. S_s = 3,705 lb/in.^{2}
11. S_s = 2,090 lb/in.^{2}
13. σ = 17,500 lb/in.^{2}
14. N = 1.4
15. N_c ≈ 1,268 rpm
    N_c ≈ 2,535 rpm
    Very approximate

CHAPTER 18
Power Screws and Ball Screws

1. a. T_{up} = 5,000 in.-lb
   b. T_{dn} = 1,540 in.-lb
   c. Yes

CHAPTER 19
Plain Surface Bearings

1. P = 40 lb/in.^{2}
   V = 210 ft/min.
   PV = 8,400
   Teflon-filled nylon
2. T = 900 hours
3. Max. = .022 in.
   Min. = .012 in.
   So start at .012 in.
5. L = .52 in.
7. Due to wear: Teflon-filled nylon (TN) or Teflon-filled acetal (AF)

CHAPTER 20
Ball and Roller Bearings

1. L_{10} = 9 m-rev
3. P_d = 1,900 pounds
5. L_{10} = 1,550 m-rev
   L = 9.8 years
7. Number 6202
8. 61,400 miles
9. Number 6207

Glossary

A

3-D modeling A computer-generated image in 3 dimensions showing what a product or machine might look like.

addendum Could be thought of as the outside diameter of a gear and relates to the clearance needed for gears or the maximum diameter of a gear system. It is the amount that is added to the pitch diameter.

alternating current The sinusoidal wave form in which most current is generated and transmitted.

angle of contact Sometimes referred to as "angle of wrap" for flexible-drive systems; the angle at which the belt or chain is in engagement with the drive pulley or sprocket.

angular deformation Twist in a round bar; usually expressed in degrees or radians.

attributes The basic properties of a solution that would not include specifics about solutions.

axial In the same direction as the centerline of a part, generally caused by a tensile "pulling" or compressive "pushing" force.

B

back driving A unique property of ball screws due to their low friction whereby rotary motion can be created from linear motion. When the nut is pushed longitudinally along the ball screw, the ball-screw assembly rotates.

backlash The clearance between gear teeth or the amount of clearance if the gears were to reverse direction.

back-of-envelope calculation A common term for an abbreviated or simplified engineering analysis.

ball screw A power screw that, instead of using sliding friction, has rotating balls that act similar to a bearing to transform rotary to linear motion.

basic research The scientific investigation of basic principles that typically does not include the application of those principles.

beach marks The result of crack propagation where, during each flexure cycle as the part recloses in a slightly different position, some wearing occurs on the part, resulting in an appearance that looks like the tide going out on a beach.

bevel gear An angular gear used to join perpendicular shafts that is similar but different in profile from a spur gear.

bimetallic Created from two or more materials with different coefficients of thermal expansion, such that a curvature or change of shape occurs with temperature changes.

boundary lubrication When there is actual contact between the surfaces but the sliding forces are minimized by the presence of lubrication.

brainstorming Free-flow thought for the creation of many ideas in order to solve a problem.

brake fade A property of brakes whereby the braking power is reduced when the brake pad material heats up with use, either because of glazing of the pad or other temperature-related property changes.

brazing A process similar to soldering that typically uses a flame and an alloying metal.

breakdown torque The point at which, when additional load is applied to a motor, the speed may decrease as opposed to increasing.

Buckingham method An alternative method for the analysis of impact loads or velocity factors in gearing systems. This method uses actual expected clearances of the gears to determine a dynamic load.

C

cal-rod heater An electrical element often used to heat large bolting or stud materials. It is similar to what would be found in a circular shape on an electric stove.

cam follower A term for a bearing mounted on a bolt or shoulder screw that was originally intended for activating a switch or valve against a cam but may be used for other purposes as well.

cap screw The technical term for what is commonly called a bolt.

capillary forces The forces that tend to hold a fluid to itself. These are the internal molecular attraction forces in a fluid that aid in lubrication.

centrifugal clutch A type of clutch in which a rotating weight is spun more and more rapidly, causing a friction material to be pushed into engagement and causing the clutch to engage.

C-face A term applied to a face of a motor system to which gearboxes or other assemblies can be bolted directly to the end of the motor.

circular pitch Relates to the physical size of the gear tooth but, unlike diametral pitch, is measured along the circumference of the gear circle.

clutch Any method used to engage or disengage one rotating shaft from another; for instance, for attaching the rotating parts of a transmission to an automobile engine.

coefficient of friction A term used for the ratio of the force required to drag an object across a surface to its weight or normal force.

coefficient of thermal expansion The ratio of the expansion of a part to the change in temperature.

computer modeling Creating a visual depiction of a proposed idea, often in three dimensions, in order to have a better feel for what the product would look like. Computer models often cost less than creating a physical prototype or a model.

conceptual design The creative process in which a concept evolves to solve a problem or a need.

conceptual testing Similar to feasibility testing except that it may involve only a specific concept; can be very limited in scope.

Conrad deep-groove bearing A type of bearing that has a radial groove cut into both the inner and outer races to guide the ball assemblies and also to allow for minor thrust loading.

constant force spring motor A spring system that uses a wound-up flat spring to pull on an object in concert with a reel to create a linear force.

consumer testing Testing of the product by potential customers; typically includes compiling their opinions.

contact pressure The ratio of force to the surface area. In the case of a round shaft, the surface area is typically calculated as the diameter times the length required to compensate for the angle of loading.

copyright The method by which ownership is established for a creative work of art, music, or text, similar to what a patent is for an invention.

corporate knowledge Specialized technical knowledge that a company may have obtained over time that is applicable to their products.

couplings Any method used for attaching two shafts; may allow for angular misalignment.

crack propagation A condition in which the microscopic flaw would increase in size during each flexure of a repeated load.

creativity sessions A more formal process of brainstorming that involves a group of people brainstorming about an idea or problem in order to come up with possible remedies or solutions.

critical speed Refers to a speed at which a part would vibrate due to its natural frequency; usually extreme care is taken to avoid operating the part close to or at its critical speed.

cross-helix A specialized form of a helix gear in which the gears are perpendicular to each other.

D

dedendum The distance from the pitch circle to the root of the gear or the base circle.

deflection analysis An analysis of the movement of a part or assembly under load.

denting The phenomenon in which, due to high static or low-speed loads, the balls in a ball bearing form a recess in the race over time.

diametral pitch Commonly referred to as simply "pitch," it is the size of the gear tooth and is measured as the number of teeth per inch of diameter.

direct current A form of electricity in which there is no sinusoidal wave transmitting the power, simply a potential between two separate conductors.

distortion energy theory A method of combining principal stresses in ductile materials to predict failure.

ductility The level of deflection before failure; often measured by the amount of energy that can be absorbed before failure.

durability testing Testing to determine whether an appropriate lifetime for a machine or part of a machine can be achieved.

dynamic braking A system that uses the magnetic field in an electric motor to slow down rotation as opposed to creating rotation.

dynamic load factor A value based on the surface speed of gears and the manufacturing accuracies; used to compensate for the clearances or impact loads inherent in gear systems.

E

eccentricity of the load A load that is applied to a column but not on the column's centerline. This could cause a bending stress along with the column loading.

efficiency The ratio of the power or torque into a system to the output.

elastic deformation The deformation of a part below the elastic limit such that the part returns to its original shape when the load is removed.

electromagnetic By imparting an electrical field, a force is created to engage a clutch or brake system. This often may be referred to as a "solenoid" brake or clutch system.

elongation The amount of stretch or change in length of a material when a load is imparted.

endurance limit If a load were applied repeatedly to a part, this is the stress level that would not cause a postulated defect to propagate.

equivalent combined radial load The method whereby the thrust loading is factored in to result in an equivalent radial load.

expansion bolts A bolting system that has an expanding outer sleeve or ferrule that can be made larger when tightened.

extension spring Very similar to a helical spring with the exception that it is made for retracting as opposed to being compressed.

F

factor of safety A margin built into a design to allow for unusual situations or factors not considered in the analysis.

fail-safe brake A term used for brakes that require power to unlock the brake system so that they are automatically set in the event of a power failure.

fatigue analysis The study of the effect of repeated loading on a part.

feasibility testing The testing of a prototype or product to determine if the principle of operation is feasible.

finite element analysis A structural analysis that identifies loads and stresses on individual sections of a part.

flexible-drive systems A term used to encompass belt and chain drives or other drives where a fixed distance between the shaft centerlines is not required.

fluid pressure When a force acting on a fluid creates internal pressure within the body of fluid.

flywheel A large, rotating weight in which energy could be stored as rotational inertia.

force critical Often referred to as the critical force, this is the force that could act on a column to the point of causing instability, at which point a lower force could continue to cause the column to fail.

forensic engineering The process of investigating what happened to a part, machine, or structure that unexpectedly failed. Typically, this process tries to determine what loads and associated stresses were acting on a part to cause the unexpected failure.

functional requirements A listing of the requirements that the product needs to accomplish, not how it would be accomplished.

functionality testing Testing to determine whether a prototype or model actually performs in the manner for which it was conceived.

G

galling A phenomenon whereby a surface gets smeared or dragged, resulting in rough surface finishes and results in poor frictional properties.

gaskets A typically soft material that is used for sealing pressure-retaining parts.

gear train Two or more sets of gears combined to result in higher ratios.

Gerber theory Similar, albeit slightly different, to the Soderberg equation, this theory of fatigue failure is sometimes used for brittle or nonductile materials.

glazing One of the properties that causes brake fade. The surface of the friction material becomes either smoothed over or, under extreme temperature, becomes semiliquid, thereby reducing the coefficient of friction.

Goodman criteria Quite similar to the Gerber theory; used for fatigue failure of brittle or nonductile materials.

H

heat-affected zone A term used when, during such processes as welding, the heat causes changes in the base property of the material or results in a residual stress.

helical compression spring A wound spring formed by bending the spring wire into a helical shape, such as the front spring on an automobile.

helical gear A gear that uses the same general form as a spur gear but is cut on an angle.

helix angle The angle at which the tooth form is oriented with respect to the centerline of the gear face. The helix angle would be perpendicular to the normal plane.

herringbone gear A left- and right-hand helical gear combined in a single assembly to eliminate thrust loads on the shaft.

hydraulic The use of a fluid; for example, water or oil.

hydraulic pressure The use of a fluid pressure to create power or work.

hydraulic tensioner A large machine that uses hydraulic cylinders to stretch a bolt for imparting an initial tension.

hydrodynamic lubrication The phenomenon that occurs when the shaft or other smooth object is supported by the lubrication properties of the fluids and no actual contact is made between the metallic surfaces.

hydroplane The process whereby a film of lubrication rides on the surface and, due to the capillary forces in the fluid, results in no direct contact between the surfaces.

hydrostatic The phenomenon of creating a pressure in a fluid in order to physically lift the object; similar to, for example, the puck in an air hockey game.

I

impact load A sharp load that is typically created from a clearance or a gap, such as a space in a machine part or an automobile hitting a guardrail.

impulse load A force applied rapidly or of short duration, typically at less than one natural frequency of the receiving part.

inactive coils In a helical spring, an often flattened coil on either end that, in some cases, is ground to result in a square end. The inactive coil does not add to the springiness of the system.

incremental control When small changes in position are achievable.

inductive motor The most common type of electric motor in which inductive forces rotate the motor by means of a magnetic field.

inefficiency The opposite of efficiency or simply the measure of wasted energy.

intellectual property The term that is applied to an idea or a creation, based on the fact that by being its creator, the individual should have rights to that idea or creation.

interference fit As it applies to bearings, this occurs when the total of the dimensions for the inner race, the balls, and the outer race are greater than the space available. This results in a residual load in these parts, but it makes for a tighter bearing assembly.

internal gear A gear profile made on the inside of a circle.

involute curve The special shape of a gear profile that allows for rolling action when mated with a similar involute shaped gear.

isotropic Having similar properties in all directions. For example, steel is generally considered to be an isotropic material because it has the same stress allowables regardless of direction of stress.

J

journal bearing Another term for a plain-surface or sleeve bearing.

K

Kevlar A term used for reinforced strands of materials that are made from a carbon fiber possessing high strength and flexibility.

keyway A slot cut into a shaft and a mating hub that uses a square or other-shaped key to transmit rotational motion.

kinetic energy The energy associated with velocity, as in a moving automobile.

L

leaf spring A spring system that uses the principles of a simply supported beam as a spring assembly. Some automobiles and many heavy trucks use spring systems of this type.

Lewis form factor A factor used in the design of gears to compensate for the sharpness of the inner root of the teeth, the shape of the teeth, and other factors.

linear bearing A bearing assembly that allows motion axially along a shaft as opposed to radially; in some cases may be combined with a radial bearing.

locked-rotor torque A term that applies to the torque of a motor to start rotation.

M

major diameter The larger or outside diameter of a thread, sometimes referred to as nominal diameter.

market-pulled When a perceived need in the marketplace encourages the creation of new products.

metric module A very similar characteristic to circular pitch; used in the metric SI system for specifying gear-tooth size.

microscopic flaws Defects in a part typically between grain boundaries, usually visible with a microscope but not with the naked eye.

minimum film thickness The terminology used for determining the thickness that many lubricants try to maintain due to their internal capillary forces. It can also be thought of as the natural thickness of the fluid on a perfectly flat and smooth surface.

minor diameter The smallest diameter of a thread, sometimes referred to as the root diameter.

miter gear A bevel gear in which both gears have the same number of teeth and, hence, a one-to-one ratio.

mitigate defect propagation A term for different methods used to reduce the effect of a postulated defect on a part. These methods may include leaving residual compressive stresses, changing radiuses, and changing dimensions.

modulus of elasticity The relationship between stress or load and the deflection of a material.

Mohr's Circle A graphical method for combining stresses. In Mohr's Circle, the stress values are added based on their directions graphically, and then the combined stresses are determined from this graphical combination.

moment of inertia Property of shape relating to the distance that the material resisting bending is placed from the neutral axis.

multiple-start worm A worm gear that contains two or more concentric thread faces resulting in a progression of a similar number of teeth on the main gear; often resulting in slightly higher operating efficiencies.

N

natural frequency The same as critical speed; the speed at which the part has an inherent tendency to vibrate.

needle bearings Roller bearings that lack an inner race; the shaft is used as the inner race in order to save space.

needle valve Used in pneumatic or hydraulic systems to control the flow rate.

neutral axis An axis at which, under bending or torsional stress, no stress exists. For a symmetric part, this is the centerline.

Newton The unit of force in the metric SI system. This is a similar quantity to a pound in the U.S. customary unit system.

noncompressible A term that is used to descibe a fluid that has a high bulk modulus such that little or relatively little change in volume occurs with pressure.

normal plane diametral pitch The diametral pitch in an imaginary plane perpendicular to the face of the gear teeth, which would be used in the manufacture of helical gears.

normal stress Stress caused by an axial or bending load that tends to pull the molecules apart.

P

Pascal A unit of stress or pressure in the metric SI system similar to a psi or pound-per-square-inch stress or pressure unit in the U.S. customary system.

patent pending A term that can be applied to a product in order to inform others that a patent has been applied for and is expected to be received.

patentable A process in which the right to use an idea is protected by law, discouraging others from using the idea.

patent A form of protecting legal rights that is created by a government to insure that someone creating an invention can have the sole right to its use for some period of time.

pillow blocks A term designating a bearing mounted in a housing.

pinion The term that is applied to the smaller gear in a gear-mating system, as opposed to the larger gear, which is referred to only as the gear.

pitch diameter The point on a sprocket or gear at which the load is assumed to be applied. This is sometimes referred to as the "running circle."

pitch or diametral pitch An index of the size of the teeth that is found by dividing the total number of teeth by the diameter of the gear.

pitch-line speed The same value as surface speed or surface velocity. This would be the speed at which the gear teeth come into engagement or the rotational speed for a wheel or belt speed for a belt system.

plain surface bearing A term that refers to a smooth-surface bearing that relies on sliding between the shaft and the bearing, typically with some form of lubrication.

pneumatic The use of a gas; for example, air or compressed air.

pneumatic pressure The use of air or other gas pressure for creating power or work.

polar moment of inertia This is a similar property to the moment of inertia. However, it is the moment of inertia that would resist a torsional load.

polar section modulus This is a similar property to section modulus. However, the polar section modulus would be the ability of a material to resist a torsional load.

postulated defect An assumed microscopic flaw that could propagate when subjected to repeated loads.

potential energy The energy associated with elevation such as that of a falling object.

power The rate at which the work is accomplished.

power accumulator A term used for multiplying the power, typically for braking systems, in order to minimize the amount of force necessary to activate the brakes.

power screw Often referred to as a "lead screw"; a threaded fastener-type assembly used to convert rotary motion to linear motion.

pressure angle A property of the gear-face profile related to the involute angle.

pride of ownership Refers, in the engineering world, to the sense that one's own ideas are inherently better than those created by others.

prior art Refers to an idea that may have been created previously. In the patent process, it would prevent a patent from being issued because the idea is not unique.

problem statement A description of the problem to be solved, often detailing only attributes of the specific problem.

proof strength A stress value below the yield where bolting material would not typically fail; somewhat similar to the elastic limit.

proportional limit A point similar to yield strength but often slightly below where the stress-versus-deflection of a material is linear.

prototype A partial or full-scale model of a product that is usually used for testing.

pullout torque When a synchronous motor needs a greater torque than can be created this is the value at which the motor falls out of synchronization or stops.

Q

quarter-turn fastener A type of latching system that, when turned, pulls parts into engagement.

R

rack-and-pinion gear Uses a conventional pinion gear in concert with a straight gear to convert rotary to linear motion.

radius of gyration A measure of the stability of a column found from the square root of the ratio of the moment of inertia to the cross-sectional area.

raveling A similar process to galling; the difference is that in raveling the material may be pushed or displaced and in general can damage the bearing race by roughening the surfaces.

reciprocating loads A term used for either repeated or repeated and reversed loads, implying that the load is applied over and over again in some manner.

rectifier An electrical device that converts alternating current into direct current.

regulator Typically, a spring-operated bellows or diaphragm assembly that allows pneumatic or hydraulic fluids to pass through at a set pressure.

repeated and reversed stress A stress that varies from tensile to compressive or reversing direction shear that can exacerbate the effect of a postulated defect on a part.

repeated stress A stress that is repeated over and over again but does not go through a reversal cycle.

residual compressive stress A compressive stress that is induced in a part such that it remains when no load is acting on the part.

residual stress A stress left over in a part due to such processes as heat treating or welding. This is a stress that exists in the part even in the absence of an external load.

rod end bearing Similar to a pillow block bearing but mounted in an assembly that would typically be attached to the end of a bar or shaft.

roller bearing Utilizes the same principles and is similar to a ball bearing, except by using a roller the contact area is increased.

rolling contact bearings A bearing that, instead of having sliding friction, contains balls or rollers that roll like wheels.

S

seals Similar to a shield but typically made of elastomeric material to try to prevent, like a shield, the entrance of contaminants. May aid in keeping lubrication in the bearing assembly.

section modulus Similar to moment of inertia, but for a uniform concentric shape. It is the moment of inertia divided by the distance from the neutral axis to the outermost fiber.

self-activating A property of certain styles of brakes whereby a slight force causes the brake system to pull itself into harder engagement.

self-locking A property of certain mechanical elements, often applied to worm gears, when the output shaft will not rotate except in the presence of rotation of the input shaft. Can be used as a built-in brake system.

serpentine A term used for a complicated belt or chain drive system that may go around many components and in some cases uses both sides of the belt or chain.

service factor Similar to a factor of safety but based on specific usage of systems, such as whether the loads would be smoothly applied or other factors inherent in the end usage of gears or other products.

service temperature The temperature at which a system operates. This needs to be taken into account in order to determine preload temperatures for tensioning bolting or stud systems.

servomotor A term that applies to a motor that has feedback loops built into the system for telling a control system the actual motor position and/or speed.

setscrew A cap screw–type mechanism mounted perpendicular to a shaft that, due to the friction in the tip of the screw, can hold a hub in position on a shaft.

shear modulus of elasticity This is a similar property to the modulus of elasticity except that it is the relationship between force and deflection from a torsional load.

shear stress Stress caused by a torsional or shearing-type load that tends to make the molecules slide by each other.

sheave The term for a pulley used in a belt system.

shield A type of sealing mechanism for a bearing that uses a small cover plate attached to one or the other of the races to prevent the entry of contaminants. In some cases it is free floating.

sleeve bearing Another term for a plain-surface bearing in which a thin cylindrical material is used to allow for shaft rotation.

snap rings Commonly referred to as "spring rings." Small components that can be used to keep a component from sliding along a shaft. They require a groove or other recess.

Soderberg equation A method for evaluating the effect of different types of repeated stresses on a part.

soldering A process for joining parts by use of an alloying element, typically at a significantly lower temperature than that used for welding.

solenoid valve An electrically operated valve in which the valve is opened and closed by an electrically operated magnet system.

specific heat A property of material that is the amount of energy required to raise its temperature. It is usually expressed in terms of Btu's or calories per unit weight or mass.

spline A device that allows a change in length of a shaft while transmitting rotary motion.

spring constant The equivalent stiffness of a part, similar to a spring rate for a spring, that can be determined based on the stiffness of any potential load-receiving part.

spur gear The most common form of gearing; uses a shaped tooth to transmit and sometimes change rotational speed.

strain gage Used to measure stress in a part. A strain gage is essentially a miniature resistor that changes resistance as it is stretched and can measure strain or unit deflection when epoxied to a part with the appropriate instrumentation.

stress concentration A location or point on a part at which stresses may be higher than average due to changes in shape or other factors.

stress risers Often referred to as discontinuities or stress concentrations at a point where due to shape changes in a part where stresses are concentrated when subject to a load.

stud Similar to a bolt, with the exception that there is no head and typically both ends are threaded.

surface velocity The surface speed of the gear or other rotating object at the pitch line. For example, for the tire of an automobile it would be the speed that the car is moving.

synchronous generator A common type of generator used in a power plant that typically rotates at the synchronous speed or, on some occasions, a multiple thereof to create (in the United States) 60 cycle or (in Europe) 50 cycle power.

synchronous speed A ratio of the speed that is generated by the power company.

T

tapered roller bearing A roller bearing made in a conical shape, so that it can support thrust loads as well as radial loads.

technology-pushed A term that is used when advances in engineering and technology encourage a company to upgrade its products.

telescoping Describes a series of tubes of different sizes that slide within each other to make a longer assembly.

three-phase power Electricity that is generated with three sets of windings in a generator system, such that each revolution adds part of a sine wave cycle to the available power. Three wires are needed to transmit the power.

thrust bearings Bearings that are intended primarily to resist thrust loads unlike most ball or roller bearings, whose primary purpose is resisting radial loading.

torque A rotational force created by a force applied at some distance from the centerline that tries to cause rotation.

torque coefficient A value similar to the coefficient of friction used for determining the torque required for turning threaded fasteners.

torque tube A round spring system that relies on a torsional deflection of a solid or hollow tube to act in concert with a lever arm in a spring system.

torque wrench A wrench with a built-in measuring system that measures the torque that is imparted to a bolt or cap screw.

torsion bar suspension A torque-tube assembly that also uses the deflection of a bar to create a suspension system, frequently used in the rear wheels of an automobile or truck.

torsion spring A spring that looks very similar to a helical spring but whose purpose is to resist rotary motion or to create rotary motion.

torsional loading A load that is created by a torque, such as in an axle, due to a force being applied at some distance from the centerline that causes a shearing-type, twisting stress.

trademarks A name or symbol used frequently to identify a product. A trademark allows its unique use by the creator under certain conditions.

transformer An electrical device that changes voltage and amperage to different levels.

transmitted force A force that would act tangent to the running circle for gears, sprockets, or pulleys. This is the value that would be used in the power analysis.

transverse pressure angle The corrected pressure angle for use in helical gearing that corrects for the helix angle.

transverse shearing stress A stress in a beam parallel to the neutral axis caused by the difference in deflection longitudinally between the tensile and compressive stresses acting upon a beam.

U

ultimate strength The stress level at which failure occurs.

unit analysis The analytical process of checking the units in a calculation. It can also serve to verify that the inputs to the formula are correct.

universal joints A flexible-shaft attachment method that allows for angular misalignment between shafts.

V

viscosity A measure of the flow characteristics of a fluid. With bearings, it is often used as an indication of the fluid's lubrication properties.

W

welding A process whereby materials are raised to above their recrystallization temperatures and

bonded together under high heat, with or without an additional filler element.

work The result of a force moving an object a certain distance.

worm gear A gear system that uses a lead-screw–type shape to turn a gear, which results in high ratios.

Y

yield strength A stress level at which permanent deformation of the part occurs.

Bibliography

Faires, Virgil M. *Design of Machine Elements*, 4th ed. New York: Macmillan, 1965.

Faires, Virgil M., and Wingren, Roy M. *Problems on the Design of Machine Elements*, 4th ed. New York: Macmillan, 1965.

Hall, Allen S. Jr., Holowenko, Alfred R., and Laughlin, Herman G. *Theory and Problems of Machine Design*. New York: Schaum, 1961.

Kolstee, Hans M. *Machine Design for Mechanical Technology*. New York: Holt, Rinehart and Winston, 1984.

Mott, Robert L. *Machine Elements in Mechanical Design*, 3rd ed. Upper Saddle River, New Jersey: Prentice Hall, 1999.

Oberg, Erik, Jones, Franklin D., and Horton, Holbrook L. *Machinery's Handbook*, 20th ed. New York: Industrial Press, 1979.

Parr, Robert E. *Principles of Mechanical Design*. New York: McGraw-Hill, 1970.

Schertz, Karen A., and Whitney, Terry A. *Design Tools for Engineering Teams: An Integrated Approach*. Albany, New York: Delmar, 2001.

Shigley, Joseph E., and Mischke, Charles R. *Mechanical Engineering Design,* 6th ed. New York: McGraw-Hill, 2001.

Index

A

Abbreviations used in text, xix
Acme thread, 395–96, 401–3
Addendum (gears), 211, 229
Alternating current (AC) motors, 172–89

B

Ball bearings, 426–37
Ball screws
    efficiency, 403–5
    torque, 403–5
Beam
    deflections, 42–44, 48, 467–68
    moments, 42–44, 467–68
    shapes, 473–80
    stress, 40–44, 48
Bearings
    ball, 426–37
    boundary lubrications, 412
    cam followers, 441–43
    full film, 413–21
    hydrodynamic, 413–21
    hydrostatic, 413, 423
    journal, 411–23
    life, 418–21, 428–37
    linear, 442–43
    needle, 438–39
    pillow blocks, 440–41
    plain surface, 411–23
    pressure velocity factor, 415–18
    rod end, 441–42
    roller, 437
    sleeve, 411–23
    tapered roller, 437
    thin film, 411–23
    thrust, 423, 440
Belt and chain drive loads, 305–8, 377–78
Belt drives, 305–17
    flat belts, 305–8, 315–16
    timing belts, 316–17
    V-belts, 311–15
Bevel gear loads, 285–87, 375–76
Bevel gears, 211–13, 233–37, 285–87
Bibliography, 519
Bolted connections, 112–28
Boundary lubrication (bearings), 412
Brainstorming, 453–56
Brakes, 349–66
    disc, 350–51, 355
    drum, 351–53
    friction coefficients, 354–56
    power, 357–58
    self-energizing, 354
    self-locking, 354
    temperature rise, 364–65
Buckingham method, 268–71
Buckling of columns, 48–56

C

Cable chains, 323–24
Cam followers, 441–43
Cast iron properties, 487

Index

Chain drives, 316–23
    center distance, 308–10
    forces on shafts, 305–8, 377–78
    lubrication, 318–19
    pitch, 316–20
    sizes, 316–20
Circular pitch (gears), 226–27, 229, 290
Classes of gears, 260
Clutches and brakes, 349–66
    band brakes, 352–53
    coefficient of friction, 354–56
    cone, 350, 359–61
    disk, 350–51, 355
    drum, 351–54
    energy absorbed, 363–66
    plate, 350
Codes and standards, 17
Coefficient of thermal expansion, 11–12
Coefficients of friction, 354–56
Columns, 48–56
    buckling, 48–56
    effective length, 50
    end connections, 50
    Euler formula, 51–53
    Johnson formula, 51–53
    slenderness ratio, 51–54
Combined stresses, 65–75, 382–84
    distortion energy theory, 74–75
    principal stress theory, 70–73, 283–84
    shear stress theory, 74, 382–83
Compressive stress, 39–41, 48
Constant velocity joints, 339, 341
Contact stress, 334–37
Conversion factors, 494–96
Coplanar shear stresses, 67–68
Copyrights, 458–59
Couplings, 337–39
Creative process, 452–58
Creativity sessions, 456–58
Critical speed, 386–88

D

Dedendum (gears), 211, 229
Deflection
    axial, 40–41, 48
    beam, 42–44, 48, 467–68
    spring, 161–63
    thermal, 11–12
    torsional, 46–48
Design calculations, 12–15
Design process, 3–7
Design stresses in shafts, 373–86
Diametral pitch (gears), 226–27
Direct current (DC) motors, 172, 179–80
Disk brakes, 350–51, 355
Distortion energy theory, 74–75
Drum brakes, 351
Ductility, 11
Dynamic load factors (gears), 260–61

E

Efficiency
    power screws, 397–99, 400–405
    worm gears, 293–96
Elastic analysis (bolted joints), 122–26
Elastic limit, 9
Electric motors, 172–89
Elongation, 40–41
Endurance limit, 85–88, 384–86
Energy analysis, 134–42
Energy absorption (springs), 166–67
Euler formula, 51–53

F

Factor of safety, 7–8, 56–57
Failure theories, 70–75
    distortion energy, 74–75
    principal stress, 70–73
    shear stress, 74
Fasteners, 112–28
    grades, 115–18
    head markings, 118
    proof strength, 115–17
    thread standards, 113–18
    tightening, 119–22
    torque coefficients, 119
Fatigue, 82–104
    crack propagation, 82–84
    endurance limit, 85–88
    failure theories, 90–92
    Gerber theory, 92
    Goodman theory, 92, 97
    life, 101–3
    strength, 85–88
    Soderberg theory, 90–97, 99–101
    stress concentration factors, 99–101
Flat belts, 305–8, 315–16
Force and power, 22, 373–74
Formulas from text, 497–502
Friction coefficients, 354–56
Full-film lubrication, 413–21

G

Gasketed connections, 126–27
Gauges of wire, 488
Gears, 208–302
    addendum, 211, 229
    bevel, 211–13, 233–37, 285–87, 375–77
    Buckingham method, 268–71
    circular pitch, 226–27, 229, 290
    classes of, 260
    dedendum, 211, 229
    diametral pitch, 226–27
    dynamic load factor, 260–61
    forces on shafts, 253–55, 279–81, 285–87, 375–82
    helical, 211–12, 232–33, 279–84
    involute tooth form, 223–25
    Lewis form factor, 258–60, 290–91
    metric module, 228–29
    miter, 211–13, 233–37, 285–87
    pitch diameter, 223–25
    pressure angle, 223–25, 230, 279–80
    rack-and-pinion, 214
    self-locking worm, 238, 294
    spur, 210–11, 220–31, 252, 375
    strength of, 257–60, 282–84, 290–93
    thermal rating (worm gears), 239–40, 295–96
    trains, 216–20, 239–44
    velocity ratios, 216–20
    wear factors, 272–73, 295, 493
    worm, 214–15, 237–40, 287–96, 377–78
Gerber theory, 92
Goodman theory, 92, 97
Glossary, 509–18

H

Helical compression springs, 151–64
Helical extension springs, 151–53
Helical gears, 211–12, 232–33, 279–84
Helix angle, 279–80
Horsepower relationships, 24–25
Hydraulic cylinders, 193–201
Hydrodynamic lubrication, 413–21
Hydrostatic bearings, 413, 423

I

I-beam shapes, 475–76, 480
Impact
    analysis, 134–45
    beams, 141–42
    stress, 134–44
    velocity, 138–42
Inertia (moments of), 31–32, 42, 46, 469, 473–80
Initial tension, 119–22
Involute shape
    gears, 223–25
    splines, 334–37

J

Johnson formula, 51–53
Joints
    bolted, 119–26
    gasketed, 126–27
    preloaded, 119–26
    universal, 339–41
Journal bearings, 411–23

K

Keys and keyways, 328–34
Kinetic energy, 135–36

L

Lead screws, 394–405
Leaf springs, 165–66
Lewis form factor, 258–60, 290–91
Line of action (gears), 223–25
Linear bearings, 442–43
Lubrication
    bearings, 410–11, 414, 444
    chains, 318–19
    worm gears, 295–97

M

Mass, 22–23
Material properties, 8–12
Maximum shear stress, 74
Metric module, 228–29
Metric units, 16
Mises-Hencky theory, 74–75
Miter gears, 211–13, 233–37, 285–87
Modulus of elasticity, 10–11, 162
Modulus (section), 31–32, 42, 469, 473–80
Mohr's Circle, 70
Motors
    electric, 172–89
    enclosure types, 181–87
    hydraulic, 202–4
Moments
    beams, 467–68
    of inertia, 31–32, 42, 46, 469, 473–80

N

Natural frequency, 386–88
Needle bearings, 438–39
Neutral axis, 42
Non-ferrous metals, 485–86
Normal stress, 39–44, 65–67
Notch sensitivity, 99–101

P

Patents, 458–61
Pillow blocks, 440–41
Pitch diameter, 223–25
Plain surface bearings, 411–23
Pneumatic drives, 193–203
Polar moment of inertia, 46, 469
Polar section modulus, 46, 469, 478
Potential energy, 135–36
Power, 23–29, 373–74
Power screws, 394–405
    acme thread, 395–96, 401–3
    back driving, 404
    ball, 396–99, 403–5
    efficiency, 400–405
    lead, 395–97
    pitch, 395–97
    self-locking, 399–400
Pressure, 29–31
Pressure angle (gears), 223–25, 230, 279–80
Pressure velocity factor (bearings), 415–18
Principal stresses, 70–73
Problem solutions, 503–8
Proof strength (bolting), 115–17
Proportional limit, 9

R

Rack-and-pinion (gears), 214
Radius of gyration, 49, 469, 474, 477–80
Repeated loading, 82–104
Residual stress, 103
Retaining rings, 341–43
Rod-end (bearing), 441–42
Roller bearings, 437
Roller chain, 316–23

Rotational inertia, 363–66
Rotational speed, 24–29

S

Safety factors, 7–8, 56–57
Screw threads, 113–15, 394–96
Section modulus, 31–32, 42, 469, 473–80
Self-locking worm gears, 294–95
Service factors, 262
Set screws, 341
Shaft design and loads, 372–89
Shape properties, 469, 473–80
Shear modulus of elasticity, 46–47, 162
Shear strength, 39, 44–45, 485
Shear stress, 44–48, 382–83
Sleeve bearings, 411–23
Slenderness ratio (columns), 51–54
Snap rings, 341–43
Soderberg equation, 90–97, 99–101
Solutions to problems, 505–8
Specific heat, 363–66
Splines, 334–37
Spring
    buckling, 163–64
    deflection, 161–63
    energy, 166–67
    helical, 152–64, 490–92
    index, 156, 160
    leaf, 153, 165–66
    rate, 155, 162–63, 490–92
    torsion, 153
    torsion tube, 46–48
    Wahl factor, 159–61
    wire, 157–58, 162, 488–89
Spur gears, 210–11, 220–31, 252
Square threads, 394–96, 400–401
Stainless steel (properties), 486

Steels (properties), 470–72
Stiffness, 123–26, 137
Strain, 9
Strength of gear teeth, 252, 257–60, 282–84
Stress
    axial, 39–41, 48
    beams, 40–44, 48
    bending, 40–44, 48
    columns, 48–56
    combined, 65–75
    compressive, 39–41, 48
    concentration, 99–101, 482–84
    contact, 334–37
    fatigue, 82–104
    impact, 134–44
    keys, 331–34
    normal, 39–44, 65–67
    principal, 70–73
    repeated, 82–104
    shafts, 372–74, 379–89
    shear, 44–48, 382–83
    springs, 150–67
    spur gears, 210–11, 220–31, 252
    tensile, 39–41
    thermal, 121–22
    torsional, 46–48, 382–83
    ultimate, 10
    vertical shearing, 48
    welded connection, 128
    yield, 10
Structural shapes, 473–80
Surface speed, 255–57, 416–18
Symbols used in text, xix

T

Tapered roller bearings, 437
Tensile stresses, 39–41
Thermal expansion, 11–12

Thermal rating (worm gears), 239–40, 295–96
Thin film bearing, 411–23
Threads, 113–15
    fasteners, 113–26
    power screws, 394–96
Thrust bearings, 423, 440
Timing belt, 316–17
Torque, 25–26, 373–74
Torqueing methods, 119–22
Torsion springs, 153
Torsional deflection, 46–48
Torsional shear stresses, 46–48, 382–83
Trademarks, 460

U

Ultimate strength, 10
Universal joints, 339–41
Unit systems, 16
U.S. customary units, 16

V

V-belts, 311–15
Velocity ratios, 216–20
Von Mises theorem, 74–75

W

Wahl factor, 159–61
Wear of bearings, 418–21
Wear of gears, 272–73, 295, 493
Welded connections, 128
Wide-flange beam shapes, 475
Wire for springs, 157–58, 162

Wire gauges, 488
Woodruff keys, 330
Work, 23–24
Worm gears, 214–15, 237–40, 287–96
    efficiency, 293–96
    loads, 287–93, 377–78
    mechanical ratings, 287–93, 295–97
    self-locking, 294–95
    thermal ratings, 295–97

Y

Yield point, 9
Yield strength, 10